THE WORLD WAS MY GARDEN

Allison Armour and the Author (*left*) beside a giant Aroid which they brought from the Cameroon to the Harvard Arboretum at Soledad, Cuba.

THE WORLD WAS MY GARDEN

Travels of a Plant Explorer

By
DAVID FAIRCHILD

Assisted by
Elizabeth and Alfred Kay

THE BLACKBURN PRESS

Reprint of 1938 Edition by CHARLES SCRIBNER'S SONS

The World Was My Garden
Travels of a Plant Explorer

ISBN-10: 1-932846-30-1
ISBN-13: 978-1-932846-30-0

Library of Congress Control Number: 2017963144

THE BLACKBURN PRESS
P. O. Box 287
Caldwell, New Jersey 07006 U.S.A.
+1 973-228-7077
www.BlackburnPress.com

Foreword

WHEN DANGER threatens a friend, there comes to us a sudden realization of the breadth of his attainments and of our depth of affection for him.

Several years ago, during an expedition to the tropics, David Fairchild contracted a serious infection. Months of illness followed, and at last he sank into a coma. During the dark hours of waiting, we came to a realization not only of our personal loss should he go, but also of the vast store of knowledge and experience which he had accumulated and which would be lost with him. We comprehended how stupid we had been that we had not charged ourselves with the task of preserving this knowledge.

So, when the hand of death withdrew and David's strength returned, we inveigled him to peaceful Hidden River Farm in New Jersey and provided a stenographer to take down the story of his life—which is also the story of plant introduction into this country. When he wearied, we drove him on and, with whatever counsel we could give, we have tried to help him form his multitude of experiences and myriad of facts into readable book form.

But the most difficult feat has been to persuade David Fairchild to include himself in his autobiography. He can understand that readers might enjoy the home life and ultimate purpose of plants, but, because of his sincere modesty, it has required steady insistence to constrain him to permit his readers a glimpse of himself as well. We hope that many will read between the lines, and envision the charming character of this delightful man whose blue eyes sparkle with a deep, unquenchable interest and enthusiasm for all things. His is a nature devoid of selfishness or the slightest tinge of pettiness. Each day brings to him new wonder and a new challenge to learn and to give to others. In his company one gains new eyes to see, new ears to hear, and a new comprehension and love of the world we live in.

May the readers of this book discover that the world is their garden too!

ELIZABETH D. KAY
ALFRED G. KAY

Contents

Contents

Illustrations

With a few exceptions, the photographs in this book were taken by the author. Many were made in the course of his early explorations. Those taken during his association with the Department of Agriculture are reproduced by courtesy of the Department. The picture of the Langley trial flight is from a copyrighted photograph, and is included by courtesy of the National Geographic Society. The photograph of the Langley machine on the roof of a house-boat is by courtesy of the Smithsonian Institution. Several photographs were taken by the late Frank N. Meyer, agricultural explorer for the Department of Agriculture. The photograph of the beautiful blossom of the Japanese Mume is from Lee S. Crandall's camera.

ix

Illustrations

Illustrations

Illustrations

THE WORLD WAS MY GARDEN

CHAPTER I

Background

I N ANY ESTIMATE of human life there are two factors, both of which are extremely difficult to weigh—the factor of inheritance and the factor of environment. Furthermore, it is perhaps an academic question whether they are distinct, or whether they are merely convenient expressions which we use to describe a background, the beginnings of a human experiment.

When I settled to the task of this biography, before my story could reach the exciting years of exploration and plant introduction, it seemed necessary to account for my being myself, and for "myself" being a plant explorer. How had it come about?

Gazing back through the mists of sixty-odd years, I realize that I both had the suitable heredity and was born into an environment adapted to the development of a naturalist or horticulturist. In other words, my path was almost predestined at my birth. I do not believe that I consciously chose its direction, but rather wandered down its attractive way unconscious where I was going.

My forefathers were early settlers in Ohio and Michigan, but were intellectual pioneers in the field of education rather than men who actually cleared the land.

My grandfather, Grandison Fairchild, was one of the founders of Oberlin College in 1833, and his son James was president of Oberlin for a quarter of a century. Another son, Henry Fairchild, was president of the college at Berea, Kentucky, and devoted his life to the work among the mountain whites; and a third son, my father, George Fairchild, was president of Kansas State College of Agriculture for many years.

Once established, Oberlin College became a religious center of importance, and my father grew up in an atmosphere saturated by the strictest Puritan dogma and doctrine. All lectures and classes were opened with prayer; while dancing, smoking, swearing, drinking, cards, and theatres were taboo.

Intolerant in matters of religion, Oberlin was remarkable for its liberalism when, in 1841, it was the first college in this country to give degrees to women, and later was the first American college to admit Negroes as students.

In consequence of Oberlin's encouragement of women students, my Quaker mother, Charlotte Halsted, journeyed west to enroll there, and promptly fell in love with my handsome father, and he with her. Father did not have a penny to his name, but this appears to have been the usual condition of bridegrooms in those days.

Fortunately for the young couple, an agricultural school, something then almost unheard of, had been started in the wilds of Michigan, and Father was offered the position of Professor of English Literature in this pioneer institution.

When one considers the number of students graduating today from the many agricultural schools throughout the world, it is hard to realize that there was not a single such school in existence before the middle of the last century. As far as I know, the first school was in Gembloux, Belgium. In the United States, I believe that the pioneering effort begun in the forests of Michigan in 1850 was the first of its kind. It was called the Michigan State College of Agriculture. Through the efforts of a few remarkable men, it soon became an outstanding and successful school, and from it there went forth an ever increasing number of exponents of the theory that the farmer should be educated. These men, in turn, helped to create other schools throughout the country.

Though not a scientific man himself, Father soon recognized and appreciated that a new era was beginning to alter the rigid dogma of the time, and his ability as a conciliator made his presence extremely valuable. "New Thought," as exemplified by the controversy between science and religion, became a burning question among the students, while the Temperance movement reared a crusading head under the banner of the Blue Ribbon Pledge.

Background

We may perhaps regret the intolerance of those old Calvinists, but it was that very quality which made them the empire builders that they were. In the same way, their analytical soul-searchings frequently developed in them the qualifications for scientific research. Their rigid discipline was applied first to themselves, and they demanded definite standards of character and achievement of themselves as well as of others.

In my day, we did not think that society owed us a job. We went out and made one for ourselves. Today, when times have made so many men soft, the scientists still carry on with single-minded purpose.

Had I had the choice of a place to be born, a family to be born into, and an environment with which to surround myself, I could hardly have chosen more wisely than Fate chose for me in 1869 when I was born at Michigan State College. My parents belonged to a class to whom the intellectual future of this country meant more than anything else in the world. As a boy, I heard countless discussions in my father's house concerning some of the most fundamental changes which have taken place in the education of the youth of America; discussions among pioneers who have left their mark on much that is fine and splendid in the civilization of our country.

There were five of us children, and the setting of our childhood was quite ideal. The campus of the college was really only a clearing in the great Michigan forest and it contained all of the elements necessary to develop happy, healthy boys and girls. There was a brook teeming with water life; sugar maples to tap when the sap ran fresh in the spring; and great black walnut trees which stood here and there on the campus and were, in some way, my childhood companions. My earliest recollection is of myself as a toddler searching for their nuts in the frosty grass.

A country road passed behind the row of professors' houses where our playmates lived, and under it the brook had been diverted into a large culvert. Wonderfully mysterious was that dark tunnel with the icy stream flowing through it, and there we found lively crayfish and caddis fly larvæ among the pebbles. It took courage to crawl farther and farther into the dark culvert, but in return we were rewarded by experiencing the thrill of exploration.

Being the youngest of the group of boys and girls on the campus, I was, so to speak, dragged about by them, and bullied into doing utterly futile

3

things that robbed my early childhood of many impressions which, had I been more by myself, would have left deeper marks on my memory and personality.

There was one good thing, however; my childhood was more casual than that of most children today, who are forced into some form of regimented play. I can see no advantage (particularly to a young naturalist) in this over-organization of childhood. But then I am old-fashioned. I believe that a large amount of unorganized time is valuable in life.

One feature of our home life which might shock the good housewives of today was the trundle bed. My two brothers and I slept in one. It was called "trundle" because it was on wheels, and in the evening was pulled out from under my parents' large bed. Next morning it was trundled back again.

My father had a library which was never closed to us. We would spread the big volume of Wood's *Natural History* on the floor and lie on our tummies while we studied the many pictures. We used to play games of recognition, holding our hands over the descriptions of the animals while we guessed their names.

During the long winter evenings, my mother read to us Abbott's biographical histories, *Robinson Crusoe* and *Swiss Family Robinson,* and our daily play was much influenced by the romantic stories of people marooned on a tropical island. Sixty years later, when attempting to land in a small boat, I was upset in the surf on the very beach which Defoe selected as the scene of Robinson Crusoe's shipwreck. Alexander Selkirk's shipwreck on Juan Fernandez first gave Defoe the idea for his story, but, finding that this island was bleak and barren, he transferred his hero to the tropical island of Tobago in the West Indies, and there I landed in the foam.

The unfortunate sequel of this story is that, after visiting Tobago and other tropical islands, I tried to read *Robinson Crusoe* again, only to find it so fantastically exaggerated that I had to give it up.

Living as we did on the campus, we children grew up among the professors. Occasionally they even joined in our investigations of brook and forest.

W. J. Beal, a gentle Quaker, was Professor of Botany and possessed amazing patience and thoroughness. To determine their length of life, he buried seeds of cereals and vegetables in labelled bottles, digging them up every ten years to test their power of germination.

Occasionally one reads newspaper accounts of "mummy wheat" taken from Egyptian tombs three thousand years old. According to these tales, when it is planted today it not only grows but produces extraordinary yields. These stories are without foundation. Yet on some such yarn William Jennings Bryan based one of his most popular sermons on immortality. His daughter, Ruth Bryan Owen (now Mrs. Rohde), told me that the sermon was even reproduced on a Victrola record. Be that as it may, Professor Beal's prolonged experiment determined that grains of wheat lose their power of germination in less than twenty-five years, and many other seeds lose theirs in a much shorter length of time.

Professor Beal was a born investigator and was one of the first men to stress the advantage of planting only selected strains of trees and vegetables. It is hard to realize that the idea of plant selection was so novel sixty years ago that he wrote a paper on the subject for one of the early horticultural meetings in Michigan.

One of the most outstanding students at the College in those early days was Mother's brother, Byron D. Halsted, who later played an important part in my life. He came often to our house, and I remember his excitement over his new microscope as vividly as I remember the thrilling new world visible through its lens.

Uncle Byron left Michigan to work under Doctor William G. Farlow of Harvard, studying microscopic fungi. Doctor Farlow had but recently written the first accurate description of a destructive plant disease to be published in this country. His subject was the "Black Knot" of the plum and cherry, which has been largely responsible for the destruction of sweet-cherry orchards in America.

Another student who stimulated my imagination was a young Japanese named Tomari who came to Michigan all the way from Japan—an inconceivable distance to me then. On departing, he gave my father a set of crystal studs in a charming box of colored straw which fascinated me so much that I had to be forbidden to handle it. Twenty-five years later I visited Tomari in Japan and found him a learned professor, one of the founders of the agricultural school of Komaba.

Two other graduates of the Michigan State College travelled westward to the Land of the Rising Sun: Edward M. Shelton in 1870 and

Professor C. C. Georgesson in 1880. Shelton took with him pure-bred cattle for the Japanese Government—which were, I believe, the first blooded stock introduced into Japan—and organized an Animal Husbandry Department in the first agricultural college established there. Both Shelton and Georgesson returned to this country to become professors in the Kansas State College, and subsequently Georgesson went to Alaska and has performed signal service at the experimental station there.

Liberty Hyde Bailey was another of those early Michigan graduates. It would be hard to overestimate the influence of his publications on the development of intelligent gardening in this country. For over sixty years his stimulating and forceful personality has been outstanding in the horticultural world, and his *Standard Cyclopedia of Horticulture* has been an authority frequently consulted by all serious-minded gardeners.

Quite as great in his way was Eugene Davenport, whose influence created the University of Illinois, which is now a teeming center of agricultural research and education.

When the formative years of one's life are spent among men such as these, it is little wonder if one becomes "agricultural-minded." Personally, I cannot imagine existence in a family where the parents are interested only in a social life, but I feel sure that it would be very boring. Maybe that is why some boys get into hot water. When absorbed in the world of growing things, even a boy can be too busy to make trouble.

CHAPTER II

Kansas

W HEN I WAS ten years old, my father accepted the presidency of the little agricultural college in Kansas which had been founded in 1863.

I would give anything to feel again the excitement of those moments. The farewells, which meant so much to my mother, had little effect on me. I was all enthusiasm at the idea of emigrating to the "far West."

We children were fascinated by the Pullman car with its polished, inlaid woodwork, and we also enjoyed the difficulty of eating in the dining car, hard to appreciate nowadays when the roadbeds are smooth and it is no longer a perilous adventure to drink a cup of coffee.

Of course the greatest thrill of the journey was crossing the Mississippi River. In the moonlight the train crept onto the trestle, and we peered at the muddy waters through an intricate pattern of bridgework until it seemed as though the crossing would never end.

The second day we reached the little town of Manhattan, Kansas, which in 1879 was a cheerless waste of treeless, muddy streets lined with brick and wooden stores. After an unsatisfactory breakfast in the dirty little hotel, the family drove out across the prairies, through muddy sloughs where the bones of buffalo were bleaching in the sun. (After all, this was only three years after Custer's "Last Stand," and but ten years after the first Union Pacific "iron horse" ploughed its way west through Indian country.) Several miles from the town we saw the college—four stone buildings and the president's house.

At first my mother felt that our move was disastrous. I think that she nearly died of homesickness. When winter settled down, with temperatures of twenty below zero, and the wild winds howled around the little house and drifted the snow in upon our beds, it was almost impossible to keep warm. These winds were also hard to bear in March and April when they carried such clouds of dust from the ploughed fields that we could not see across the road.

However, my father found the situation extremely interesting. Here was a college of two hundred students so intent upon securing an education that many of them were working their way through, and most of them economized by cooking their own meals.

My father had, of course, brought with him his religious convictions, and we were soon surrounded by what we considered the only right atmosphere—all others being more or less criminal. I mean the religious atmosphere exemplified by the hymns of Moody and Sankey.

In moments of solitude, when I was milking the cows or herding the cattle, the "life hereafter" took preference over the actual life around me. The many religious songs which I knew by heart tended to make the actual world seem somehow inferior to the one beyond. Prayer meetings at the College occupied a prominent part in the activities. There were no dances, of course; the nearest approach to frivolity was an occasional "sociable," where boys and girls promenaded, chanting the "Dusty Miller."

It was all rather a man's world. There were many more men in the community than women, and every girl had a choice of escorts. So appreciative were we of our womenfolk that I never heard of a chaperon.

Up to this time, death had not entered into my life. No one whom I knew had passed beyond the veil, although I had been terribly shocked by the demise of an injured horse belonging to our neighbors. When the poor beast was led away and shot, I stole out to see it lying there so silent and still, and the sight haunted me for days.

It is curious that size should have so much to do with our emotions regarding death. I could casually squash flies on the windowpane, and kill snakes, frogs, and even squirrels, although these began to have emotional value. But the death of a living thing as large as a cat or rabbit and, of course, a dog, upsets the emotional centers of even the least sentimental of us. Certainly the death of that horse was a terrible incident to me.

Suddenly one of my playmates, a boy of my own age, broke his leg

Kansas

while riding in the buggy with his father. His foot slipped from the dashboard and caught in the wheel. It was a compound fracture, and our family physician shook his shaggy head as he said, "I fear that he cannot live." The boy's leg was amputated immediately. Later word came that gangrene had set in. And then the funeral.

To the medical profession of those days, a fracture which broke the skin, technically a compound fracture, meant almost certain death. Modern methods of disinfection were still unknown. In fact, it was not until seven years after this that I first heard the word "bacterium," when my classmate Swingle painted for me a world filled with bacteria, floating particles in the air, microscopic plants. Only those of us who lived before the days of bacteriology can realize what an amazing thought it seemed when first presented to the world.

We had arrived in Kansas in 1879, just after the grasshopper year, when the sun was darkened by clouds of grasshoppers which consumed everything green, even, so 'twas said, the green paint on the window frames. This plague left the farmers head over heels in debt to the eastern bankers. And although the grasshoppers had passed, other enemies had begun to appear, among them the chinch bug and the wheat smut, which between them threatened the wheat crop.

At the College, an experimental patch of alfalfa caused a stir because of its long-rooting habit and its resistance to drying winds and the cold of winter, but it was years before the introduction of alfalfa transformed eastern Kansas and the bottom lands of the whole Middle West into veritable gold mines.

In the orchard, I knew all the main varieties of apples. There were scores of them, but their names seemed no more difficult to remember than the names of my playmates. I was equally at home in the vineyard on the hillside, where I used to fill my hat with grapes—Catawbas, Concords, and Delawares.

The occasional groves of cottonwoods, box elders, and maples were a poor substitute for the great forests of Michigan. But their scarcity increased our interest in each individual tree, and gave us an appreciation of trees in general, in contrast to Michigan where trees were unhesitatingly cut down if they were in the way. The hours under those Kansas cottonwoods later gave me real understanding of the shelter-belt experiments in the Upper Mississippi Valley and Canadian prairies, and added zest to the quest which we made in China and Siberia for the hardiest trees and shrubs for the treeless wastes of the Dakotas.

The Kansas State College of Agriculture was one of the "Land Grant Colleges." In 1862 a bill known as The Morrill Act had been passed by Congress, and considerable tracts of public land were given to the various States for these institutions. Land over ten miles from a railroad was not considered valuable in those pre-automobile days, and more acreage was given the western States in consequence. Therefore, when those States prospered and the land increased in value, their "Land Grants" proved a great asset if properly handled. Kansas State College had been given a sizable grant, and my father was able to sell the land to good advantage and thus insure a substantial endowment.

However, it soon became evident that the colleges would need more funds for scientific experimental work, and, in 1887, an Act was proposed in Congress by William Henry Hatch, Congressman from Missouri, to provide Federal assistance for the agricultural colleges. This enterprising man had established the Bureau of Animal Industry in 1884, and due to him the Department of Agriculture gained its seat in the Cabinet in 1889.

However, when the Hatch Act was first proposed, it met with strong opposition from a powerful organization called "The Grange," composed of conservative and retroactive farmers who did not want their sons exposed to fancy "book learning." They feared that additional funds would be used by the colleges for classical "trimmings."

Father went to Washington and managed to conciliate the Grange by having inserted in the bill the stipulation that the appropriation must be used exclusively for experimental work in agriculture. With this guarantee, the Grange withdrew its objections, and the bill passed, creating Federal Agriculture Experiment Stations in connection with agricultural colleges.

The federal funds thus appropriated created many positions for research men and there was a demand for young fellows who had studied the new world revealed through the powerful microscopes of Carl Zeiss of Jena.

All my life it has seemed strange to me that the vast majority of human beings are content with only hearsay accounts of the wonders found "through the microscope." It is a breath-taking world, filled with myriads of strange and fascinating objects which the naked eye could never see. Any one who has never looked thus into the heart of a flower has not fully lived.

The human mind prefers something which it can recognize to something for which it has no name, and, whereas thousands of persons carry field glasses to bring horses, ships, or steeples close to them, only a few carry even the simplest pocket microscope. Yet a small microscope will reveal wonders a thousand times more thrilling than anything which Alice saw behind the looking-glass.

In the eighties, natural history societies sprang up here and there, and at our college the rather meagerly attended meetings were held in the chemical laboratory. They were eerie affairs as they occurred at night and the room was dimly lighted.

At one of the first meetings, a young classmate of mine, W. T. Swingle, amazed us by delivering a scholarly address on the fungi which cause such diseases as the wheat rust. It was an entirely new subject to us, and we sat spellbound while he presented his discourse. Entranced, we watched Swingle's long arms wave about and his piercing gray eyes dart from one to the other of us. Delivering an address was a praiseworthy feat for him, as he was easily embarrassed and inclined to stammer.

His subject was the life history of *Puccinia graminis,* and he spoke at length and with great detail in order to define clearly the distinction between the three types of spores or carriers of the fungus which he said caused the rust on wheat. The first type, the teleutospores, are the resting spores which carry the disease through the winter. These can be seen by the naked eye only when they are present in masses resembling amber streaks on the wheat stems. The second type, the uredospores, are the summer spores which form the orange powder on the leaves, known to every boy who has wandered through an infected wheat field.

But most amazing was Swingle's description of the third type, the æcidospores. These, he said, are found in the pretty little red spots which appear in early summer on the barberry leaves. However, they are not capable of growing on the barberry, and are carried by the wind from there to the wheat plant. The question then presented itself as to whether

The World Was My Garden

the wild grasses were also host plants for this stage of the rust disease, and what could be done about it.

It was fitting that such an occasion should give me my first insight into the great intellect of Swingle, who was to be my friend and associate for fifty years. This same subject has fascinated many other of my later friends, and involved me in its problems when searching for rust-resistant wheats and cereals. (A quest which deserves a book of its own.)

Mark Carleton was a classmate, too, and probably sat listening to Swingle that night. We would all have been astonished could we have realized that Carleton's name would go down in history as the man who explored the "Black Lands of Russia" and sent home the Durum wheats.

In fact, we were all to be actors and supers in a gigantic drama. Its importance has meant little to the press which sees news only in the startling headlines of the day, but I believe that history will evaluate our work more highly, for historians view life in terms of related instead of unrelated days.

At this time I was deep in the rather tedious discipline of quantitative analysis of feedstuffs, learning from Doctor J. T. Willard the importance of a fraction of a milligram and the difficulties of getting a beaker chemically clean. One day my father suggested my taking a course in Botany under Professor W. A. Kellerman, and this suggestion was a turning point in my life, for I soon discovered that the field of botanical observation was vastly more to my liking than the meticulous accuracy of Organic Chemistry.

Under the enthusiastic guidance of Professor Kellerman, I became immersed in the problem of tumbleweeds. The barbed-wire fences—rather a new invention in those days—were piled high in the autumn with a great variety of dried weeds which scattered their seeds as the continual winds rolled them across the prairies.

Doctor Kellerman's assistants were incredibly busy. We collected every leaf-spot and parasitic fungus which we could find, and a world formerly filled with innocuous green leaves suddenly became a place full of dangerous spots and discolored surfaces. Each speck was immediately under suspicion as a possible indication of some new leaf-spot fungus which, under the microscope, might reveal characters as interesting and definite as those of the tree from which the leaves had been picked.

Kansas

Gradually it dawned upon my young mind that I had stepped into the beginnings of a new science, the science of the diseases of plants—Plant Pathology as it later came to be called.

About this time, some military-minded Congressman conceived the idea that the Land Grant Colleges should be centers of military training, affiliated with West Point Military Academy. A graduate of West Point was therefore sent to teach military tactics to our farmer students and to drill them in the arts of infantry maneuver.

My father, although not a pacifist, was opposed in principle to anything militaristic. He objected to the presence of military uniforms on the campus, and forbade their use except during the hours of actual maneuvers. The college had no dormitories as yet, so we used an old stone barn in which to change our clothes before and after drill.

From this first prairie squad to skirmish in the meadows, came three major-generals of the Great War—Major-General Helmick; Major-General Coe, who was for many years a teacher of Mathematics at West Point; and Major-General James G. Harbord, who might have been Pershing's successor had he not decided, almost on the eve of his appointment, to become president of the Radio Corporation of America instead.

How long ago it all was! How far away war seemed one summer afternoon when Coe and I drove across the Kansas River! As I gathered sawfly larvæ in the scrub, he told me of his ambition to go to West Point.

And I can see Jim Harbord, librarian of the college library, seated at the typewriter cataloguing books and humming "The Gay Cavalier." I can hear him slam the typewriter and say in a decided voice,

"Dave, I am going to West Point!"

"How can you?" I remember saying. "Aren't you too old?"

"I don't care," he answered. "If I'm too old for West Point, I will enlist as a private." And he did.

In those days the ranks of the regular army were not considered the place for an educated man, so Jim had a hard time. But the Spanish War came ten years later, and in the Philippines unusual opportunities developed for advancement. Thus, in Manila, during the "Days of the Empire," these three cadets from the Kansas Agricultural College rose to positions of responsibility and later all three reached real prominence in the Great War. No other college, I believe, outside of West Point furnished as many major-generals for the conflict—the conflict which was supposedly fought to end war!

I earned my spending money in the carpenter shop, where I soon discovered that time flies fast when one's hands are occupied. It was my father's belief that education which does not include training of the hands as well as the head is very one-sided indeed. It is a great experience for a boy to work out a problem like the cutting and fitting of a door frame, where even a slight mismeasurement, or failure to saw to a knife-mark, spells disaster.

In those days the white-collar workers were by far the better paid. Through the years I have watched the scales tipping in favor of the worker who deals with concrete things, and have looked with skepticism upon the attempts man has made to clarify and evaluate the different kinds of labor in a world where such things as abstract values are well-nigh impossible to define.

Great was my excitement when I heard that the famous naturalist, Alfred Russel Wallace, was to lecture at the college during his visit to America.

When Wallace came he stayed at our house, and charmed us with his simplicity. At the same time, he overawed the audience at his lecture by his masterly presentation of the subject which was nearest to his heart, the theory of natural selection—known today as Darwinism.

Not many years after that, Wallace's *Malay Archipelago* was my guide-book to that vast region on the other side of the globe, and I visited Dobo in the Aru Islands, where Wallace spent the weeks of his illness. As he lay helpless in a native hut, in his brilliant mind were forming the conceptions of natural selection in the struggle for existence. These thoughts he later communicated to his friend Charles Darwin in a now historic letter which reached Darwin soon after he had reached the same conclusions himself. Darwin showed Wallace's letter to his friends and they urged him to publish his own conclusions with regard to evolution jointly with Wallace's letter—conclusions which Darwin had reached during his years of exploration as a naturalist on the *Beagle*. This Darwin did, and the observations of these two men have dramatically altered the trend of human thought.

Soon after my graduation in 1888, Uncle Byron Halsted brought his wife for a visit. By this time Uncle Byron was quite well known through his botanical and horticultural writings, and had become Professor of Botany in the Iowa State College of Agriculture at Ames.

The visit must have been a trying experience for Aunt Sue. The weather was exceedingly hot and there were few "things to do." She had travelled and was extremely musical, having the works of Chopin, Schubert, and Beethoven at her finger tips. Literature also appealed to her, and she had a particular love for Dickens.

Like many other religious people of the day, my father felt a reservation with regard to Dickens. He frequently deplored the undercurrent of caricature in Dickens' books, which he believed was not Christlike. Therefore, when my aunt proposed reading *David Copperfield* aloud to us youngsters, I knew that Father did not approve, but he did not know how to forbid it without offending his guests.

I was a prairie boy who had never seen the waves of the sea, nor heard the whistle of a ferryboat; Europe was as far away as the Mountains of the Moon; and music was limited to selections by the little church choir. Naturally, I was completely captivated by my young aunt from the East. In rapt attention, I absorbed the story of David Copperfield and believed it to be the greatest romance ever written.

Forty-seven years have passed and I have never been able to make myself read *David Copperfield* again. I feared to break the spell. However, recently the story appeared on the silver screen and I watched that delightful child growing before our eyes from childhood to manhood and understood (if one can understand these things) what a tremendous influence the reading of that book has had upon my life.

One day, as Uncle Byron and I walked together along the banks of Wild Cat Creek, he asked if I would care to go back with him and study in his laboratory in Iowa. I had always loved Uncle Byron and was overjoyed by the invitation, and my parents were delighted to see me started on a scientific career.

Uncle Byron's laboratory was full of fascinating paraphernalia in that disorder which seems inevitable in a busy laboratory. He was studying the contrivances which flowers develop to insure their cross-pollination by bees and other insects, and, like all great teachers, he fascinated his pupils by his own interest in a subject. The days were all too short, and the evenings were filled with my aunt's lovely music and books which Uncle

Byron read aloud to the assembled family. It was a delightful atmosphere, warm with friendship.

Iowa resembles Kansas, and its spectacular career as a farming state was then beginning. An experiment station had been started and Uncle Byron greatly enjoyed his work because the plant problems were all new. However, Aunt Sue was not particularly happy so far away from her New Jersey home, and was delighted when my uncle received a call to Rutgers College and its newly formed experiment station on the Raritan. When Uncle Byron suggested that I accompany him to Rutgers as a postgraduate student, I eagerly accepted.

New Brunswick, New Jersey, was the oldest town I had ever seen. Its Dutch architecture, open drains and sloping roofs, from which rain poured on to the street, all amazed me. The town seemed full of graveyards.

On an unforgettable day, Aunt Sue took me to see the ocean. Truly those who are born within the sound of the sea can have little idea of the profoundness, the overwhelming sensation which its immensity, its ceaseless motion, and its powerful breakers have upon a person who has never seen more than a ripple on a mill pond.

No coast of the Seven Seas ever impressed me as did my first sight of the Atlantic Ocean. First impressions always have meant much in my life. I wonder why so few people distinguish between their first sight of anything and the hundreds of later impressions which may follow in jumbled confusion.

In the laboratory I became deeply interested in a collection of plants which I was making, and I began to think of a future career. One day my uncle brought Beverly T. Galloway, of Washington, home to luncheon. This long-legged, thin man, with kindly brown eyes behind big spectacles, had a smile which won him friends everywhere. I already knew his name, having seen it in the agricultural journals.

I do not remember much about the luncheon, but before it ended he threw a bombshell into the family circle by inviting me to join his staff in Washington and become an employee of the Government at one thousand dollars a year.

Aunt Sue strongly opposed the move, for it interfered with her plans for me. She was very anxious for me to study in Europe before settling down anywhere. Uncle Byron tried to reassure her, explaining that mine would be a scientific position, associated with educated men, where I could continue my studies. From Kansas my father wrote urging me to accept the offer. So, in July 1889, after an emotional farewell at the New Brunswick station, I left home again, this time for the city of Washington, where I knew no one but Mr. Galloway.

As reading matter for the train, I had a highly technical description of a facultative parasite known as Macrosporium. This organism, under proper conditions, could cause the death of onion plants. My appointment in Washington was contingent on my passing a Civil Service examination, and I believe that I thought facultative parasitism would be a question in the examination.

CHAPTER III

I Enter Government Service

How splendid the broad avenues of Washington seemed after the narrow streets of New Brunswick! It was July, but after the heat of Kansas I did not mind torrid temperatures.

In 1889, all that existed of the Department of Agriculture was housed in an ugly old building with a mansard roof topping its red-brick walls. It was situated in a park south of Pennsylvania Avenue, just beyond one of the most disreputable quarters of the city.

I had no difficulty in finding Mr. Galloway. His new creation, the Section of Plant Pathology, was crowded into two or three rooms, one of which was an attic under the broiling mansard roof. The personnel consisted of Mr. Galloway and five assistants. Three worked with microscopes, and two were learning to write fairly presentable letters on the noisy typewriters of the time. The letters were then considerably blurred in wet-press copy books, carbon paper being as yet unknown. Even the White House sent out letters corrected in pen and ink.

The friendliness with which this little group greeted a nineteen-year-old boy comes back to me after all these years.

There was Erwin F. Smith, busy studying peach yellows. This disease eventually destroyed the delicious Crawford peaches of the eastern shore of Maryland, and has made all but some strains of Chinese peaches too difficult to grow with profit.

Merton B. Waite was there, fresh from the laboratory of Doctor Burrill of Illinois. Doctor Burrill discovered the first recognized bacterial disease of plants, proving that a rod-like, microscopic organism, *Bacillus amylovorus,* caused the destructive fire-blight of pear trees.

Mr. Galloway himself was particularly interested in the diseases which threatened the grape-growing regions of this country.

I Enter Government Service

I felt immediately at home, for I realized that these men were pioneers in the study of plant diseases, like those with whom I had been associated in the laboratories of Doctor Kellerman and Uncle Byron.

It was Saturday, and Mr. Galloway took me on the horse-cars to his little house in the sparsely settled suburbs. He had been married recently and was waiting for his wife to join him in the autumn. We sat down after lunch to discuss the work for which he had brought me to Washington. I think he recognized signs of homesickness in me which corresponded with his own mood. As evening fell, he strummed on his guitar and sang "Love's Old Sweet Song."

This was almost half a century ago, but I still beg him to sing the song again when we are together. We have been close friends during all these years, and we have seen the little Section of Plant Pathology become the center of a vast Bureau, filled with scientific men and women who have probably done more for the plant growers of the world than any other group of people.

I became a regular visitor at Mr. Galloway's house. One day Mrs. Galloway complained that some sweet potatoes were rotting in an appalling manner. I immediately set out to discover the trouble. After weeks of investigation, I found one of those facultative parasites, a mold—a beautiful mold (*Rhizopus nigricans*), one of the most interesting of the host of common molds which make their homes in fruits and roots. This study of the soft rot of the sweet potato awakened my interest in other diseases of this important food plant which can be so easily grown in southern climates, and is second to none in the quantity of starch and sugar it produces.

Notwithstanding my interest in the experiments of those first months in the Capital, I greatly missed the affectionate home life which I had had with Uncle Byron and Aunt Sue, and when my vacation came I spent it with them. I returned to Washington and, two weeks later, received a frantic telegram from Uncle Byron saying that Aunt Sue was dying of pneumonia.

The message reached me at my microscope and I ran for my bicycle, left it in a drug-store near the station, and caught the first train for New Brunswick. I arrived, terror-stricken, to find that she had gone two hours before, Aunt Sue whom I adored, leaving me conscious of a debt of gratitude which I shall never forget. She it was who brought into my life new ambition and a love for music and literature. Best of all, she had given me the necessary faith in myself, without which no mortal can go on.

For years the influence of Aunt Sue deepened the experiences of my life, and perhaps prevented me from caring for many of the more transient things which absorb the average young fellow. It was her influence which eventually made me break away from Washington to augment my meager education in the gardens and laboratories of Europe, because this was the ambitious dream which she had planted in my soul.

Throughout that winter I strove to bury sorrow in work, continuing my investigations of the various rots of the sweet potato. The colored plate and drawings I made were considered good enough to publish in the Year Book of the Department of Agriculture. I have them still.

The approach of spring in 1890 found me completing elaborate wall charts on starched cotton cloth, to be used in an illustrated lecture on plant diseases which I was to make before the Horticultural Society of Western New York. This would be my maiden effort, and I was so apprehensive that I lost a good deal of sleep over it, as I do 'to this day when faced by the necessity of making any public address.

After elaborate preparations, I left for Rochester, New York, with the roll of charts under my arm. On the way, I stopped overnight with my classmate, Ernest Fox Nichols, whose brilliant career as a physicist included the invention of an instrument for measuring the pressure of light from the fixed stars. Ted Nichols was later president of Dartmouth College, and had been elected president of the Massachusetts Institute of Technology not long before he died on the lecture platform while presenting a paper before the Academy of Science in Washington. Ted's career, like those of our collegemates, the three major-generals, reflects great credit on the little college on the plains of Kansas.

Ted and I talked far into the night, relying on his alarm clock to wake us in the morning, for I was to catch an early train. But the clock failed us, and a frightened young man arrived in Rochester barely in time for the lecture.

It was evident that the gray-bearded horticulturists were somewhat shocked when a mere youth appeared and began hanging a lot of charts on the platform. Possibly the charts were rather a shock too, for they contained Latin names and sketches of strange objects which were completely unknown to most of the audience. However, I had my facts pretty well in hand, and, although many in the audience doubted whether these "parasitic fungi" could have anything to do with mildews, rots, and black knot, some men there realized that I was telling them something new and important, and they kept me on my feet for an hour after my lecture was over, answering their questions.

Exhausted, I at last sat down, still only half realizing the rare situation which made it possible for so young a man to talk authoritatively about matters of such importance to horticulturists. It was indeed a new science which I represented.

After my return to Washington an S. O. S. came from the grape growers of western New York State. A curious disease had appeared in the vineyards, a disease which baffled us, but which today would probably be classed among the virus leaf diseases of plants. At that time, even the word "virus" had not been coined. It was only when the mosaic disease of the tobacco plant appeared that pathologists recognized the existence of a multitude of diseases caused by organisms too tiny to be visible under even the highest-powered microscopes.

I was ordered to Lockport, a center of the new Niagara grape industry, to demonstrate methods of spraying. Some of the earliest and most effective results obtained by Doctor Galloway and his staff were in the use of Bordeaux treatments as a control for diseases of the American grape.

The discovery of Bordeaux mixture is a romantic story in itself. In the French province of the Gironde it was the custom to spray the grapevines along the roadways with a brilliant blue mixture of bluestone (copper sulphate) and lime, in order to discourage thieves. The fear of being poisoned acted as a deterrent and also the color adhered to clothing and helped to convict the culprits. M. Millardet, Professor of Viticulture, noticed that vines thus sprayed were not attacked by the downy mildew, an American fungus which was then devastating the vineyards of his country. He thereupon developed Bordeaux mixture, which immediately became popular in France. A special nozzle, called a cyclone nozzle, was also invented to provide the proper spray.

Doctor Galloway and I had experimented in a vineyard in Virginia. The place was run in the careless manner of those days, and black rot had completely ruined the grapes for some years past. It was a hillside vineyard, ideal for our experiments. Both leaves and fruit were thoroughly infected with the characteristic spots of the microscopic fungus.

The World Was My Garden

To make a demonstration, we segregated the four corners of the vineyard, separating them from one another by untreated areas which crossed in the center.

Doctor Galloway and I had spent many days and nights devising an apparatus better suited for our purpose than the knapsack sprayer imported from France. I drew the sketches of our invention and made the models for the castings at a little foundry on Pennsylvania Avenue.

The experiment in the Virginia vineyard was a great success. I remember my amazement when I saw a full crop of fruit in the corner blocks which I had sprayed with Bordeaux mixture, less perfect crops on the corners sprayed with other mixtures, and a complete failure—a mere handful of grapes which had escaped the rot—on the Greek cross of unsprayed vines down the center. The success of this experiment added impetus to the enthusiasm which spread through the country, and Doctor Galloway skillfully made use of it to secure an addition to the modest appropriation of $35,000 a year allowed to the Section of Plant Pathology.

Doctor Peter Collier, one of the first American chemists trained in Germany, was director of the State Agricultural Experiment Station in Geneva, New York. He had transformed the lower stories of a large house into a laboratory and lived with his family on the top floor.

To this experiment station I was sent in both '91 and '92 to test the spraying methods which Doctor Galloway and P. H. Dorsett had found successful in the treatment of the leaf diseases of nursery stock. We hoped to demonstrate that Bordeaux and other poison mixtures could be so applied that they would protect seedling fruit trees from leaf fungi and thus prevent the falling of the leaves. The seedlings could then be budded successfully in midsummer. However, if the leaves fall, the bark becomes tight and buds cannot be inserted. The potash disappears from the leaves into the bark and stiffens it, for there is a complicated but fascinating side to the phenomena of leaf fall.

Nothing that I have ever done was more educational than the planning and carrying out of that experiment on the Geneva Station grounds. I had at my disposal thousands of little seedlings, and arranged them in rows, duplicating and triplicating my controls until I thought that I had eliminated all chance of error from the influence of the soil. But, when I began to check the results, I found untreated seedlings which had not dropped a leaf, and treated ones from which the leaves all fell in early summer. The hereditary natural resistance had been greater in the one case than the extreme susceptibility in the other, even when sprayed.

In the field of medical science, the advent of the guinea pig and white rat have discredited the "case history" methods of a generation ago which lacked the factor of controls. The treated patient got well, but where was the untreated one? Maybe that case recovered also. And how about the hereditary set-up of resistance? The value of identical twins as offering material for control in medical experimentation is just beginning to be appreciated.

At Geneva I reaped great benefits from the technical training which I had received from Merton B. Waite. I was very thankful for the hours I had spent helping him to cultivate fungi in test tubes on culture media made from the tissues of the diseased plant which he was investigating. In a corner of Doctor Collier's library I set up a little cubicle of boards and canvas, in which I could make my transfers and keep my sterilizer. There I carried on investigations of the various plant diseases which came tumbling in upon me from many parts of the State. It was new and exciting work, so exciting that even Miss Amy Collier's Beethoven Sonatas, floating down from the floor above, failed to distract me.

As I did not dance, play cards, or care much for society, I spent long hours in my little makeshift laboratory, or in my boarding house struggling with the German language.

During the summer, I received many calls for aid, accompanied by specimens of diseased plants. One day a letter arrived from a certain Mr. John Burroughs, of Esopus-on-the-Hudson. He sent a packet containing some currant stalks, with the statement that his currant bushes were slowly dying. I was delighted to find that the dark, sunken areas on the specimens contained a fungus which I succeeded in growing in my test tubes.

After a successful inoculation of some currant bushes on the station grounds, I longed to see the dying bushes of this Mr. Burroughs. Early one morning I arrived enthusiastically at his door but received such a chilly reception from Mrs. Burroughs that I turned to beat a hasty retreat. Fortunately, Mr. Burroughs arrived on the scene and so charmed me with his interest in the currants that I spent a delightful morning discussing plant diseases with him. He showed me his study, which he called "Slabsides," and I departed feeling that Mr. Burroughs of Esopus was one of the most interesting men I had ever met.

Upon my return to Washington, Doctor L. O. Howard, the entomologist, remarked in his teasing way,

"I hear you've been hobnobbing with John Burroughs."

"Do you mean the man at Esopus who has diseased currant bushes?" I inquired.

When Doctor Howard realized that I did not know of John Burroughs, and had never read his books, he repeated the story so often that I have never been able to outlive this joke on myself.

The following winter John Burroughs, on a visit to Washington, took the trouble to look me up. As he greeted me he said,

"You probably don't remember me. I'm the man at Esopus with the dying currant bushes."

Later, I not only came to know him better but read and enjoyed all that he had written.

There were strange contrasts in the little microcosm which then constituted the Department of Agriculture. Some of the men seemed to be relics of a former era, still unaware of the tremendous strides which had taken place in the use of the microscope. In vivid contrast to these fossils, were such men as Theobold Smith, whose laboratory adjoined mine up under the old mansard roof.

Late one afternoon, long after most of the Department had gone home, I heard Theobold Smith's light step behind me and his enthusiastic voice calling, "Fairchild, would you like to see the cause of Texas cattle fever?" After months of work, he had just discovered the parasite in a drop of steer's blood which he had taken from a cattle tick. It was a momentous discovery, the first of its kind.

I had heard much about the terrific losses of cattle on the plains. Whenever herds of domestic cattle were driven from Texas to the slaughterhouses in Chicago and Kansas City, they died by the hundreds if their paths happened to cross a trail made by the longhorn Texas cattle of the plains. Apparently the native Texas cattle were not susceptible to the fever themselves, but were passing it on somehow to their less fortunate brethren.

From his investigations, Theobold Smith had felt confident that ticks must be the carriers of the disease, but he could not be positive until the day when he actually found the microscopic parasite in a drop of steer's blood extracted from a Texas cattle tick.

The idea that an insect could convey a blood disease was a new one, and Theobold Smith's conception and proof of the theory startled the scientific workers of zoölogy. It was a great privilege to have known him and to have heard from his own lips the account of this important discovery in the field of medicine.

The idea of insect carriers, "vectors" as they are now called, made a tremendous impression on me. Three years later, I asked the research man at the new bacteriological laboratory in Java why he did not study the mosquito. He was very proud of his method for collecting dew on chilled tumblers set out in the evening in the low swamps around Batavia. He believed that, by using the new bateriological methods of Koch, he would find in this dew the cause of malaria—an imagined bacterium of some kind. He shrugged his shoulders at my account of Theobold Smith's discovery and gave me a hundred reasons why mosquitoes could not possibly carry malaria.

Again, in 1898 I was fleeing at sundown from Rio, as all foreigners did in those days, to escape from the yellow-fever area. On the ferry-boat crossing the harbor, I met a young German who was studying mosquitoes. He listened with the keenest interest to my story of the cattle tick. He had noticed that the mosquitoes in the town were particularly active about dusk, and he had vaguely begun to suspect that they might transmit the disease in some way.

However, it was not until many years afterwards that I met the real discoverer of the carrier of malaria. I dined with Sir Ronald Ross in Ceylon, and heard him describe the thrilling moment when the thought first came to him that certain glistening bodies in the salivary glands of the mosquito might be forms of the malarial parasite. Theobold Smith's discovery antedates that of Sir Ronald Ross, but in no way detracts from its greatness.

The momentous consequences which resulted from the work of these men have always seemed to me among the greatest miracles of science—miracles which confuse and confound the flat theories and dogmatic pronouncements of the obvious-minded. I sometimes wonder whether the marble palaces of research, which today line the Mall in Washington, will ever produce any greater or more profound discovery than this one which Theobold Smith made in the broiling attic of that old brick building way back in the ignorant nineties. Men, ideas, tools and quiet—not desks and the noisy machinery of correspondence—are the building stones of scientific discovery.

That winter, Waite and I lived in a boarding house just off Fourteenth Street. Our evenings were passed in study, and we vied with one another in buying scientific handbooks with which to fill our shelves. Neither of us could compete with young Swingle, my old classmate from Kansas, who had recently joined the force bringing with him a library in five or six languages. In fact, he started a perfect frenzy of enthusiasm among us all to acquire any language which had literature on plant pathology.

Waite's profound research into the nature and cause of the fire blight of the pear was spectacular and had far-reaching results. He had an incubator full of the pear blight organism, *Bacillus amylovorus,* and could produce the blight at will by dipping a needle into the culture and inserting it into the growing tip of a pear branch. He believed that bees carried the blight from infected flowers to healthy ones. Doctor Maxwell, a little country doctor of Still Pond, Maryland, challenged Waite to prove this on his trees. The results were disastrous! A short time after Waite inoculated the flowers on a few trees, Doctor Maxwell suddenly realized that the bees had spread the blight all over his orchard. He sent a frantic telegram to Washington, but there was little that could be done, for the disease had made such headway that Waite could not stop it.

Waite next went South and there made another far-reaching discovery. In an immense orchard of Bartlett pears, the trees mysteriously failed to bear fruit. Waite managed to solve the problem, and returned to Washington in a great state of excitement. His discovery was that the Bartlett pear flower is practically sterile to its own pollen. Hence he found that the only fruit was on trees along the outer edge of the big orchard. He interpreted this phenomenon as indicating that the bees from near-by orchards of other varieties had brought foreign pollen and pollinated the first trees they came to. Basing his experiments on this assumption he proved that it was correct.

It was a very real discovery, a precursor of what has now become a generally accepted principle of horticulture, the principle of mixed plantings.

Walter T. Swingle in his attic laboratory, engrossed in studies of plant diseases.

Personnel of the Section of Plant Pathology in the early 90's:

| Joseph James | Theodore Holm | Merton B. Waite | P. Howard Dorsett |
| | David Fairchild | Beverly T. Galloway | Walter T. Swingle |

Merton B. Waite discovered that the Fire Blight of the pear is spread by bees, and that the Bartlett pear is sterile to its own pollen.

Beverly T. Galloway, Chief of the Section of Plant Pathology, in his attic laboratory in the old Department of Agriculture Building in 1893.

I Enter Government Service

When Waite returned from this Southern orchard, I was about to leave for my second season in Geneva, New York, and undertook to make some bagging experiments for him. He wanted to find out whether clusters of blossoms could pollinate themselves if enclosed in paper bags to exclude all bees and other insects, and, if so, whether they were able to produce fruit.

The first weeks of that summer were hectic ones, as I busied myself feverishly tying thousands of bags on pear trees around Geneva.

The results proved quite conclusively that the Bartlett was sterile to its own pollen. Their bags contained only dried blossoms. With some other varieties this was not the case, and the bags contained young pears, although definitely different from those produced on the same trees without the bags. The fruit in the bags was not only different in form but also, as a general rule, was seedless.

I fear that I rather annoyed dear Doctor Collier that summer when I proposed to keep artificial rain pouring over one side of a plum tree as it came into bloom, leaving the other side exposed to the natural conditions of drought and rainfall of the region. The rôle which rainfall plays in the orchard was suspected, and, to my delight, my experiment corroborated the general belief that exceptionally heavy rainfall during the blooming period will prevent the proper fertilization of the flowers and the setting of the fruit. Further than this, heavy rain may cause such fruit as does set to rot.

In the fall of 1892, with a suitcase full of photographs and records, I returned enthusiastically to the Washington laboratory under the rafters. There I found that P. H. Dorsett, a classmate of Doctor Galloway, had joined our little group. During the ensuing winter we embarked upon a perfect whirlwind of interesting activities. An atmosphere of scientific Bohemianism prevailed among us, and we organized a Botanical Society which met at each member's quarters in turn. The meetings always ended with some original collation which, although temperate, was generally a real test of one's digestive abilities.

About this time, people from all over the country began sending specimens to us for identification and information. The steadily increasing flood of correspondence quite overwhelmed us and soon necessitated an additional number of clerks.

The World Was My Garden

Among the books in our library was one which contained beautiful illustrations of the spice trees of the Malay Archipelago. An ambition to go to Java grew in me and undoubtedly had its inception in the pages of this book. I had saved what seemed to me a great deal of money out of my salary, and I wrote to Java asking how cheaply I could live in Buitenzorg. However, the prices were entirely out of my range, and reluctantly I gave up the idea for the time being. Instead, I had a long talk with C. Wardell Stiles, a brilliant young zoölogist just back from Germany. At his suggestion I applied for the Smithsonian Table in the then famous Zoölogical Station at Naples. I was anxious to find a means of studying abroad, and this would be the equivalent of a scholarship if I could secure it.

At this moment there opened before us a great opportunity to publicize the results of our experiments. The plans for the Chicago World's Fair were announced and an appropriation was made by Congress for an Agricultural exhibit of colossal dimensions. We immediately decided to demonstrate the various plant diseases, what damage they were doing, and what we had done to control them. As exhibits, we prepared bottles of fungicides, wax models of diseased leaves and fruits, herbarium specimens, carefully worded labels, and a set of microscope slides. We also planted a box with tiny pear seedlings. These we planned to inoculate with the germs of pear blight so that they would die before the very eyes of the visitors at the Exposition. Beside this, there was to be a revolving case of shelves in which to exhibit fresh specimens of the whole range of plant diseases as we then knew them. Accompanied by an assistant, I was sent to Chicago to set up the exhibit.

The Chicago World's Fair of 1893 was the greatest dream-city man had ever built. It burst upon the consciousness of Middle Westerners with a force which it is impossible for the generation of today to understand. There was comparatively little foreign travel then, and few had seen the beauties of European architecture. Electric lighting was new, and the lagoons with gondolas, new electrically driven motorboats, and magnificent fountains lighted by colored electric lights, were all bewildering.

The exhausting activity of those three months at the Fair nearly broke my health. I arrived in early spring. Great sheets of ice had been floored over under the Agricultural Building in the rush to complete the building for the opening. Bitter winds from the Lake swept through the open doors. I stood day after day in the cold building, explaining plant diseases to a crowd of visitors, depleting my strength and enthusiasm to a degree which I had not dreamed possible. An added strain was my determination to see every exhibit in the vast array during my free time.

Our exhibit at the Fair attracted a great deal of attention. There were always an extraordinary number of people there whenever I demonstarted the way pear blight killed one seedling after another in our box of starved-looking little pear trees, etiolated by the lack of sunshine.

About the time I went to Chicago, I began keeping little red notebooks (a habit which I still continue). Those first ones are full of names and addresses of interesting people who chatted with me at the Agricultural exhibit. I recall in particular the Swedish plant-breeder, Hjalmar Nilsson, who was one of the first men to realize that a field of barley is not just a field of barley, but is instead a collection of widely different, individual barley plants. In consequence, a field contains many weaklings, some rather mediocre plants, and here and there unusual plants with possibilities in the way of production, strength of straw, and time of ripening. Until our day, this idea had not been realized by any of the barley growers who had cultivated this cereal from the time of its domestication in Asia Minor thousands of years ago.

While I was at the Chicago Fair, Doctor Stiles secured for me the Smithsonian "working table" at the Naples Zoölogical Station, which was the outstanding place of its kind in the world. In conception it resembled the present summer Biological Laboratory at Woods Hole, Massachusetts. The allotment of a table carried with it laboratory equipment and all expenses involved in experimentation.

CHAPTER IV

I Meet Barbour Lathrop and Reach Naples

I HAVE NEVER quite understood how I had the courage to resign from the Department. All the older men, including the Assistant Secretary of Agriculture, advised me strongly not to do so. They told me of young men who had gone to Europe and returned filled with newfangled scientific notions which were of little use in practical American life. However, I always remembered Aunt Sue's desire that I should study abroad, and in the end it was her ambition for me which proved the deciding factor.

Once the matter was settled, great was my excitement at the thought of going to Europe. The North German Lloyd had recently inaugurated their first Mediterranean cruises, and I secured an inside room on one of the largest steamers afloat, the *Fulda,* a vessel of 7000 tons.

Finally, in November, 1893, I set out on a stormy voyage across the Atlantic, the roughest and most uncomfortable of my life. The gale blew us south through the Azores, swept the decks, wrecked all sorts of things; even the table racks were torn loose and the dishes smashed on the floor.

When I wrote my mother a lengthy letter describing this eventful crossing, it did not occur to me to mention an incident which proved one of the most momentous of my whole life.

The storm eventually subsided, and I amused myself by photographing the gigantic waves breaking over the bow. One morning I was amazed to see a man standing in the doorway of the Second Officer's cabin, clad in wrapper and pajamas. It was the first pair of pajamas I had ever beheld; heretofore, every one I had seen attired for slumber had worn a nightshirt. Somewhat stunned, I stared at this tall stranger in open-mouthed astonishment. I realized that he must be the important occupant of the Second Officer's cabin, the star passenger of the ship and a friend of the Captain. He was very good looking, standing there in the sunlight, quite the distinguished man of the world. It did not occur to me that I might ever meet him; still less did I dream that we would become intimate companions and friends for life.

Professor Raphael Pumpelly, the great geologist of Harvard, was also on board. During one of our chats, I told him of my longing to go to Java. The Professor said there was a Mr. Barbour Lathrop on board who had been in that fascinating, far-away place, and added that he would introduce me if I would accompany him to the smoking room. Thus came about my meeting with the man who was to "direct my destiny."

We stepped over the threshold into the tiny smoking room, and there I beheld the handsome, rather severe-looking gentleman of the pajamas. This time, however, he was formally attired. He was seated as I was to see him so many hundred times, with a novel in one hand and a cigarette-holder in the other. He received me cordially enough, and seemed interested that a young man should want to hear about Java. He recounted some of his early experiences there, including a rhinoceros hunt in the western part of the island, where he had come down with a fever and had been very ill.

I remember his telling us that he considered the three most beautiful scenes in the world, the view from the hotel in Buitenzorg, Java; the view from Damascus; and the harbor of Rio. He had already been eighteen times around the world, so I felt that his opinion was worth hearing.

During the conversation, he would pause, blow the stub of his cigarette out of its Turkish holder, take a fresh bit of cotton from his case, tamp it down into the holder with his pencil and, with the precision which comes from long habit, fit a fresh Egyptian cigarette into the holder and light it. This performance was as much a part of him as his walk or voice.

When I told him that I had been forced to give up my idea of going to Java because it was too expensive, he said, "Why don't you collect plant specimens for the Smithsonian Institution and pay for your trip that way?"

The World Was My Garden

I explained that I was not that kind of a botanist, and that if I went I wanted to study living tropical plants in the Botanic Gardens, not collect herbarium specimens.

The interview was rather a short one, and I left quite awed, feeling that I had met one of the most widely travelled men in the world.

When the *Fulda* neared Gibraltar, a committee arranged the usual farewell concert. Mr. Barbour Lathrop, of course, was master of ceremonies. Ladies in low-necked dresses and gentlemen in Tuxedos filled the dining-room. Having no dress clothes, I took refuge behind one of the pillars of the saloon and from there enjoyed Mr. Lathrop's amusing introduction of each performer. I thought him the wittiest man I had ever heard.

In the confusion of my first landing on foreign soil and the fascination of Gibraltar, the first old town which I had ever seen, I soon forgot Mr. Lathrop. He left the ship there to visit the Foreign Legion which was in action somewhere in Morocco. Naturally, I never expected to see him again.

On the boat I had also made the acquaintance of a delightful Italian portrait painter, Signor Mariotti, and his wife. They were returning from the World's Fair, where he had exhibited some of his pictures. Signora Mariotti, a Boston woman, was a great admirer of everything Italian. She was in immediate sympathy with my desire to increase my slight knowledge of the language acquired at the Berlitz School in Washington, and suggested that I live in an Italian family. She thought that her brother-in-law in Naples would accept me as a boarder.

In those days, Italy swarmed with pestiferous ragamuffins who made the lives of visiting tourists miserable, fighting with each other over every piece of baggage. In this respect Naples was beyond words. The high-pitched, screaming voices, the continual gesticulations, and the insistent demands for more money, made it a nightmare to me. I threw myself helplessly on the Mariottis' hands, and was thankful to climb up with the driver of the little carrozza and drive off with them through the narrow streets to the apartment of Signor Mariotti's brother.

As we dined in a little café with a stone floor covered with sawdust, I had a chance to become acquainted with the little railroad functionnaire into whose family I was coming as a boarder.

My diminutive bedroom looked toward Vesuvius. There had been a light fall of snow on the volcano and Vesuvius was in slight eruption. At night, a stream of living red ran down the white cone. It was a soul-stirring sight for a boy from the plains of Kansas.

Although the view from my window might be fascinating, my breakfast the next morning was disappointing. I found the hard bread and coffee with goat's milk most uninspiring. The spoon and knife impressed me as having been only sketchily washed. On the other hand, I was glad to be settled, my expenses were well within my means, and I would not hear anything but Italian spoken about me. Furthermore, the eldest son took a fancy to me. He and I used to wander around the streets at night and visit marvellous shows where "Due soldi" was the modest price of admission.

The little theatres, ringed round with boxes, were miniature affairs, but oh, what tragedies they portrayed amid a torrent of language and marvellous gesticulation! In one play, which I have never been able to forget, the thirteen characters were done away with in quick succession as scene followed scene. Each succumbed to a more ghastly and violent death than the preceding one, until eventually only the hero and heroine remained. Then, as a grand finale, the lover stabbed his sweetheart and hurled himself into the sea!

A dirty little street car, drawn by pathetic horses, took me from the railroad station to the Stazione Zoologica, an impressive white building in the Villa Nazionale.

I presented my papers to Doctor Anton Dohrn, to whose farseeing mind the conception and creation of the Biological Station were due. In this international rendezvous, scientific men were welcomed and provided with facilities for the study of the great biological mysteries of life. The laboratories contained the best equipment obtainable, and scientific men from the great universities of Europe worked there in monastic peace.

Into this environment I arrived, a youth of very fragmentary education, coming as the first representative of the Smithsonian Institution, to occupy a reasearch table and work on equal footing with these great European savants. Fortunately, both Professor Dohrn and his associate, Paul Mayer, seemed willing to overlook my lack of training, and I soon found myself installed under the supervision of Doctor Mayer, whose devotion to science was a real inspiration. Science was his religion—his very life.

His contributions and discoveries in the field of microscopic stains were so fundamental that today his carmine alum stain is found in every biological laboratory in the world.

I loved Doctor Mayer the moment I saw him and therefore readily accepted his suggestion that I study the karyokinesis of the nuclei of a strange, deep-sea alga which grew in places on the sea bottom of the Bay of Naples. These were the early days of the study of the living cell —the nucleus of the cell in particular with its chromosomes and centrosomes. The entire, fascinating phenomenon of nuclear division had recently been given the name of karyokinesis.

Lo Bianco, the head boatman of the Station, dredged near the island of Capri in many fathoms of water for the little green balloons, iridescent in the sunlight, which he put in the tank beside my chair. *Valonia utricularis* was their name, and for five months, for many hours a day, I turned the highest objective of my microscope on the nuclei-filled protoplasm of this fascinating denizen of the sea.

Unlike almost all other plants, Valonia is composed of a single mass of protoplasm spread out in a thin layer inside a tough-walled balloon the size of a man's thumb. Scattered through this protoplasm, without a semblance of cell walls to separate them, were thousands of nuclei, seemingly as independent of each other as are those which compose the root-tip of an onion—although those of the onion are separated by cell walls, as are the tissues of most living substances.

Using bits of woolen thread, Paul Mayer demonstrated for me the selective action of his newly discovered carmine alum stain. With the help of the stain, I proceeded to study the naked protoplasm filled with nuclei, hoping that the alum would stain the nuclei, and that I should discover something new and interesting with regard to their method of division. Despite my anxiety to contribute to the new accumulation of knowledge about the organization of cells, I did not prove much, for the nuclei were small and there were many chromosomes. What were then considered to be the most important organs of the cell, the centrosomes, were beyond the visibility of my microscope. Nevertheless, I spent happy months living, in my imagination, inside those fascinating little seaweed balloons, and I gained an insight into the mechanism of heredity and the beginnings of individual life.

Also, Naples in the nineties was an interesting place outside as well as inside the Biological Station. There are no more picturesque spots in all the world than Capri and Posilipo, while Vesuvius with the buried cities of Pompeii and Herculaneum at its base, the museums of magnificent bronzes and other archeologic treasures make the region of Naples utterly fascinating.

Conscientiousness has its disadvantage at times, and my absorption in my studies kept me indoors when the street-cries, the heavenly blue of water and sky, and the outline of Capri across the bay offered constant temptation to "play hooky," as the boys would say.

One day, after I had been there about a month, Michaele brought in two calling cards. I believe that I have them still. Engraved on them were the names of Barbour Lathrop of Chicago, and Professor Raphael Pumpelly, of Cambridge, Massachusetts. That these distinguished men were calling upon me quite set me in a whirl. I hurriedly threw some rotting seaweed into the wastebasket, straightened up the room a bit, and met them on the stairs.

I was impressed anew by their striking personalities—Professor Pumpelly with his long, flowing beard, erect stature and piercing eyes, and Mr. Lathrop, almost military in appearance, with gray mustache and hair parted up the back as was the custom in those days.

"The Professor and I thought we'd come to see how you were getting along, Fairchild," said Mr. Lathrop. "I've been to the Aquarium below here many times, but never knew there was anything like this in the upper story."

My guests seated themselves and Mr. Lathrop unlimbered the paraphernalia of his cigarette-holder.

"One is allowed to smoke here, I suppose?" he said, and then inquired what I was doing.

"I am studying the karyokinesis in Valonia," I replied, throwing at him the newest words in my vocabulary.

"The what?" he demanded.

I tried to explain the deep significance of my search for the nuclei in the protoplasm of my little green seaweed from the Bay of Naples. I could see that my explanation did not impress him.

"You told me on the boat that you wanted to go to Java. Have you given up that idea?" he continued.

I told him yes, I had given it up, for it was beyond my means.

"Well," he said, "I've decided to give you a thousand dollars with which to go to Java. I want you to understand that I look upon this thousand dollars as an investment, nothing more. I have had you looked up, and you seem to be all right, and when you are ready for the money I will send you a letter of credit for that amount. This is no personal matter; I want that clearly understood. It is merely my idea of making an investment in Science."

And then, even before I had time to thank him, he rose and said, "I'd like to have you lunch with me at the Tivoli Hotel at one o'clock. There will be a pretty girl there, too, I warn you."

I do not know what I said. There is no record of it in my notes of the time, although I recorded his conversation. Uncle Byron and Aunt Sue had once given me fifty dollars as a Christmas present and I thought that I had fallen heir to a fortune. A gift of one thousand dollars sounded too fantastic to be true, and the assurance of seeing Java seemed like a fairy tale. My head was in a whirl.

I was still in pretty much of a haze when I joined Mr. Lathrop at luncheon, and sat beside a pretty girl with red hair, with whom I did not get on very well. After lunch I drove with Mr. Lathrop to the great Don Carlos Theatre to get tickets for the evening performance of "Carmen."

"I suppose you speak some Italian, don't you? You can buy the tickets," he said.

He soon discovered that I did not know the difference between orchestra seats, box seats, and seats in the balcony.

"Your Italian doesn't amount to much, Fairchild. I can't speak the language but I can buy tickets for the theatre," and he swept me aside, much to my mortification.

The show began at nine o'clock and ended at two in the morning. Mr. Lathrop of course knew the story. For my part, I found my fragmentary Italian of no assistance in following the plot. Besides, I was overawed and frightened by Mr. Lathrop's dynamic personality.

We drove back to the hotel in the moonlight while Mr. Lathrop recounted his experiences in Naples during his previous visits. When we arrived at the little street which led up to my lodgings, we stopped. I thanked him and said good-bye.

I Meet Barbour Lathrop and Reach Naples

He waited for a moment, and then said, "Well, don't you think you'd better have my address?"

Confused by my stupidity, I took out my little red notebook. Of course I must know how to write to him when I was ready to go to Java! And there in the moonlight I inscribed on the back cover,

> Barbour Lathrop, Esq.,
> Bank of Scotland,
> 19 Bishopsgate Street Within,
> London, E. C.

I looked at the faded writing again just the other day.

Then he said, "Good-bye, Fairchild," and called "Avanti" to the coachman. With a crack of the whip they disappeared into the night.

The happenings of that day passed so swiftly, and proved so dramatic, that they left me dazed. I was not as yet accustomed to the sudden occurrences in life which completely change the course of events, and it was some time before I could concentrate on my Valonia again. But, after a bit, the influence of scientific men around me, and their deep knowledge, once more held me enthralled. More important still, associating with these men convinced me of the abysmal depths of my own ignorance. How foolish I would be to go to Java without a better training! I must acquaint myself in some degree with the great botanical laboratories of Germany before journeying farther afield.

I had realized from Mr. Lathrop's conversation that he expected me to sail for Java in the near future, and, as the months passed, I received brief letters from him which indicated as much. The postmarks of these letters fascinated me: Singapore, Hongkong, Yokohama, Hawaii, even Cape Town. He was always on the move.

I wrote to Mr. Lathrop and explained my feeling of inadequacy. It would not be fair to accept his money until my scientific knowledge warranted his "investment" in me. Consequently, I decided to spend the next two years preparing myself for the promised trip to the East Indies.

My associates in the Stazione Zoologica were zoölogists, with the exception of two botanists, Doctor Golenkin from St. Petersburg, and Doc-

tor Kline from Budapest. Also, they were all Europeans until William
Morton Wheeler arrived. Wheeler was already known as a distinguished American investigator in the field of cellular morphology and
cytology.

Thomas Huxley's brilliant assistant, young Moore, was at Naples
hunting for centrosomes in the sperm-cells of the dogfish. He was an
erratic fellow. I remember his telling me that he had climbed the Marble Arch during a royal procession and spat upon Queen Victoria when
she passed below. He seemed rather proud of this, which was a shock-
ing offense to my way of thinking. His quarters were across the funi-
colare from mine, and on the occasion of my first invitation to dinner
at the Meuricoffres (the leading bankers of Naples), Moore loaned me a
pair of what he called his "down belows" and a bandana handkerchief
with which to hold them up. All evening in the brilliantly lighted salon,
he took delight in calling attention to the fact that my trousers were
coming down.

Years later, when I visited him at the School of Science in London, he
told me two anecdotes about Huxley and Darwin which illustrate the
difference between the personalities of those two intellectual giants.

The elevator in the School of Science had a habit of sticking between
floors. The first day Moore was there, the corridors resounded with loud
and violent profanity. Seeing his amazement, the watchman said, "Oh,
that's just Professor Huxley stuck in the lift."

The contrasting story was of a day when Moore and Howell, another
of Huxley's assistants, were sitting in the laboratory, feet on the table.
Some one rapped gently on the door.

"Come in," said Howell. No one appeared, but another timid rap was
heard.

"It's that stupid servant who washes the glassware. Come in!" yelled
Howell, louder than before. Still nothing happened.

"Damn it," Howell shouted, "Can't you come in without my opening
the door for you?"

At that, the door timidly opened and there stood Charles Darwin, the
greatest scientist in England.

"Is Professor Huxley here?" he asked gently, making no reference to
his rude reception, to the immense discomfiture of the two young men.

Poor Moore! His last letters were from the Mountains of the Moon, near the headwaters of the Nile, where he went to study the strange fishes in the tropical lakes of that region.

My friendship with William Morton Wheeler, the other American at the Biological Station, continued until his death and grew with the years. But in those early days of our acquaintance, he suspected me, as a product of the Western plains, of knowing little about marine biology. One day in Naples he put me in my place quite brutally when I mistook some tiny rotifers, which had gorged themselves on chlorophyll granules, for swarm spores of my pet Valonia.

"Oh, rats!" he said. And I fled in disgrace.

But Wheeler's teaching instinct was too great to allow him to remain scornful. The next day he called me in to show me a sea urchin's egg lying in a watch crystal under his microscope, and explained the phenomenon of fertilization in the life of the sea urchin. It was an enormous egg surrounded by the flagellated swimming sperms, a sight which every schoolchild should see, the beginning of an individual existence, a marvel which never fails to thrill me to this day.

Wheeler knew Dante almost by heart and also introduced me to the melancholy poems of Count Giacomo Leopardi, noted for their perfection of style. We climbed Vesuvius together, and frequented the curious cafés which used to line the Via Roma in those days.

A true spirit of research and peace presided over the Stazione Zoologica. We had a complete library, efficient servants, and a devoted staff. Indeed, it was a scientific monastery, the only one in which it has been my good fortune to occupy a cell. No click of typewriters or noisy conversation disturbed the quiet of the laboratories.

My study of Valonia was taking me a long way from Agriculture; so for relaxation I used to visit the Botanical Gardens, and take notes on the many subtropical plants. Naples also introduced me to a fascinating new menu. For luncheon I would order dried figs, the overripe fruits of "il sorbo" (*Sorbus domestica*), and vermicelli with "pomo doro." It is hard to realize that in those days a dish of spaghetti was rarely seen on an American table.

I collected over sixty forms of "pastas" (products made from hard wheats) and sent them to my boarding house in Washington. When I returned there I demonstrated to the "boys" how macaroni should be cooked and served, and the difference between macaroni, spaghetti, and vermicelli.

39

The World Was My Garden

Too close application to my microscope, too many hours searching for dividing nuclei, brought on a state of nervousness which temporarily forced me to abandon the quest.

Harassed by the specter of a "nervous breakdown," I took to the hills back of Naples and found my way to Camaldoli, where, in the monastery, were many white-robed Capuchin monks. Soon I became quite well acquainted with one of them, Brother Paolo. The peace and quiet of the place appealed to me, as did the thought that this old monk had chosen to spend his days in such a lovely spot.

Eventually I became so captivated by the monastery that I asked Brother Paolo how I might leave the world behind and enter the Order of the Capuchins. The old monk seemed to like me and was much interested in my conversion until I inquired whether I might bring my microscope and continue my search for the secrets of life buried in living substances. He shook his head sadly, and said that he feared this would not be possible. Just why, he did not know.

He showed me his own cell with its narrow, hard bench, tiny wash basin, and dirt floor, and said quite simply that he felt it was too luxurious. He contemplated going to some other monastery where he would have to sleep upon the floor. Little by little I became analytical, critical, and puzzled by his motives. The monastic life began to seem only sentimental asceticism, and my enthusiasm cooled from white heat to zero. Why become just another monk in a monastery?

One afternoon as I stood in deep conversation with the old, white-robed monk, he said, "Here comes the Father Superior."

A younger man was approaching. He lacked the charm of Brother Paolo—I could never lose my heart to him. When he climbed the narrow trail and stood beside us, he opened his hand and in it was a little song bird which he had killed. With the ardor of a connoisseur, he felt its tiny breast, and I realized with horror that his mind was absorbed by thoughts of how the bird would taste "en casserole."

A feeling of revulsion swept over me. The thought of these white-robed monks aiding in the extermination of the song birds of Italy was too much for me. I turned and left Camaldoli and have never cared to see the place again.

CHAPTER V

Breslau, Berlin, and Bonn

THE WEEKS PASSED swiftly, and, as I planned to matriculate for the summer semester in the University of Breslau, I must leave Naples.

My route lay through Rome, and I spent a day of sight-seeing with the Mariottis, my friends of the steamer.

Being completely ignorant of Roman history, I was entirely unprepared for the overwhelming significance of Rome and left there with a most confused recollection of the Coliseum, Vatican, picture galleries, Catacombs, St. Peter's, and masterpieces of Michelangelo—all jumbled together.

On the Kansas prairies, everything made by human hands was new, and only a single generation had separated us from the savage days of the tomahawk. In consequence, Rome was beyond my immediate comprehension. The magnificent examples of Art frequently depicted naked forms—things taboo in my childhood——and left me somewhat aghast. My mind was in a whirl as I departed for Fiume.

In the nineties, few Americans thought of entering Germany through the back door, via Budapest. I felt quite an explorer when I landed in the capital of Hungary and called upon the famous botanists there. They in their turn had never seen an American, much less an American botanist, and I found myself telling them more about my own country than I managed to make them tell me about theirs.

The thing which impressed me most in Budapest was the fact that it had cable cars instead of horse-drawn vehicles. It was not until years later that the cable-car system was laid in Washington.

I could not understand a word of the language, but I carried away the impression that the Hungarians were a very emotional people. I gathered this from a performance in the National Theatre, where I found myself in the balcony surrounded by weeping and embracing couples. Their emotion was undoubtedly a reflection of what was happening on the stage, but it seemed to me unnecessarily exaggerated!

I have often been asked why I went to Breslau to study Botany when there were many German laboratories much better equipped. I am not very sure about this myself, but I believe that I was led astray by Theodore Holm, a Danish botanist, who had joined our staff in Washington. Holm had impressed us all with his thorough knowledge of systematic botany and the report of his arctic explorations on the *Fram*. To him, a classification of sedges and grasses by means of their tissues indicated the greatest scientific acumen, and his pen-and-ink sketches quite overawed me. Ferdinand Pax was his ideal in systematic botany, a field in which my education had been much neglected. Holm therefore strongly advised me to spend a college term with Professor Pax at Breslau.

Breslau in the nineties seemed a long way from home. It was a different world. Situated on the dull, flat plains of Silesia where tiny rivers meandered sluggishly, the town consisted of cobblestone streets lined solidly with apartment houses and restaurants where beer, and yet more beer, was served. *Sauerbraten* or veal cutlet and boiled potatoes composed the usual menu.

My lodgings consisted of the standardized combination of living and bedroom inhabited by the German students of those days. The room contained a short, narrow bedstead with an enormous featherbed as coverlet; a hair sofa in front of which stood the inevitable green baize-covered table; and, in the corner, a white porcelain stove for burning peat briquettes.

Almost the first words from my landlady startled me: *"Haben Sie angemeldet?"*

I had no idea what she meant. When she realized this, she sent me off posthaste to the Police Station with whatever credentials I had. It took me some time to realize that the people of Silesia felt themselves, even then, under the shadow of an ever-present Russian menace. The restaurant conversations, the military discipline, the discussions among the students, had an undercurrent which I came to associate with the psychology of race hatred and national antipathies. Until then the fear of war had never meant more to me than distant history. Yet here I found myself living with people in daily dread of their neighbors across the border.

At the old botanical garden my reception by Professor Pax could not have been more friendly. I was his first American student and he seemed proud to have me. The garden was fascinating with its neat beds and carefully labelled plants. It was indeed a strong contrast to the ill-kept, badly labelled Italian gardens I had visited.

The greatest scientific botanist of Breslau was not Pax, under whom I had chosen to study, but Ferdinand Cohn. Cohn was considered by some to be the father of the science of bacteriology. It was to him that Robert Koch, the greatest figure in this new science, owed his first inspirations. Cohn aided Koch in publishing his famous work on Anthrax.

Professor Cohn was more than cordial to me. He was a rather short, thickset man, with two pairs of glasses on his nose. I found him in his laboratory, drawing in colors some microscopic algæ grown in the same lily tank in the garden from which years before he had taken those drops of water in which he first saw swarms of bacteria. He found that I was familiar with the paraphernalia of bacteriology and also with the name of Koch, his pupil, and his intelligent, mobile face lighted up as he showed me some mounted slides of the bacillus of cholera. They were slides given to him by Koch after his return from India.

It is my understanding that it was Cohn who first conceived of the possibility that somewhere among these organisms in the drops of water there might be some which caused disturbances in the intestines, and that even cholera, one of the greatest scourges of the world, might be associated with one of these water-inhabiting organisms. This was some time, of course, before Koch had made his historic visit to India and discovered that a bacillus caused Asiatic cholera.

How carefully Professor Cohn put away those precious, microscopic slides! How proud he was of his star pupil!

I was working one morning in Cohn's laboratory near a young German at a neighboring table. Suddenly I realized that the young man was standing close beside me, clicking his heels and bowing repeatedly: "*Gestatten Sie mir. Ich bin Herr Fedde.*" I did not know it, but he was introducing himself in the approved German fashion. I rose clumsily and told him my name. Friedrich Fedde later was editor of the most famous of all the yearly summaries of botanical information, *Just's Botanisches Jahresbericht*.

Fedde and I became good friends, and eventually he invited me to see a students' duel. So one morning I went with him to a rather dreary duelling place outside the city limits. Duelling was already under some kind of a ban, although the police never molested this function of the German students, and in Breslau duels were a common occurrence. I was eager to know what the duel was all about, and was informed by one of the participants, a friend of Fedde, that his antagonist had elbowed him off the sidewalk, which was classified, so I understood, as one of the minor insults.

The other day I ran across the duelling card which I kept as a souvenir of this affair. It has ninety-nine pencil marks made by the seconds, one for each of the sword strokes of the combatants.

The duellists were buckled into leather neckshields, armshields, goggles, and earshields, as the nervous tension of the crowd increased. To see two men with long, thin weapons as sharp as razors stand face to face and slash viciously at each other is an exciting experience. The swords were flat and very flexible.

After some moments of combat, the seconds rushed in with their swords and knocked up the weapons of the contestants, calling "Halt," so that the doctor might inspect a gash across the forehead of Fedde's friend. The blood gushed forth, and I thought our duellist must be nearly dead when I saw the doctors lay him out and then deliberately stand him on his head. Soon, his pale face grew ruddy again, whereupon they stood him on his feet once more and he fought the duel through. I cringed a bit when his antagonist scalped him and a lock of hair shot past me and landed in the sawdust.

I did not remain to see the saber duel which was to follow, although Fedde thought that I should; I had had enough. I later heard that one of the participants had his skull badly cracked.

A. Weberbauer was Pax's assistant, and we soon became fast friends. His face was scarred from a duel, and I am sure that he considered me rather effeminate when I told him I did not care for this sport. However, one morning in the garden, I heard him yell as though he were being murdered, and discovered that he had been stung by a honeybee.

I had the laugh on him then. In my boyhood, on the plains, we used to hunt bumblebees and fight them with bats made of shingles.

Weberbauer was working on the order Rhamnaceæ for that monumental book on plants, *Die Natürlichen Pflanzen Familien* (The Natural Plant Families). This publication was subsidized by the Government and directed from Berlin.

I disappointed Professor Pax by my ignorance of Eichler's floral diagrams, in which he took a keen delight. He soon taught me to have great respect for them, although I must confess that they have played a very small part in my botanical existence.

If Professor Pax was shocked that I knew nothing about floral diagrams, I was equally surprised that he had only once seen, and had never tasted, a tomato. It is amazing to realize how restricted was the food distribution before the days of improved refrigeration. Later, while in Bonn, I disgraced myself by laughing at a young German back from Brazil describing to his audience how a banana should be peeled.

I applied myself assiduously to the highly technical process in which Professor Pax and Weberbauer took such delight. They soaked up herbarium specimens in potassium hydrate and teased them to pieces, a process which seemed crude to me after my long experience with the microtome. I think that Professor Pax must have considered me particularly stupid, for I had not the requisites of a systematic botanist. However, he saved my face by asking me to prepare a paper for the Agricultural Society of Silesia on fruit-tree diseases and the methods of combating them in America. This experience made me realize how aloof the scientific German botanist of that day held himself from the practical problems of the peasant and his crops. Almost nothing in the way of actual help had as yet come to the peasants from the highly technical scientific studies of the botanical institutions of Germany.

Just as I was leaving Breslau, a letter came from W. A. Taylor, Assistant Pomologist of the Department of Agriculture, suggesting that I make a visit to Corsica and send him authentic cuttings of the citron. In the meantime my old chum, Ted Nichols, had married and was coming to Germany accompanied by both his wife and her twin sister. Naturally, I was keen to see him.

I wrote accepting Taylor's offer to visit Corsica, and asked for a cablegram of definite authorization from the Department. Leaving my forwarding address with Professor Pax, I pushed on to Frankfurt-am-Main, but evidently I did not make adequate arrangements for forwarding a cable by wire, and I dearly paid for this oversight.

The reunion with Ted Nichols in Frankfurt was delightful. Ted and his wife planned to remain in Germany, while his sister-in-law and a friend were to cross the Alps into Switzerland. It was therefore decided that I pilot the two ladies through Switzerland on my way to Corsica.

In those days the magnificence of glaciers and snow-capped peaks was enhanced and accentuated by the simple peasant life of the country. Tinkling cow-bells and yodeling herdsmen created a colorful atmosphere completely lacking in the modern tourist-beset Switzerland of today.

At Locarno, I said a regretful farewell to my delightful companions and pushed on to Florence. No news had come from Washington, but since I had left my address at each hotel, I felt, with the confidence of youth, that a message would catch up with me. So I spent happy days in Florence, where I annoyed the attendants of the art galleries by the clatter of my hobnailed shoes over the marble floors.

I could not help worrying about the authorization from the Secretary of Agriculture which should come with a letter and draft from Taylor and was terribly distressed when I reached Livorno and the Consul had nothing for me. He questioned me rather closely with regard to my object in visiting Corsica, but I had been cautioned to keep my mission quiet lest the native growers be disinclined to provide cuttings for a rival industry in the United States. I therefore merely tried to secure what information I could with regard to the shipment of citrons to America in brine, and the type of salt-water tanks in which the Corsican rind is first rotted before it is candied.

For the benefit of those who know citron only as they meet its fragrant fragments in a slice of fruit cake, I should say that these fragments are the candied rind of a very thick-skinned citrus fruit (*Citrus medica*) which looks like a rough lemon. It is a species quite distinct from all the other citrus fruits. Candied citron forms one of the most delicate spices used by cooks in making Christmas plum pudding and the rich wedding cake which wedding guests take home to dream upon.

I had of course heard of the Corsican vendettas, and I knew that the mountainous parts of the island were not considered safe because of bandits. But these rumors merely made me keen to push on, and, moreover, this was my first commission to work for my Government in a foreign land.

It was an ill-fated expedition from the start. The sea was rough and I crawled off the miserable little boat the next morning in the disgustingly bilious mood which a nasty night at sea never fails to produce. I had been accustomed to a certain degree of dirt but, seen through my "green spectacles," the town of Bastia appeared unbelievably filthy, with mangy dogs wandering about the unappetizing hotel where I put up.

I was in great anxiety lest the letter containing my commission had not arrived, and it seemed hours before the post office was open. Finally, to my delight the postmaster handed me a much redirected envelope from Breslau. In it Professor Pax enclosed a cablegram which he had received and mailed to Frankfurt. From there it had chased me over the Alps to Corsica.

I opened it and read: "Secretary refuses authorization."

I still recall the shock, the feeling that the ground had crumbled under me. But, mixed with my surprise and disappointment, was the indefinable exhilaration which one feels in an emergency. I had spent more of my money than I could afford to reach Corsica. But there I was, with an adventure on my hands, and I enjoyed it.

I hired a cab and drove out along the coast, wondering just what to do. We stopped at a tiny village, Erba Lunga, one of those lovely spots which so enhance the Mediterranean islands. The little café by the roadside was run by a sad and disappointed Corsican who had once been in the States. As I ate my lunch, his life story thoroughly distracted me from my own problem. He took me into his unkempt garden to taste his figs and, as he plucked them from his trees, a drop of honey hung on each blossom. Such figs, such sweetness—shall I ever taste their like again?

The hours in Erba Lunga gave me time for reflection. I realized that I could not afford to stay in Corsica but, before leaving, I might as well try to obtain the citron cuttings.

I had noticed a little macaroni factory by the roadside, and could not resist stopping to see how Corsican macaroni was made. The owner of the factory gladly showed me through his mill and sold me some new forms of "Pasta" for my collection. He was a friendly soul and I felt that I could safely ask him about the citron groves—where they were and how to get there.

"I have a friend," he said, "who is Mayor of Borgo, a village in the center of the citron district. You can reach there by donkeyback from the railroad. I'll give you a note of introduction to him."

The following morning I travelled across the island, left my bags at the little station, and hired a donkey to carry both myself and my camera and tripod.

Up the steep trail we climbed to a picturesque village on a tiny plateau overlooking deep valleys on either side. I had not yet seen the hill towns of Italy, and thought this a most delightful spot, dirty and dusty though it was.

Armed with the macaroni-maker's card, I inquired for his honor, the Mayor. To my astonishment, I had to go through a pigsty to reach the Mayor's "palace," but was cordially greeted when I climbed the rickety stairs.

The Mayor, a ruddy-faced bandit of a fellow, spoke little Italian, but managed to explain that, unfortunately, his best friend's daughter had died and he was just starting for the funeral. Pouring out a glass of wine, he offered me the hospitality of his house, and said that he would be back in the afternoon.

In the wine, a patch of mould floated casually. The Mayor tossed off his drink and disappeared down the stair, leaving me wondering what I was to do next. Through the cracks in the floor I could see the untidy pigsty. I stooped and poured the wine into the mud below. After that there seemed nothing to do but wait.

The scenery from the window was superb, and I decided to pass the time by taking photographs. Stepping gingerly to avoid the pigs, I returned to the main street of the town. A small crowd gathered around me, chattering an unintelligible dialect. There was one old man who persistently repeated an Italian word that sounded like *"Pianura."* I was photographing the quaint houses and I thought that he wanted to know why I did not take the great plains below which were too misty and far-away to make a clear picture. He was rather annoying, and I probably was a little sharp with him, for he hurried off up the street. To please an old woman in the crowd, I agreed to photograph her daughter,

A Citron Grove on the West Coast of Corsica showing the trees trained in vase form to give the branches the right spread.

48A

Photograph taken by the Author just before his arrest in Borgo, Corsica.

48B

who was soon to be married. While I stood with my head under the camera cloth, some one touched me on the elbow:

"Vos papiers, s'il vous plaît."

A man in uniform, a gendarme of some sort, stood beside me. In the emergency, my Rutgers French failed me utterly, and I could neither understand him nor make myself understood. He motioned me to follow and, pushing our way through the crowd, we entered his house and he shut the door. Seated in the corner of the room was an evil-looking woman with a baby at her breast. I can see her still. Her presence seemed a menace.

I had never been searched before, but the gendarme did not leave much doubt as to what he wanted me to do. So I laid everything on the table. My little red notebook was filled with closely written notes in Professor Mayer's favorite crimson ink, giving my observations from day to day on the marine algæ of the Bay of Naples, together with the scientific names of plants I had seen in Sicily, Germany, and in the Alps. The pages were a hodge-podge of German, Italian, and English, and I realized, as this ugly-looking officer thumbed through the book, that it must seem suspicious to his illiterate mind, although I doubt whether he could read much of any language. He scowled and looked from me to his wife, who kept firing questions at him in Corsican dialect.

I wondered whether I should have to spend the night in the gaol, which I was sure would be neither a comfortable nor a sanitary place. By the time I had become thoroughly scared, the gendarme picked up my pocketbook and there, to my delight, I saw a United States Treasury check for fifteen dollars, sent me as a reimbursement. I had already tried to explain that the valise with my papers was at the railway station.

I pointed to the steel engraving of Ulysses S. Grant on the check. *"Io sono Americano,"* I shouted at him, waving the thing in his face.

But why had I come through to Borgo? And why those suspicious-looking notes? And why my camera? All this time his unpleasant wife was talking to him in an undertone. I could not explain my mission. Neither could he believe that I was just an American tourist taking pictures of his little town. He was too familiar with the game of espionage. Corsica had been a warlike land for centuries, and wars have always bred a general hatred and suspicion of all strangers.

49

Finally, convinced by Grant's rugged features, he reluctantly let me go, but ordered me to leave the town at once.

I did not wait to thank the Mayor for his glass of wine, but was soon trotting down the mountainside on my little donkey. I had noticed groves of citron trees on my way up and, finding that I was not pursued, I cut bud-sticks from some trees beside the path and hid them under my coat. Needless to say, I did not linger to take photographs. I was far too scared for that.

In Ajaccio, a very different place from Bastia, I found a nurseryman in whose back yard I packed my handful of cuttings with some others which he helped me secure from trees along the terraces. I knew little about the safest way to send cuttings through the mail—in fact, little of such work was being done anywhere. I therefore adopted a method I had read of, and stuck the end of every cutting into a raw potato, hoping in this way to supply the cuttings with the necessary moisture for the long trip to Washington. I am glad to report that they reached home successfully and proved of real value to our citrus growers.

As I planned to spend the winter with Ted Nichols in Berlin, I crossed to Marseilles and hastened on to Germany after a few days in Paris.

I do not recall exactly where I lived in Berlin that winter. My room was only a place in which to sleep. Ted, with his bride, Kate, and her twin Carrie, were living in style on the Pariserplatz near the great Brandenburger Tor, and I spent most of my evenings with them, often at the opera listening enthralled to the wonder of Wagner's music. I can still remember the swelling volume of the orchestra and feel the strict discipline and almost religious awe of the audience during the intermissions. No disturbing noises, no loud talk, and no late arrivals pushing past one, as is so common in Anglo-Saxon theatres.

D'Albert was in his prime, but I remember one concert which was considerably marred by his uncontrolled shirt-front. I believe the contraptions are called dickies. His insisted upon slipping out of his waistcoat, and required the constant attention of one hand to keep it even partially in place.

To give anything but the most fragmentary sketch of the winter of '94 and '95 in Berlin is impossible. I met many dynamic personalities, and tried to assimilate much scientific knowledge. A great deal of time seemed to be wasted on street-cars, and there was so little daylight that I could not accomplish half that I had intended to do. Yet I assimilated much of botanical interest which has played a considerable rôle in my later life.

I was systematically preparing for my trip to Java, and greatly broadened my horizon under the stimulus of a group of men who have all left their mark on the science of botany. The lecture system was at its height, and I arranged a schedule which taxed my energies to the limit. Taking notes in German from the rapid presentation of the professors was a difficult feat, particularly as I often knew comparatively little about the subject under discussion.

Professor Simon Schwendener, who lectured on the mechanical principles of plants, was nearing the close of his career. He would step out on the platform, shake his large, shaggy head, and begin: *"Meine Herren. Heute haben wir die Aufgabe—"* ("Gentlemen. Today we have the problem—") He then plunged into his subject, enthralling his audience with his theories of the mechanics of plant structure.

Professor Schumann augmented his discourses by use of the first substantial textbook on paleobotany. Through his presentations, we glimpsed the vegetation of the globe as it was millions of years ago.

I gained the friendship of Professor Otto Warburg and have kept it through all these years. His lectures on tropical plants gave me my first acquaintance with the wealth of species I was soon to meet in Java. "After all," he used to say, "flowering plants are nearer to us humans than your beloved fungi." I still find his *Pflanzenwelt* the most useful botany in my library. Why it has never been translated I cannot understand. It assembles a much wider range of knowledge and is more beautifully illustrated than any other book I know. He told me how he happened to write it. At breakfast one day his little boy asked his father what pearl barley was. Professor Warburg began telling his son stories of the various plant products which appeared upon the table. Later he expanded these talks into a three-volume work.

Because of my interest in fungi, I wished to meet Lindau and Hennings. I found Lindau busy with the mushrooms and large, fleshy fungi, and Hennings studying the microscopic forms.

When I met Professor Paul Magnus, he invited me to his home, and I found myself in an uncomfortable situation. Magnus was a Jew and not persona grata with Lindau and Hennings. If I associated with him, I would not be acceptable to them. Yes, there was, even forty years ago, a distinct antipathy in Berlin towards the Jews.

I tried to carry on some laboratory investigations under dear old Professor Julius Klein, who had succeeded Nathaniel Pringsheim, the famous experimenter. Klein's institute necessitated a long, uninteresting trolley ride and I did not often have time to go. But, although I did not accomplish much in physiological botany, I acquired a great respect for the old laboratory and what it stood for. I think that some amœbæ which appeared as contaminations in my agar cultures were perhaps the first amœbæ to be cultured in agar. Later the method was much used in the investigations of amœbic dysentery in the Philippines.

At Christmas, Professor Klein invited me to tea, and I had a glimpse of Christmas Day in a German family. The tree was draped with strings of popcorn much as trees had been trimmed at home. Professor Klein explained that these strings had been sent to him from America and were kept from year to year, and I was surprised to realize that neither popcorn nor any other corn was grown in Germany.

In Berlin there was great botanical activity connected with the German colonies in Africa, particularly the Cameroons. This work, dominated by Professor Adolf Engler, had been for several years concerned with the publication of a monumental encyclopedia on the plant families, arranged according to their natural affinities. I therefore spent much of my time at the botanic gardens, where Professor Engler and his associates had their headquarters and where the immense herbarium was housed. Work was going on at full speed on the flora of West Africa. The glamour of the Cameroons pervaded the place and gave a real objective to the work.

Thirty years later, I visited the Cameroons. By that time, it had become a British mandate and, to my disappointment, I found at the botanic garden that the only men who knew the plants or cared anything about them were two half-castes. To make matters worse, some drunken sailors, as a practical joke, had mixed up the labels. It was disheartening to realize that the scientific work and accumulation of knowledge about the vast plant-world of the Cameroons, which I had seen in Berlin, had been interrupted by the insanity of a general European war. To me there is nothing more tragic than the scattering of workers and the end of investigations which come about as a result of war.

Breslau, Berlin, and Bonn

Like most Americans, I chafed at first at the regulated city life of Berlin, the strict police surveillance, the signs "Verboten" on every hand. In fact, an attitude of personal criticism seemed almost universal. I saw little "social life"; the family life of the professors was in striking contrast to the home life of the average American college professor. The German wives played no rôle such as do women in our American universities.

To be sure, Professor Engler did give an entertainment at his home, a so-called costume ball. Ted Nichols had been told that it was up to me to make a call on Frau Geheimrath Engler before the event. Ted said that this call had to be made at noon in full dress. I was in a quandary for I had never owned a dress suit. The twins dressed me up in Ted's coat one morning and started me off. I could feel the people staring at me in the street car as I rode out to make my call. Frau Engler emerged from her kitchen and looked at me in such open-eyed wonder that I doubted then, and still do, whether any of the other botanists had ever called upon her in a full dress suit at high noon. However, she was very cordial.

While I was rounding out my botanical education, Ted Nichols was studying a fascinating problem in the basement of the Physical Institute. He succeeded in making what was then probably the most sensitive instrument in the world with which to measure the pressure of light. With this instrument he explored the unknown region between the long waves of electricity and the very short ones which we see as light. I used to stop there on my way home from the Agricultural Institute. He gave me my first glimpse into the world of electrical phenomena, but in his wildest dreams I am sure that he never foresaw the radio of today.

As the winter wore on, I became restless and dissatisfied with my progress. Just about that time Alfred Moeller came to Berlin. I had known Moeller—a fascinating person—in Washington when he stopped there on his way back from Blumenau, a German settlement on the coast of Brazil. At that time he had shown us his material which he believed proved that the leaf-cutting ants of southern Brazil were "mushroom-growers and mushroom-eaters," and had probably been so for millions of years. He worked out the microscopic details of the fungus gardens of these ants of the genus Atta, following a suggestion made by the brilliant engineer, Thomas Belt, one of the early naturalists, whose book *A Naturalist in Nicaragua* attracted the attention of Charles Darwin.

53

The World Was My Garden

I was fascinated by Moeller's researches, and, as no one had ever done anything in the Oriental tropics in this field, I made up my mind to look into the subject during my prospective stay in Java.

"Why don't you go to Professor Brefeld and study fungi with him?" Moeller asked me.

The idea appealed to me, and the opening of the spring semester found me in Professor Oskar Brefeld's little laboratory at Münster. Moeller had been Brefeld's star pupil, and when I presented a letter from him, the Professor welcomed me with open arms.

An accident had cost Brefeld the use of his left eye, but his right one seemed to have the strength of ten and enabled him to spend endless hours at the microscope, studying the fascinating spores of the various species of fungi which he collected in the forests of Westphalia. His long, gray hair was brushed upwards from his massive forehead in true German style, and his appearance was impressive, even in the ancient gray wrapper which hung dejectedly from his shoulders. As the Professor walked before me to show me the table where I was to work, I had my first view of the back of this garment and was amused to see that, where the seat had become worn during his long vigils at the microscope, his wife had attempted to patch it. The patch, however, hung by the top edge only and flapped in the breeze.

In the nineties, American students were rare, and German laboratories opened their doors to them with a hospitality which was remarkable considering the meager funds which the laboratories had at their disposal. I shall never forget the courtesy with which Professor Brefeld initiated me into a technique which he had developed during his investigations into the life histories of innumerable molds and other fungi. I was charmed both by him and by the simplicity of the surroundings, which were delightfully refreshing after the confusion of life in Berlin.

Professor Brefeld promised me some material on which to begin work in the morning, and I departed to find lodgings. Two old ladies who lived on what seemed to be a quiet street were glad to take me in. They both were charming and tried hard to convert me to Catholicism. When they finally succeeded in persuading me to go to the Cathedral on Easter Sunday, they divided their forces and stood guard at the two entrances to make sure that one of them would see me come in.

But I was mistaken about the quiet street. What a noisy place my little room proved to be! The clatter of wooden shoes on the cobblestones was even more disturbing than the horses' hoofs in the early mornings when I lived in Washington.

In time I became so well acclimated to German customs that I even bought myself a Schnurrband, an arrangement which young men wore at night to train their mustaches in imitation of the Kaiser's. I remember this incident because it indicated the widespread influence of the young Emperor. A chromo of him hung in every home, and German babies in their cradles contemplated a picture of the "War Lord." People spoke carefully about the Kaiser, and *lèse majesté* was a serious offense.

The morning after my arrival, I was unlimbering my microscope when the Professor's "Diener," a devoted servant and assistant, came marching in. In his hands, held as proudly as a flaming, Christmas pudding, was a plate heaped high with horse-dung. With a flourish, he placed it on my table, covered it with a bell jar, and assured me that it was completely fresh and absolutely perfect in every way. Incredible as it may seem, after my first feeling of revulsion had passed, I spent three of the most entertaining and instructive weeks of my life studying the fascinating molds which appeared one by one on the slowly disintegrating mass of horse-dung. Microscopic molds are both very beautiful and absorbingly interesting. The rapid growth of their spores, the way they live on each other, the manner in which the different forms come and go, is so amazing and varied that I believe a man could spend his life and not exhaust the forms or problems contained in one plate of manure.

Brefeld was the first person to demonstrate the fact that bacteria have resting spores inside of them—spores which are resistant to boiling temperatures. Yet he considered the entire technique of the solid media used in test-tube cultures or Petri dishes as unnecessary nonsense. Brefeld had no test tubes, no sterilizer, no Petri dishes, and had no use for plate cultures of any kind. In order to discover the resting spores, he had sat at his microscope for twenty-four hours with his eye riveted on a single bacterium within which he could clearly see a dark body. No one knew the function of this body, but he was determined to discover its secret, and his vigil was finally rewarded by the sight of a tiny germination tube issuing from it. He then realized that the dark body was a spore, a resting spore. This was a spore which was not killed by boiling, and his discovery explained why fruits or vegetables may ferment ("spoil" as we say) even after they have been boiled, due to the presence of these spores which have survived boiling temperatures. Because of this discovery, the method of repeated boilings was developed, allowing for incubation periods between.

His success in making this discovery had convinced Brefeld that the surest way to understand the behavior of these tiny organisms was to watch a single individual closely for a long period of time. After all, there is much to be said for the principle of watching the hole and seeing the rat come out of it, instead of deducting by some roundabout reasoning that the hole might be a rat hole.

Brefeld had promised to show me the "secrets" of his technique, and I was considerably surprised to find that his entire equipment consisted of solutions of beerwort and horse-dung extract. This was kept in little flasks which were boiled every day by the Diener, who also cleaned the glass microscope slides with hydrochloric acid and alcohol, and I must say they were clean slides. When a drop of horse-dung extract was put on a slide with the little glass stick which remained in each flask, the drops spread evenly over the surface and, with careful technique, it was possible to place in the center of this drop the single, minute spore which one wished to study. Once sown with their respective spores, these slides were placed on metal trays under bell jars in a saturated atmosphere, and were examined through the microscope every day.

Another fascinating organism which I studied was a singular fungus which lives in the intestines of the frog. This microscopic fungus universally inhabits the intestines of frogs and lizards all over the world, and has a most ingenious method of scattering its spores. It shoots them from the tips of special hyphae or fungal threads which raise themselves above the surface of the excrement and can be seen with the naked eye. This fungus has, I believe, the shortest nuclear history of any known—three cell divisions. I devoted six months to the *Basidiobolus ranarum,* as this tiny denizen of frogs' intestines is named, and if I had time today, I could take up the study where I left off, and spend many exciting months experimenting with it.

One warm summer day, while I was immersed in the delightful occupation of fungus gardening under a microscope, weeding out with a sterile needle any stray spores which had intruded, a tall, broad-shouldered

Breslau, Berlin, and Bonn

American walked into the laboratory. He announced himself as R. A. Harper and was the first American botanist I had met since leaving Washington. I liked him immediately for he had a merry laugh and kind, gray eyes.

Harper, too, had come to study Brefeld's mycological technique and systematic classification of the fungi. This classification was based on the theory that the higher fungi, such as mushrooms and toadstools, unlike the rest of the plant world, were completely sexless organisms. The sexuality of living organisms, as exemplified by the union of two nuclei, a male and a female, was a burning question in the nineties, and Harper soon showed me discoveries which he himself had made. These discoveries eventually proved Professor Brefeld's theory to be wrong, for they proved to every one's satisfaction that, so far as the nuclei of the higher fungi are concerned, there is a complicated union of what may be called a male and a female nucleus, which initiates the new generation as it does in almost all the other plants except the bacteria.

Professor Brefeld was not willing to face the refutation of his theory of the non-sexual origin of the higher fungi, for this destroyed the value of his classification which was based upon it. However, before many weeks had passed, he realized that his two American students, instead of studying along the lines of his ideas, were trying to disprove them.

The dandelions in the meadows were covered with a curious fungus, Synchitrium. By putting dandelion stems in water, I secured myriads of flagellated swarm spores, and developed a technique which enabled me to prove the conjugation of the minute swarm spores which represented the male and female phases of the organism. Until then, the sexuality of this fungus had not been proved.

From then on, Professor Brefeld found one pretext or another to keep away from us, and never visited his laboratory. We saw practically nothing of him, and eventually decided that we would formally call upon him, secure his blessing, and go up to Bonn. He gave us a rather sharp lecture, but years afterwards, as the truth of Harper's discoveries were recognized, he was rather proud to say that we had been his pupils.

Harper and I found Swingle at Bonn and we settled to a great winter of work in Professor Strasburger's botanical institute in the old Schloss. In my memory, I treasure vivid pictures of the moat, the ducks we used to feed, the lecture room and the brilliant lectures by this great master of Botany. Protoplasm and Irritability, "Das Protoplasma und die Reitzbarkeit"; Strasburger's discoveries in the field of cytology; his conception of the chromosomes as the carriers of heredity; his mastery of the then new science of the cell, all gave to the laboratory an atmosphere quite unlike any other.

The Professor's total budget for laboratory supplies and maintenance was only six hundred dollars a year. The straight, wooden chairs had seen better days, and collapsed, one after another, under the rough treatment of the pupils. But the Herr Professor's morning greeting, *"Haben Sie etwas neues?"* was in itself a keen encouragement to discovery.

A distinguished group of investigators were gathered there, including five Americans as well as several Britishers. I say distinguished advisedly, for Harper later became Professor of Botany at Columbia University; Osterhout is one of the research staff of the Rockefeller Institute; Mottier was Professor of Botany at Indiana State University; my Kansas classmate Swingle became one of the most brilliant investigators of the Department of Agriculture in Washington; V. H. Blackman became Professor of Botany at Cambridge University, England; and Koernicke succeeded his father as Professor of Botany in Bonn.

We spent both days and nights together. Each microscope had its round Schusterkugel (shoemaker's bulb) filled with copper sulphate to concentrate the light of the Welsbach burner on the mirror of our microscopes.

The results of that winter's work are recorded in the standard publications relating to the nuclear history of plants, and, without boasting, I maintain that no better microtome slides revealing the structure of the plant cells had ever been made than those which we fellows prepared, each in his own field, during those long, gray winter days in the old Schloss. Those slides represent actual contributions to the sum of human knowledge, and have a more or less direct bearing on the superstructure which we now call the theory of heredity—or at least on that part which has to do with the cellular mechanisms of heredity.

One morning in November, as we were settling down to our microscopes for the day, the Professor threw open the door with a flourish and walked rapidly into the room. He held a photograph in his hand and we could see that he was very excited.

"Gentlemen," he said, "my friend Röntgen has just made an amazing, unprecedented discovery. It will be of the greatest use to surgery."

He then showed us that first X-ray photograph of the key, the ring, and the purse, which has now become historic. We were among the very first to look at it, and we sensed the importance of the moment. As he described it all, Professor Strasburger made us feel as though we had been with him in his friend Röntgen's laboratory on that great day of November 8, 1895.

It is overwhelming to contemplate the complete change of diagnosis ushered in by that one discovery which has materially changed our conception of the universe, as well as mitigated untold human suffering.

One extracurricular activity of mine that year bore pleasant fruit. I had the thrill of making a successful mechanical invention. In the preparation of any living material to be sectioned with the microtome, it is necessary that it be fixed with some fluid such as picric acid or corrosive sublimate, and then later this fluid must be washed out—a long, tedious process which involves changing the water every few hours. I became thoroughly "fed up" with the drudgery of washing my material, and devised a little porcelain cup, not much larger than a thimble and full of holes, which could be supported by a cork and would float like the bobber on a fishing line. When used in the laboratory sink, with water running through it, the material in the "cup" would be washed out automatically. A manufacturer of laboratory apparatus in Bonn made a lot of these for me, and they were very successful. In fact, Professor Strasburger included a picture of the device in his book.

Thirty years afterwards, when I supposed that this invention had long ago been discarded for something better, I was equipping the laboratory of Mr. Allison Armour's yacht in Genoa, and, in an establishment there, I found hundreds of these "Fairchild's porcelain washing thimbles." When the dealer realized that I was their inventor I rose amazingly in his opinion!

Every few months a letter would reach me from Mr. Lathrop asking why the devil I had not gone to Java. In the tone of these communications I detected a growing impatience of my wanderings in Germany.

During this time I had been corresponding with the botanical gardens of Java, and early in 1896 I received a letter from their Director, Melchior Treub, saying that he was coming to Holland for a visit. He suggested that when he returned to Java in April I should go with him. This put me in a quandary. I had to decide whether to remain in Bonn for my doctorate, or accompany Doctor Treub to the Dutch East Indies. The importance of a doctorate in those days was not perhaps as great as it is now. Anyway, it did not take me long to decide.

When Treub reached Holland, I visited him in Leyden. He was different from the scientific men I had known—an aristocratic man of the world, and an accomplished linguist. His waxed mustache, eyeglasses on a ribbon, together with his fashionably tailored clothes and spotless linen, made me a little afraid of him at first. Even when I came to know him well, he still rather kept me at a distance.

Treub was already becoming an important man in Holland, and he was organizing the Dutch botanic gardens into a great agricultural department. During his stay in Holland, he was conferring with Dutch owners of coffee and tobacco plantations and was arranging for their support of laboratories of entomology and pathology which would study the diseases of these crops.

Doctor Treub and I arranged to meet in Genoa and settled the details of our trip. My long-cherished dream of seeing Java was really to come true, and I returned to Bonn with my head in a whirl.

When the time for my departure drew near, I had not yet finished my paper on *Basidiobolus ranarum*. Harper, born teacher that he was, began to worry the life out of me to complete it, and I think that the last words were written just before I left my lodgings for the train. Harper sent it to the Journal of the German Botanical Society, where it was published.

CHAPTER VI

Java Ho!

O N A BEAUTIFUL spring morning I walked into the Hanbury Botanical Institute in Genoa. Its director, Professor Otto Penzig, was anxious for me to meet a young fellow named Rapps who worked in his laboratory, because Rapps expected to go out to Java in the near future. This young man had just come through a horrible experience. While on his honeymoon, he and his bride had stayed in a hotel lighted by gas. It was Rapps' first experience with illuminating gas, and he had blown out the light. His poor bride was asphyxiated and he himself barely escaped death. I mention this as partial explanation of certain incidents which occurred when he reached Buitenzorg.

The great day finally arrived and I sailed off for Java under Doctor Treub's chaperonage on a boat of the Netherlands-India Packet Boat Company.

That trip to Java in the nineties held more of interest than any ocean voyage I have ever taken. When I boarded the boat in Genoa and saw the turbaned Sundanese stewards, the Malays in their blue costumes, and the children's Javanese nurses, or "baboos," in their sarongs, I realized that I was entering a new world. By the time we reached Port Saïd, I had acquired a considerable acquaintance on the boat.

Entering the Red Sea was a great event. When I came on deck the first morning "east of Suez," the ladies lay in the chairs all along the deck clad in strange costumes or negligées—I was not sure which. They were nearly all barefooted and the sarongs which they wore were much shorter than the dresses of those days. I took a hasty look around and decided that I had mistaken the hour; that men were not supposed to be on deck so early. Turning abruptly, I fled from what seemed a definitely boudoir atmosphere. But Treub reassured me, explaining that the women had all donned native Javanese costume simultaneously because it was a fixed rule of the Captain that no lady could appear in native dress west of Suez.

From hot Aden, we sailed directly across the Indian Ocean to the west coast of Sumatra, an island which I believe holds more tropical beauty and mystery than any other spot in the world. The bewildering romance of the arrival in Padang and my first evening among bamboos, palms, and the noisy night insects, comes back to me often. I do not understand how any one can be content until he has experienced the wonder of the tropics.

Round-the-world tourists of today who find the streets of Batavia noisy with motor traffic, policemen and buses, can have no conception of the sleepy atmosphere which characterized that old Dutch town in the nineties. Its great open square, the Konings Plein, was surrounded by enormous Ficus trees and, in their shade, turbaned Javanese wandered barefoot, swinging their beautiful bamboo hats, or carrying on their shoulders long bamboo poles with baskets at each end. Tiny ponies trotted along, pulling some white-clad official sitting back to back with the driver in one of the two-wheeled carts called dos-à-dos.

In the hotel patio, the glare of midday was tempered by the shade of gigantic, overarching banyans, and an idyllic leisureliness pervaded everything everywhere. The "boys," in beautiful, handmade sarongs of brown and indigo, came and went noiselessly across the patio and along the verandah, speaking softly to each other in a language as musical as the speech of an Andalusian. Every afternoon after luncheon, a hush descended during the siesta hour. We drowsed happily, listening to the chatter of small parrots in the branches of the trees and the occasional thud of a coconut as it fell to the ground.

The delightful village of Buitenzorg, where I spent eight months, lies in the saddle between two smoking volcanoes, the Salak and the Gedeh. A little railway train, manned by turbaned natives, took me there. Joyfully tooting its toy whistle, the train crossed the lowlands and wound up the mountainside through the most fairylike and utterly delightful scenery. The swampy plain was filled with giant ferns and Nipa palms, those stemless Oriental palms whose fronds resemble giant cycad leaves. A strange mist hangs over these lowlands, making it an unhealthy place to live.

Soon the hills were reached, with their kampongs composed of pretty, little bamboo houses thatched with palm leaves and shaded by lofty clumps of feathery bamboo gently waving in the breeze. Clean-swept pathways led to the springs or brooks, where naked children and their mothers were taking a morning bath. Surrounding every bamboo house was a bamboo woven hedge, each one a different pattern. Hanging in little bamboo baskets from bamboo poles were cooing doves, the favorite song birds of the Javanese. It seemed a civilization dependent upon bamboo for both necessities and luxuries.

Before noon, I was installed at the one-story Hotel Chemin de Fer in Buitenzorg, and here I lived during the eight months of my sojourn. When I opened the simple wooden shutters of my room, I looked out on a thoroughfare crowded with traffic, but yet a noiseless one. For although a human being passed my window almost every second throughout the day, he passed silently, barefooted.

My broad, hard bed was spotless, and on it lay one of those curious, long, round cushions with which the sleeper is supposed to separate his knees at night for coolness. Festooned above the bed, on a metal frame, was a mosquito net, another object which I had never seen before. Although this was in the days when few men besides Ross and Theobold Smith had an idea that the mosquito could be anything but a nuisance, yet there was an undefined suspicion that the miasma—the intangible something in the air which produced malaria—was kept out at night by the meshes of a mosquito netting.

Director Treub kindly insisted upon accompanying me when I made my first visit to the Gardens. I must have delighted him with my astonishment when he showed me the great avenue of huge Canarium trees (*Canarium commune,* Java Almond) which are so dramatically tropical with their weirdly buttressed roots. The trunk and limbs of the great trees were festooned with variegated climbing aroids (*Pothos aureus* and others) over which the liana, *Entada scandens,* had climbed like some gigantic reptile. At that time, this was undoubtedly the most remarkable avenue of trees in all the world.

Doctor Treub next showed me an orchid with a thousand blossoms (*Grammatophyllum speciosum*), and then took me to a tree from Africa which bore clusters of gold-edged, scarlet flowers. They were the great, flaming, cup-shaped blossoms of the African Tulip tree (*Spathodea campanulata*). As he picked up a fallen flower, he explained that,

63

while in the bud, the gorgeous, red corolla had been held in its envelope of sepals by tough leathery bracts which contained a watery fluid under pressure. He had seen birds peck at these buds and had watched the fluid squirt from them. He thought the birds were frightened by the squirting fluid.

I was so entranced by the Spathodea that it was the first flower which I photographed. Today this tree has been imported and is growing casually in Florida, so perhaps I do not quite appreciate its rarity and beauty as I did then.

Director Treub next took me into the tangled jungle of the rattan Palms. I brushed against a swaying tip and instantly felt its grip upon my shoulder. Treub stood smiling at me as I struggled to free myself. The strength and sharpness of the clawlike spines was in surprising contrast to the delicacy and harmless appearance of the leaf tip. This tip is a climbing organ, quite as delicate as the tendril of the cucumber and other such vines.

The Director led me to the little laboratory which was placed at the disposal of visiting scientists. It was a great moment when I sat at the table designated as mine and realized that I had come to stay.

Everybody had a "boy," so Treub turned his own boy, Mario, over to me. I had never had a servant of my own, and felt myself a prince. Incidentally, I have never had a servant since who cared for me as Mario did. From mounting microtome sections on microscopic slides to managing a caravan across the mountains, he took care of everything. I shall never forget his tears as I bade him a last farewell when I left the island.

As I learned a little Malay, I was astonished when chatting with the Javanese, Sundanese, and Madurese working in the garden to realize their knowledge of the plants. Treub had recently decorated Mantri Oedam, the Javanese head gardener, with the "Silver Star of Merit." Oedam was a really remarkable botanist in many ways, familiar with every plant in the garden, and knew not only its native name but its botanical name as well. He had that rare faculty of form memory.

Everything was new and fascinating. If I had not been absorbed in my resolution to discover whether the ants and termites of the Orient were mushroom growers, I should have been distracted and probably would have done little of anything because of the wealth of novelty about me.

Each morning I rose early, as one does in the tropics, and Mario would

A typical scene in Java, 1895.

Bloom of the Spathodea (African Tulip Tree) in the Botanical Garden, Java. Today, dooryards in South Florida are gay with these great clusters of vermilion flowers.

On a Guava leaf rested a large insect of the Mantis family, so leaflike in its appearance that it was almost impossible to detect. (Page 70)

In a large Termite mound, three species were living harmoniously, each with its own type of Fungus Garden.

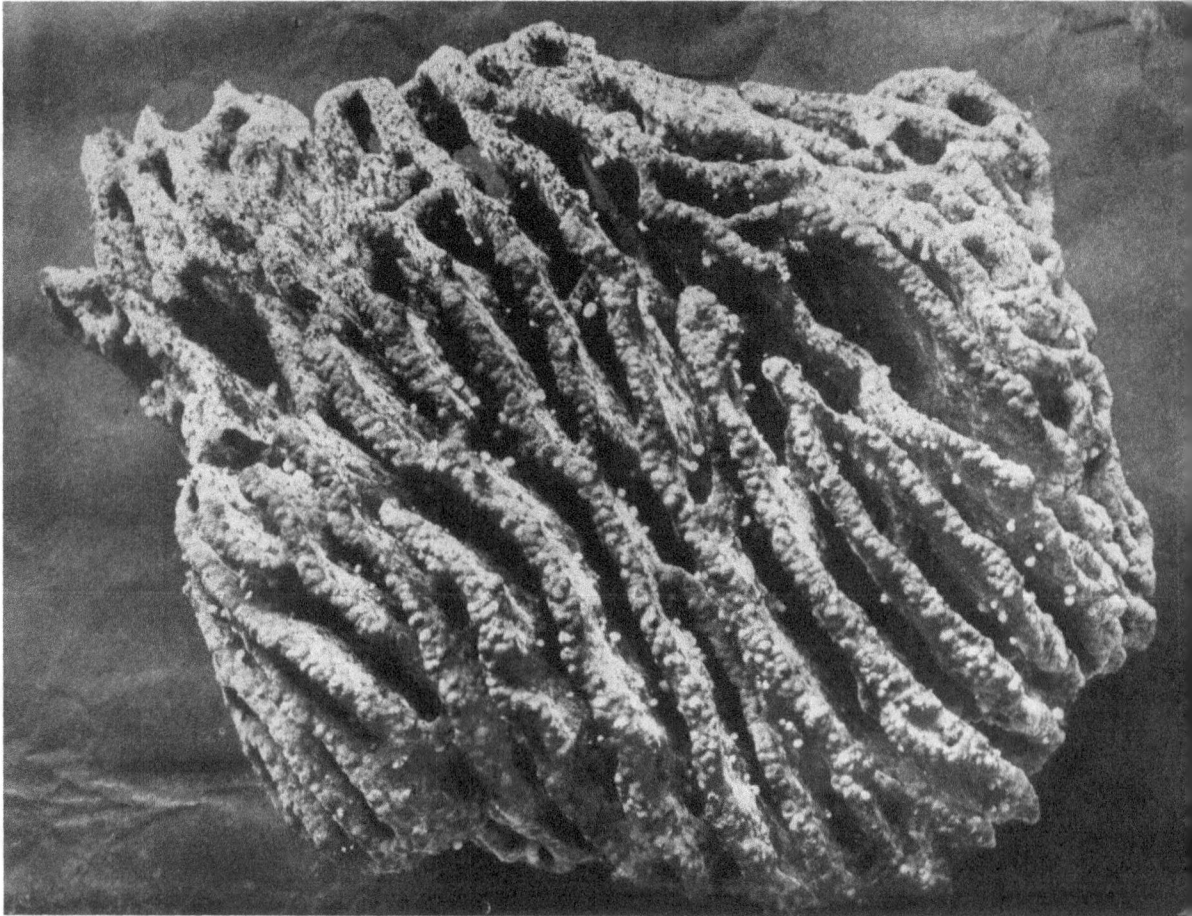

Termite Mushroom Garden with perpendicular galleries.

Mantri Oedam, head gardener in the Garden of Buitenzorg, was decorated by Treub with the "Silver Star of Merit" for his phenomenal knowledge of plants.

Cross-section showing the location of two Mushroom Gardens in one Termite mound.

A toadstool growing from an abandoned Termite garden.

64F

The forest surrounding Director Treub's mountain laboratory.

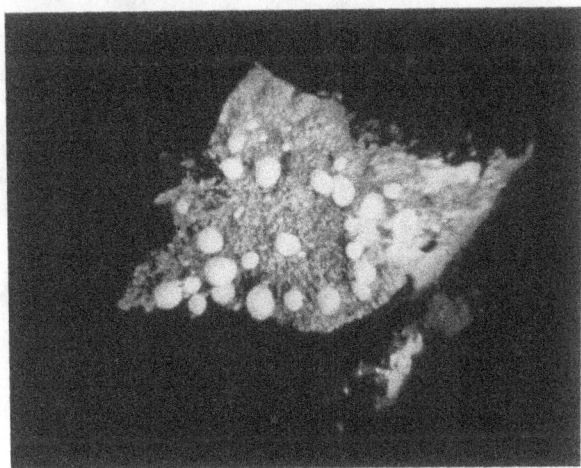

Termite Mushrooms, taken with a
high-power microscopic lens.

Great clump of Tali bamboo along a mountain road in Java. This bamboo has lately been established in the American tropics.

64H

Java Ho!

serve me with coffee. He would then run out in the street to secure a tiny dos-à-dos, and climb up in front with the driver to accompany me in the cool air of sunrise to the shady recesses of the marvellous botanical garden.

One of the wonders of the tropics is the Talipot palm (*Corypha umbraculifera*) in bloom, as it so vividly exemplifies the strange forces which move in the plant world. In the early sunlight, I stood one morning watching clouds of the small, white flowers drifting down from the enormous flower cluster fifty feet above me in the crown of the immense palm. The ground was white with blossoms, and myriads of bees were busily gathering honey. This was the closing scene in the life of this palm. Like the Century Plant, it blooms but once and dies. The creamy inflorescence weighed a ton or more, and the palm trunk supporting it was two feet in diameter.

I was not a systematic botanist, and did not perhaps fully appreciate the wealth of species which had been gathered in this Garden from the vast archipelago which stretched away across the Java Sea to that mysterious, unexplored island of New Guinea two thousand miles away.

Those were days when much attention was being paid to adaptations of all kinds of plants to climate. Gottlieb Haberlandt had just described the leaf pores which enabled plants to evaporate their moisture in the saturated atmosphere of such a place as Buitenzorg where there is over three hundred inches of rainfall annually. Hydathodes, these pores are called. Many intricate contrivances to insure cross-fertilization by birds and bees and other insects were also first described about this time.

One morning, across the street from my window, I heard the rustle of dead leaves in the native schoolyard. A giant Ficus tree had begun to drop its leaves, and a Javanese coolie was raking them up. Every day he raked them into windrows and burned them. It made me homesick for autumn days at home. But in Java the moment a leaf fell, a new one pushed forth in its place, and in another week the tree was in full leaf again.

Treub believed that tropical trees which drop their leaves had gradually worked their way northward into the temperate zone because they had at least a brief dormant period, and while dormant could stand more cold.

The Botanic Garden was full of termites. After the first bewilderment

65

of sight-seeing, I began to search for termite nests with the help of Papa Iidan, the Javanese who collected material for the scientific men working at the Garden. Papa Iidan knew these insects well, for they were at work around him everywhere. Their narrow trails of mud could be found on almost every tree in the Garden and on every wooden building, too. Once a runway was broken, out swarmed the termites by the thousands.

The termites resemble ants, although they belong to quite a different order of insects (Isoptera), an order of straight-bodied creatures, without the waistline of the ordinary ant. As a matter of fact, the ants are the natural enemies of the termites, and to defend themselves against the ants the termites have armies of soldiers. It was easy to tell the soldiers from the workers by their longish heads and the long pincer-like jaws which they were continually snapping. The soldiers were exceedingly vicious and would bury their pincers in one's finger and hold on with bulldog tenacity. Curiously enough, the soldiers were all blind, as were the workers, but the workers' heads were smaller than those of the soldiers, and their jaws, though inconspicuous, were marvellously equipped with teeth devised for sawing wood. Both workers and soldiers had long antennæ, resembling strings of tiny beads, which they waved continuously in the air.

It would take me too far afield to discuss the rôle these insects play in the economy of nature. There is perhaps no single factor more important in the complicated phenomena of the tropical forest than the termite. Every fallen tree, every dead branch or twig, almost everything made of wood, is quickly reduced to a pulp by the mouth-parts of the worker termites and passed through their digestive tracts with incredible speed. Another amazing thing about these insects is the fact that, apparently in order to digest these chewed-up fragments of wood, the digestive tract of the termite is inhabited by species of microscopic protozoans.

There is probably no more extraordinary example of architectural ability shown by any of the lower animals than that exhibited by the termites all over the world. Each species builds a distinct type of nest. In Java, termite nests, or mounds, were to be found here and there throughout the Garden and were abundant near the kampongs in the neighborhood. Before a week had passed, Papa Iidan and I had opened up many of these nests and I became engrossed in studying one of the most fascinating groups of living organisms which inhabit the world.

Almost the first mound that we uncovered contained what I at once felt must be a mushroom garden. However, it took me six months of constant work to prove beyond the shadow of a doubt that I was right, and since then other scientists have proved it even more conclusively.

A termite nest is a dwelling, a fortress, and a truck garden. Built into the complicated, mud runways of the nest itself, are the gardens in which the termites grow mushrooms to feed their young. Each species of mushroom-growing termite apparently develops a different form of garden. Like our own mushroom gardens, these are composed of dung. But the termites make their fungus gardens with an art and skill in striking contrast to our gloomy mushroom cellars. Theirs are fragile, delicate affairs, honeycombed with myriads of passageways. These pathways through the mushroom garden are separated from each other by thin partitions composed of countless millions of individual deposits of excreta worked into position by the termite workers. This material is made up of fragments of digested, dead wood which that individual has probably gathered from some decaying forest tree hundreds of yards away. On this deposit grow the mushrooms.

Travelling back and forth through the long galleries or tunnels of mud which the colony is continually constructing, the swift workers are able to secure their daily meals of dead wood from an ever-widening section of the forest without going "out-of-doors," always progressing under cover from the sunlight in their mud tunnels. Being soft-bodied insects, they do not relish the sunshine or even the outer air unless the humidity is close to saturation.

There was something utterly fascinating about these gardens built within the termite strongholds. Their interior walls were covered throughout by pearly-white filaments of fungus, which lined the passageways as though with white velvet.

I always believed that these fungus filaments were deliberately maintained a uniform length, perhaps sheared off by the saw-like jaws of the workers, just as humans keep the grass cut on their lawns. Here and there on the white velvet covering of the mushroom gardens, glistening white bodies, about half the size of the head of a pin, rose on short stems.

It was a great day when I held the first fungus garden in my hand and knew that I had been right in believing that the termites were mushroom-growers, and probably mushroom-eaters. I experienced the excitement which accompanies any real discovery, great or small. But how was I to prove the use to which these creatures put their fungus gardens? The instant a ray of light penetrated the darkness of the mushroom garden, every creature scuttled out of sight, or froze to immobility.

If you see a rabbit in a lettuce bed, you are pretty sure that he is there to eat the lettuce. But, if nobody has seen him eat it, you would be obliged either to watch him until you did see him munching the leaves, or to kill him, cut him open, and identify lettuce leaves in his stomach.

I tried my best to surprise the young termites actually feeding, but I could not get my microscope in place quickly enough. So I took the other method. I killed and dehydrated a number of the young, and embedded them in paraffin. I then sectioned them with my microtome and, by means of Doctor Paul Mayer's stains, I identified the presence in their stomachs of the sporelike bodies which composed the pearly masses of the fungus gardens. Of course, this was to be expected, for why would these creatures build and tend such complicated gardens and grow such luscious, little, white cabbages if they were not to be eaten? But just the same it was necessary to prove it.

After making this discovery I ransacked the countryside to see if there were other species of termites, and whether or not they cultivated different species of mushrooms. I found that there were at least two other species of termites which made their own peculiar fungus gardens, both strikingly different from the one I had first studied, which was about nine inches across, and resembled a birthday cake with horizontal galleries. On the other hand, the other species of termites built gardens with the galleries running perpendicularly. Many of these would be scattered in a group under a mound of clay. Sometimes two forms of gardens would be together in the same termite mound. Once I found a tiny garden no larger than a tennis ball. It was very delicate and had been built by an extremely small species of termite about one-tenth the size of the other two species.

These termite species lived side by side, apparently in peace until I broke open the nests and disturbed them. Then they attacked one another and fought to the death. I used to put the soldiers of the different species two by two under a watch crystal and study them as they killed each other. A curious feature of these combats was that, as I have said, the soldier termites were quite blind. They used to wander furiously around the arena of the watch-glass, snapping their pincer-like, sharp-edged jaws viciously while waving their antennæ in the air. When the antagonists met it was usually near the rim of the watch-glass. The antennæ and legs would begin to fly, sheared off by the snapping pincers. Invariably one of the soldiers was left dead in the arena and the other crippled.

It is only when a termite leaves his legitimate field of devouring the decaying forest trees and converting them into mould, and turns his attention to the timbers which man uses in his dwelling, that he becomes a really important and dangerous pest.

The characteristics of the termites themselves became a part of my study of their gardens. One day, while sitting on the verandah of the hotel, a swarm of fluttering, winged insects filled the air. They were termites. The Malay "boys" arrived with brooms and pails and swept up the insects by the bucketful as they landed on the floor. Thus ended the last phase of what had been a marriage flight. I watched the winged insects chasing each other along the wall two by two. With a curious jerk of their shoulders, they broke off their wings of flight and, soon after mating, each couple prepared to begin a new termite colony by digging a hole somewhere in the ground. That creatures which have once had wings should throw them off deliberately and begin a subterranean existence, seems amazing.

Next morning I sent Papa Iidan out to search for the queen and king of my termite colony; "the Rajah," he called the king. He soon brought me a piece of dark brown clay, thicker and slightly larger than my hand and with numerous little runways leading into it. It was so hard that I had to break it with a hammer. Inside, lying side by side in a royal chamber, were the queen and king. The chamber was low and broad, and the exits from it were only large enough to let the workers and soldiers pass. So enormous and "different" were the king and queen, that it was hard to believe that they had any relation to the other termites. The king was twenty times as large as any soldier, and had great facet eyes on each side of his head, in contrast to the eyeless workers and soldiers. The queen was an enormous creature, quite as large as my thumb and something the same shape. Like the king, her head had facet eyes. The pair lay helpless in this royal chamber. With my microscope I saw that, to all appearances, their legs had been chewed off; at least they were quite inadequate for locomotion.

The World Was My Garden

It seemed fantastic that this strange queen, a thousand times the size of any worker, with her great, puffy, caterpillar-like, white body, could be the mother of the myriads of soldiers, workers and possibly individuals of other castes which made up the termite colony.

I took the pains to dissect her carefully (later I even embedded another queen in paraffin and sectioned her) and found the soft mass which composed her body was almost entirely made up of eggs. Like strings of graduated beads, smaller at the top and growing large at the base, the eggs filled the entire cavity of her body. She was the egg-laying apparatus of this social organism, this collectivity, the group of busy individuals which made up the colony.

One day I timed another queen as she laid egg after egg. She averaged an egg a second—over 80,000 eggs per day. It is said that such a queen can live ten years. If she continued at this rate, she would be laying thirty million eggs a year. During this time, her consort would furnish the sperms with which certain of those eggs would be fertilized, and from them would develop the nymphs which composed a marriage flight such as I had seen that evening in the hotel.

It was with a certain sense of guilt that I spent my days—many hours flat on the ground—beside the termite nests in the Garden and racked my brains devising means by which I might see the workers feeding the queen and king. But I believed that I was discovering something new, and the termites fascinated me completely.

Recently, my friend, Morton Wheeler, published a delightful satire on human society entitled, "Foibles of Insects and Men." In it he described the termite method of practically eliminating the male from the social order, at least reducing him to a single male for the whole colony. Wheeler's charming paper was wonderfully understanding, and served to lessen my feeling of guilt that I had played with the termites instead of mastering the relationships of the hundreds of trees which had been collected in Java. Anyway, Entomology, Botany, and Horticulture are as inseparable in their interests as the Three Musketeers.

One morning, Papa Iidan brought a branch from a bush into the laboratory. There seemed nothing unusual about it, but he pointed to an amazing leaf insect nearly three inches long belonging to the family of Phasmidæ. It exactly matched the under side of a leaf, and even had little spots which looked for all the world like "leaf spots" produced by microscopic fungi. The creature crouched against the leaf and moved only when I touched it.

Life in the delightful laboratory was equalled in interest by life in the hotel. After a fatiguing morning in the Garden, I used to swing off the step of the dos-à-dos just in time for the "rijsttafel" at the long hotel table. At this amazing meal we piled our deep plates high with steaming rice, and, as the long line of turbaned waiters offered them, added in turn a bit of fried chicken, slices of egg, perhaps a sardine, a meat-ball, or a fried banana. Then came the sambalang,—a great tray containing red Macassar fish, tiny pickled ears of corn, burnt peanuts, roasted coconut, sweet mango chutney, a darker brand of Indian chutney as hot as liquid fire, a peculiarly flavored pickle made from Gnetum (a strange climbing shrub), and others which I cannot now recall. After making our choice, we poured over the heaping mass a quantity of curry sauce made fresh each day from ground-up cardamons and fiery-hot red peppers. The first mouthful of this mixture brought the perspiration to one's face and started something, whether it was digestion or not I do not know. However, in the end, one experienced a sense of well-being and drowsiness which admirably suited the tropical custom of a siesta after luncheon, and the hotel became as quiet as the grave from two to four.

It was after one of these enforced siestas, which I at first detested, that I saw my first mangosteen (*Garcinia Mangostana*). A coolie passed through the patio balancing on his shoulder two baskets hung from a bamboo pole. Seeing me, he held out a bunch of mangosteens, some dozen or more fruits tied together artistically by shreds of bamboo.

Resembling an apple of deep brownish-purple tint, mangosteens have short stems and four thick, leaflike bracts which form a rosette holding the purple fruit. The flower-end of the fruit is marked by the persistent stigma composed of seven triangular segments slightly raised above the surface.

Mangosteens have a tough, firm shell, and I had trouble breaking it open with my hands. The coolie showed me how to cut through the hard rind with my penknife and lift off the top as one would lift the cover from a sugar bowl. There, lying loosely in a pink cup, were five ivory-white segments glistening with moisture. They could be taken from their shell as easily as bonbons from a dish. Many of the segments are seedless, and the seed itself, which is also edible when cooked, is thin, brown, smooth, and flat. The meat has the consistency of a green-gage plum but a flavor which is indescribably delicious. Like many tropical fruits, there is a sprightliness of flavor, a suggestion of the pineapple, the apricot, the orange.

Of course, I immediately wanted to see this fruit on the American market, but there were many difficulties to be overcome. Java in those days was almost as distant as the moon. During the nineties, I made several attempts to get living seeds to the United States,—even coating some with paraffin, although the best method proved to be packing them in dry charcoal.

The first serious efforts of the Office of Plant Introduction were made in 1900. We soon found that the mangosteen was too tender to grow without protection even in southern Florida, and, consequently, included in our scheme the introduction of closely related relatives of the mangosteen, hoping to find hardier stocks. *Garcinia tinctoria* proved the best "relative" which was experimented with; *Garcinia binucao* from the Philippines seemed unusually hardy in South Florida and should be further tested as a stock. In fact there are many interesting, fine fruited strains of Garcinia which deserve to be studied.

We made repeated attempts to secure two species of Garcinia which are native to regions subject to frost, and were finally successful in growing *Garcinia Mestoni* from Queensland. The other species is *G. multiflora* from near Kiaying Chow.

From our experiments we soon discovered that there are definite problems to be met during the early stages of a mangosteen's existence. Not only is the vitality of the seeds extremely low, but the young plants have a very weak root system and are easily checked in growth. An ingenious method was devised to resuscitate dying seedlings by inarching them on to a vigorous, rooted seedling of some other species of Garcinia. This seemed to revive the young, dying mangosteens much as blood infusions are able to save the lives of human beings. There may be more than an analogy here.

Today there is a healthy orchard of mangosteens at the Summit Experiment Garden in Panama. But we did not manage to establish this until 1923, after an ignominious defeat at the hands of the military régime in Panama when we made a first attempt in 1905, as I shall relate.

Java Ho!

Doctor Treub told me that he preferred another Javanese fruit, the pulassan, and I must confess that it does run the mangosteen a close second. The pulassan and the rambutan are two tropical relatives of the famous litchi of South China, and are all delicious fruits which were quite unknown in the Western Hemisphere in the nineties.

The fruit of the pulassan (*Nephelium mutabile*) is the size of a plum, and has a deep pink, pebbly surface, and a fairly thick skin. The single seed, with its surrounding pulp, comes out of the shell like a grape, although it is drier.

The rambutan (*Nephelium lappaceum*) is much like the pulassan but is covered with soft, curled, tentacle-like hairs which make it somewhat resemble a chestnut bur. Unfortunately, both of these fruits proved too tender for Florida. The litchi (*Nephelium Litchi*), which resembles the pulassan in appearance, is hardier and has been successfully grown in Florida.

I did not ask the coolie why he had no durians, for I had already heard that none were allowed in the hotel. The curious odor of a ripe durian (*Durio zibethinus*) is something which few Europeans can endure. However, Alfred Russel Wallace declared that it was worth a trip to the East Indies just to taste a durian. Others, too, have told me of their passion for this fruit. To my shame I must confess that during my first stay in Java I could not bring myself to eat one. The prejudice of the people around me was so great, and the odor of rotting durians in the market-place was so offensive, that to taste the fruit assumed the proportions of a major operation which I could not force myself to undergo.

When I returned to Buitenzorg later with Mr. Lathrop, I slipped away one Sunday afternoon to the native village, and tasted the custard-like pulp of a durian. Its flavor was indescribably rich and sweet, and I enjoyed it at the moment but, like other strongly flavored foods, such as raw onions for example, its odor returned to plague me.

I can still remember the expression on Mr. Lathrop's face when I reached the hotel that afternoon.

"Fairy!" he shouted, "You've been eating durian! I smell it! You get out of here and don't come near me until that stench has worn off."

In some respects, the durian is the most remarkable fruit in the world. The head-hunters of Borneo will commit murder to possess it. The fruits

73

weigh from five to ten pounds and are about the size of a small coconut, completely covered with sharp prickles. The species most commonly cultivated is borne on a tall forest tree. Because of the weight and thorny surface of the fruit, it is dangerous to walk under a durian tree when the fruits are ripening. The seeds may be roasted and eaten like chestnuts. I have heard that the Chinese obtain an oil from the fruits which they use for washing purposes.

All this, of course, I did not know that day as I watched the coolie pick up his baskets, bow, and trot away. By this time clouds were gathering over the Gedeh. Flashes of lightning shot through the dark mass which covered the mountaintop. Rapidly the sunlight faded. Thunder rolled, and we were in for the afternoon downpour which came daily about four o'clock. The deafening beat of falling water is as much a part of the tropics as the brilliant sun and waving palms.

As quickly as it came, the storm passed. The cool, refreshed world assumed a hue of greenish-gold in the evening sunlight—a light unlike any seen in northern latitudes. Soon it was night again in that land of brief twilight.

One afternoon, I heard a curious, ripping sound and glanced toward a tall coconut tree just in time to see one of the coolies dodge a falling leaf. It was amusing to realize that my idea of a leaf, built up by exclusive association with the temperate zone, was an utterly inadequate conception. Imagine the incredulity of a Kansas boy if you told him that he would ever run from a falling leaf!

One of the delights of Java had been making the acquaintance of the coconut. It was an exciting experience to drink the milk from the nuts. Visions of Robinson Crusoe always came to my mind. How little I dreamed that I would one day own a place in the United States with coconut palms on it where I could drink coconut milk every day!

It seems almost inexplicable now that it did not occur to me to busy myself introducing the coconut into America, but in those days there were no gardens or agencies in the United States equipped to handle tropical plants. In the entire tropical regions of South Florida, there were only a few homesteaders and a meager number of winter visitors who came merely for a brief stay at the few hotels.

Fortunately, the coconut came into Florida from the West Indies almost on its own, so to speak, and there are now many thousands growing along the east coast. Their rustling fronds glisten in the sunshine and add greatly to the beauty of our streets and homes.

My fellow boarders at the Hotel Chemin de Fer were a strange but fascinating lot. My next-door neighbor on one side drank innumerable "bitterjies" and pestered me with strange inquiries about the English language, which he thought he spoke quite well.

"I am to be to are to want to go to bed. That is it English?" he once said.

My neighbor on the other side was a Belgian botanist named Cleautreau from Erera's laboratory in Brussels; a brilliant, emotional young man who wrote letters of forty pages in fine handwriting to his mother, and later, poor boy, died of a broken heart soon after his mother passed away.

DeMunnick, another hotel guest, had been a Dutch fonctionnaire. He was interested in schemes for utilizing the fibers of the Kapok or Silk-cotton tree which were then allowed to go to waste. He complained bitterly that his countrymen invested their money in American railroad stocks, but had nothing for the development of the Dutch East Indies.

This small group was augmented one morning by the arrival of Herr Rapps, the man I had met in Genoa who had blown out the gas and asphyxiated his bride. With his advent, our peaceful existence came to an end. DeMunnick soon confided to me that Rapps' baggage consisted of two enormous cases of Marsala wine and innumerable white sailor-caps. It also transpired that he had come to collect, not plants, but reptiles for a brother in Berlin.

Rapps sent word to the native quarter that he would buy live snakes and lizards, and swarms of natives appeared bearing all manner of creeping things. He soon had a dozen great, long lizards which he fastened to the legs of his sofa. When their activities kept him awake at night, he hung them out of the window, but this manœuvre was not successful, for they soon scratched the plaster off the walls and he had to cut them loose. The natives would catch them and sell them to him again the following morning. It was a mad performance from first to last.

Rapps also bought every available species of snake. We warned him about the deadly ular blang which, like the coral snake in Florida, has a mimic, a snake so nearly like it that it takes an expert to tell them apart.

I went into his disorderly bedroom one afternoon and saw an ular blang in a big candy jar with a cigarette stub holding the stopper open to give it air. I told Rapps that this was the poisonous snake, but he scoffed at me and said that he knew snakes, and this was the harmless mimic. Nevertheless I took the precaution of tying down the stopper before I left that night and made him promise that he would test the snake the next day.

In the morning, therefore, Rapps bought a young chicken. When I arrived, he had the snake tied loosely to a stanchion on the verandah and the chicken held in his bare hand as he presented the bird for the snake to strike at it. The snake most certainly would have struck Rapps' knuckles had I not pulled him back and helped him to arrange a safer test. I put the chicken in a box, covered it with a screen, and then put the snake in too. In a few minutes the chicken was dead. Needless to say, Rapps' reputation as a herpetologist sank rapidly to zero.

His behavior also went from bad to worse. He disgraced himself at Treub's immaculate retreat in the mountains by killing a wild boar, dressing it in the laboratory, and trying to make salt pork of it in a leaky flour barrel.

Eventually Rapps decided on an expedition to the little island of Nias off the coast of West Sumatra. After shipping his badly prepared specimens to his brother, he departed one day taking with him what was left of his Marsala. Somehow, somewhere, he completely disappeared on this expedition. He was not regretted, I fear, for Doctor Treub never could understand why Doctor Penzig sent out such a man.

Most of the conversation among Treub and his associates concerned the problems of the Dutch East Indian planters. Those were the early days of rubber and Sumatra tobacco. Also, at that time, Java coffee culture was being superseded by plantations of the Assam tea. There was much talk about which type of rubber tree would win out as the future source of rubber. Treub believed that the "parlor" rubber tree (*Ficus elastica*) which was native to the East Indies, would have a better chance than *Hevea brasiliensis* from the Amazon. However, the Hevea now constitutes practically all of the rubber plantations of the world.

Occasionally I would go to the Hotel Bellevue to see the sunset on the Salak. It brought back memories of the fascinating man to whom I owed the wonderful experience I was having. Mr. Lathrop had described this scene when we parted that moonlit night in Naples, and no wonder the picture lingered in his mind. The music of the bamboo flutes drifted up at twilight from the kampongs, while native men, women and children bathed in the swift stream below, with charming decorum and modesty. Rustling palms swayed above, and great, cloud-capped mountains raised their heads in the arches of the lovely rainbows which followed the afternoon rainstorms.

In the virgin forest on the slopes of the volcano Gedeh, Treub had built a small laboratory where the air was cool and fresh. He suggested my going there to carry on some microscopical studies, and Mario cheerfully packed my equipment into the Standard Oil tins which were universally used as containers. Loaded down with fifty pounds of baggage on each end of a bamboo pole, the coolies trotted off over the pass while Mario and I followed in a dos-à-dos.

The rice around Buitenzorg had been harvested, and the workers in the paddy fields were planting the next crop, wading knee-deep in the mud. The terraces extended up the mountainside almost to the summits of the hills. As we drove up the Pontjak Pass, the season seemed to change. The late summer landscape melted gradually into spring, and I saw demonstrated the relation between altitude and latitude. We were following the spring north, as tourists from Florida so often do when they leave midsummer verdure in Miami in March and find the cherry blossoms just opening in Washington.

A delusion common in those days was the belief that high altitudes in tropical mountains resemble the temperate zone, and that almost any northern plants could be grown in the tropics if planted high enough. Similarly, it was supposed that the mountain plants of the tropics could be cultivated in temperate regions. Many years and many failures finally disproved this theory. Aside from certain forms, the mountain species of the tropics cannot be cultivated in the lowlands of the temperate zone. Moreover, few of the temperate species are ever happy in the mountains of the tropics.

I received my first impressions of a virgin, equatorial forest during the days I spent alone on the steep trails of the Gedeh. Cleautreau joined me there to pursue the study of the digestive juices of the wild pitcher-plants, while I discovered some new and interesting forms of parasitic fungi. In spare moments, I proved that my beloved termites were higher in the social scale than ants. I had brought with me nests of different species of termites and put them beside nests of the same species living there in the mountains. To my amazement, individuals of the different nests did not fight each other, whereas it is well known that the members of an ordinary ant colony will at once fight the individuals of a neighboring colony. It would appear, therefore, that the ants are still in the tribal stage, whereas termites might be said to have a higher or racial organization.

I had often heard of the edible bird-nests which are so much used by the Chinese in soups, and enthusiastic when Cleautreau proposed a visit to the caves in southern Java where the swifts live and build their nests. These birds are a genus allied to the European swift.

It was quite a trip to the caves, and then we had to descend far to reach the bottom. We crawled down on long, primitive, bamboo ladders and finally found ourselves walking on a soft mass of débris on the floor of the great caverns. We could hear the swirl of the swifts above us as they escaped through an opening higher up in the cliffs.

We think of a bird-nest as constructed of sticks or mud but, when I climbed up the rock wall and pulled off one of those edible nests, it was composed of a substance as soft as some forms of mushrooms and it was almost translucent. These nests are, in fact, made of the foamy saliva from the salivary glands of many species of swifts which inhabit the caves of numerous Oriental islands and the mainland of southeastern Asia as well. Their exploitation is a profitable industry, and the right to collect nests is sold for a good price.

The caves also shelter and support a host of insects, some of which I at once began to collect. However, when I found a huge centipede within a few inches of my head, my ardor cooled momentarily. Armed with some bamboo pincers which Mario had made for me, I captured the creature, which had legs nearly two inches long and vicious-looking jaws. I also took some of the brownish, powdery material which covered the floor of the cave. When I examined it in Buitenzorg, it proved to be a writhing mass of tiny red mites!

My first information about the new quinine industry came from a lanky Hollander who appeared on the verandah of our hotel one day and settled beside me for a chat. He was a cinchona planter, one of the first to successfully grow the trees in Java, and he had built a factory to extract quinine from the bark.

Hitherto quinine had come largely from the wild cinchona trees of

Java Ho!

Ecuador, and, as is the case of all wild cultures, the supply of bark could not be relied upon. It varied with the season and the temper of the Indian bark gatherers. Seeds of the different species had been sent from Ecuador first to Jamaica and then to the Orient for trial. In the nineties, their culture was being attempted in British India, Ceylon, and the mountains of Java. Although it may have been some special climatic condition in western Java which was responsible for the success of cinchona culture there in contrast to the failure in other tropical regions, yet I am inclined to believe that it was the persistence of the Dutch planters, assisted by government investigators, which contributed to the success of the new venture.

The work of breeding the different species of cinchona and selecting the best strains was just beginning when Doctor J. P. Lotsy arrived in Buitenzorg to find out in what part of the bark the alkaloid quinine was located and how it varied in amount from month to month. I had known Doctor Lotsy in Baltimore and was much pleased when he invited me to visit him at the plantation where he was establishing a headquarters.

In a clearing in the primeval forest, the cinchona trees had been planted, and the government had built a tiny laboratory and dwelling house. Through the jungle ran crooked trails made by the small Java rhinoceros, an animal today exceedingly rare (if not extinct) in the island.

The great forest trees were laden with orchids. If any of them seemed especially interesting, in a few minutes the skilled woodsmen had the tree lying at my feet. Botanizing through the branches of a tree in a tropical forest is a fascinating experience. A wealth of beautiful forms are found among the many tiny orchids, ferns, lycopods, lichens and bromeliads. Numerous species of plants and animals live only in the aerial world of the tree-tops, neither growing on the ground nor descending to it. Like the aquatic flora and fauna of streams and lakes, they live their own peculiar existence.

After a few days in this wild and beautiful mountain retreat, I received a letter from Mr. Lathrop, mailed in Chicago and sent on to me from Buitenzorg. It brought the astonishing news that, soon after his letter reached me, he would be in Java. He was coming with his brother and sister-in-law, and he warned me to beware of Miss McCormick, a pretty American girl who would be one of the party. I gathered that they would want to see the whole of Java as expeditiously as possible.

Tumbling my specimens and baggage into the cart, I regretfully bade good-bye to Doctor and Mrs. Lotsy, whose kindness I have never forgotten. Down the mountainside I hurried, and took the train to Tanjong Priok, where I arrived just in time to meet the boat.

I was anxious to make a good impression on my benefactor, whom I barely knew, and determined to be especially helpful and attentive to Miss McCormick.

Making my way on board, I found Mr. Lathrop with his brother, Mr. Bryan Lathrop of Chicago. They were so utterly different from the men among whom I had lived that they seemed beings from another world. An atmosphere of dignity, culture and wealth surrounded them; they were obviously men who had travelled widely and knew the world, and I stood awkwardly answering Mr. Lathrop's questions, very ill at ease. Mr. Lathrop told me afterwards that he had been shocked by the appearance of his "investment." I was emaciated after a touch of fever and, moreover, was dressed in the conventional two-piece white-duck suit adopted by the Dutch, which buttoned up under the chin and was extremely unbecoming to every one.

Mrs. Lathrop soon appeared with Miss Carrie McCormick. Mrs. Lathrop could ask more questions in a minute than any one I had ever met, and her enthusiasm was unlimited. Among other things, I remember her asking me whether I had ever seen the beautiful jungle-trail in Florida between the little town of Miami and Coconut Grove. I often think of her now as I drive along Brickell Avenue, which was once this trail but today is lined with houses.

As I escorted the party to the Bellevue Hotel at Buitenzorg, they seemed interested in everything, particularly in what I had been doing. I therefore brought my microscope to the hotel, and did my best to show them the fungus gardens and the battles of the termites.

I had expected the Lathrop party to stay for a reasonable length of time, for it seemed inconceivable that people would come all the way from Chicago to spend but a few days in such a paradise. However, I soon discovered that they had only ten days to devote to the whole island. I was delegated to act as their guide, a position which I filled rather badly, for I knew little of the island outside the botanic gardens. My Dutch friends, as residents in the tropics always do, advised our seeing the waterfalls and mountains and all the things which they enjoyed themselves, forgetting that high mountain scenery in the tropics resembles that of temperate

The elaborate bamboo bridge at Bandjar Negara, Java.

Left: Fruit of the Munkonga (*Garcinia Livingstonei*), an East African relative of the Mangosteen, which has shown itself capable of living and fruiting in Florida. *Right:* Mangosteen flowers.

Garcinia Hombroniana.

The pink-fleshed Shaddock of Java.

Open Mangosteen, showing the segments ready to be eaten.

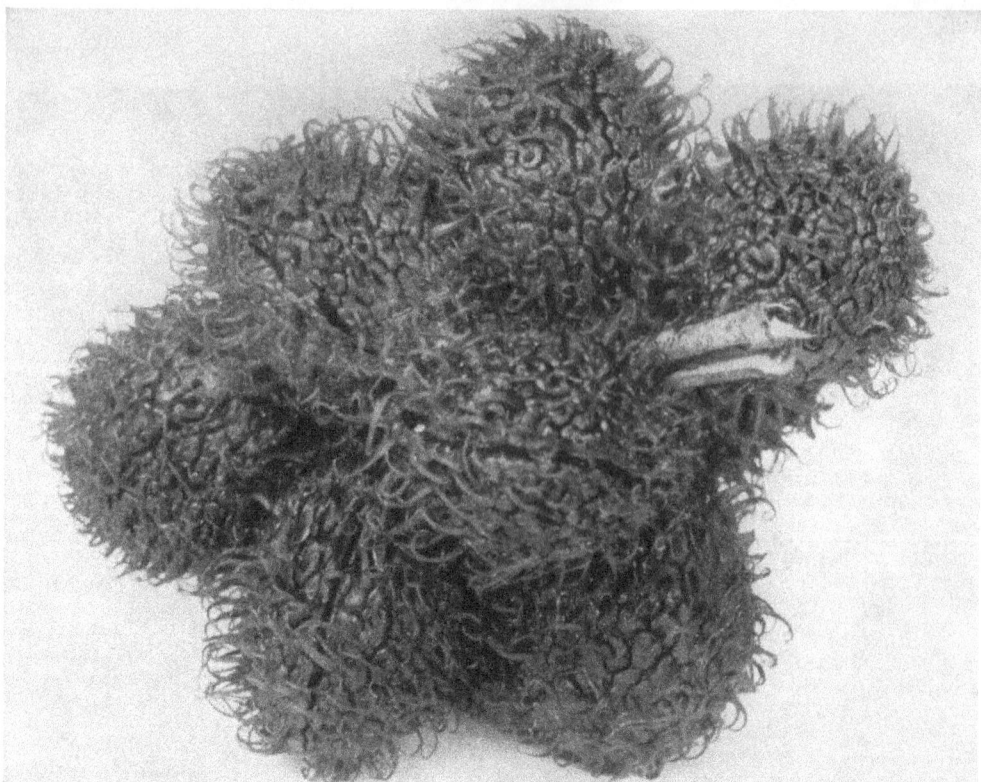

Rambutan fruits with their soft, curly covering somewhat resemble chestnut burs.

The Rambutan is a relative of the Litchi.

On our way to the Birds' Nest Caves we paused for a drink of coconut milk. (*Author seated in the center.*)

80E

The Hotel Chemin de Fer in Buitenzorg. The Belgian botanist, Cleautreau, is reclining on the left.

80F

Many of the Dutch women living in Buitenzorg adopted Javanese costume.

Dr. Lotsy's laboratory and dwelling house on the Cinchona Plantation.

regions more than do the lowlands where the palms and bamboos grow more densely.

I had heard about Bat Island, which was the roosting place of thousands of flying foxes or fruit bats, strange nocturnal creatures, larger than a crow. Rather to my surprise, the others did not care to see it and sent Miss McCormick and me to visit it alone. The tall trees on the little island were black with bats, tens of thousands of them. A Javanese coolie rowed us out in the lake and then clapped two boards together. Whereupon, the whole screaming, screeching colony lifted like an enormous black cloud and flew away. I enjoyed the sight immensely, but as we were climbing the hill I turned to Miss McCormick, who said very quietly,

"I am not afraid of snakes at all, but I simply cannot endure bats."

The trip through Java was not a great success either, and it soon became evident that Mr. Lathrop was a bit restive and obviously had not been in the habit of travelling in so large a party. One evening, in the hotel in Bandung, matters came to a head between Mr. Lathrop and his brother Bryan, and when we retired that night he said to me,

"Fairchild, I don't like this travelling with a party. I'm not accustomed to it. They don't like the same things that I do, and what's the use? If you'll come along with me, as my guest, we'll go up the coast of Sumatra and I'll show you something of the world. You're working too hard anyway. You're worn out. You need a change. We'll go back to Buitenzorg, pack up your specimens and ship them home, and make a real collecting trip up the west side of Sumatra."

I murmured protests about my unfinished termite studies; what Doctor Treub would say; and the difficulty of leaving so quickly. I had always been accustomed to doing things quietly with no hurry until the inevitable last-moment rush. However, my protests were of no avail. Although he did not say so at the time, Mr. Lathrop was not enthusiastic about the fungus gardens of the termites, nor about some of my other researches which I believed would "shed a ray of light in some dark corner in the field of science."

That evening in Bandung, my real acquaintance with Mr. Lathrop began—an acquaintance which developed into an intimate comradeship and continued to the day of his death. This friendship forms the major romance in the story of my life, for we travelled the world together for eight years.

CHAPTER VII

The Lathrop-Fairchild Odyssey Begins

Mr. Lathrop discovered that the only boat going up the west coast of Sumatra would sail two days later, and I believe that the resulting whirlwind of our departure from Java left a very bad impression on all my Dutch friends.

Mr. Lathrop expected me to take care of the baggage; I misunderstood his directions; he discovered the trunks in process of being hauled off to the wrong station. A frantic scene ensued, and when we finally reached the boat, I was completely exhausted. It was only then that I began faintly to realize with what an extraordinary, restless, forceful character I had thrown in my lot. In a vague way, I had known that Mr. Lathrop travelled fast, for he had been around the world twice while I was peacefully studying in Germany, but I had not contemplated what his speed involved. As our present trip had been proposed as a collecting expedition, I expected time to collect. However, that was not Mr. Lathrop's idea at all.

A visit to the market in Padang early our first morning showed me how many new and interesting food plants there were if only we had an established place where they could be sent. I was ready to spend a month or a year right there in Padang, but Mr. Lathrop was set for the next place.

Beside the roadway in the mountains, I saw a lovely yellow raspberry (which I finally did collect in 1924), and there I also received a lasting impression of a growing bamboo cane. A slender, jointed, thirty-foot column of deepest chlorophyll green stood beside the path. I grasped its smooth freshness and shook it; the joints separated and the whole thing came tumbling down upon my head. As I stood aghast, I little dreamed how much of my life would be devoted to the introduction of bamboos into the Western Hemisphere.

From Padang, on the west coast of Sumatra, we pushed up to Fort de

The Lathrop–Fairchild Odyssey Begins

Kock. In the Kloof von Arau, where thirteen waterfalls cascade over gray cliffs lush with tropical verdure, I lagged behind intent on my collecting. Landscapes and waterfalls would not go in my bottles, nor could I take them home to study them.

After waiting an hour for me, Mr. Lathrop returned somewhat irate. He found me where he had left me, still gathering termites. There ensued a rather heated dialogue between us which almost sent me back to Buitenzorg.

"If you're going to travel with me, I'll show you the world; but you can't stop every minute and collect specimens or you won't get any general idea of the countries we travel through," said Mr. Lathrop.

From that time on I tried to sandwich my biology between the other experiences of travel. Mr. Lathrop was not a collector at heart and remonstrated about my vials of dead insects. No one would ever need so many of a kind, he said. This sounded like heresy to me. I was a candidate for curator of microscopic fungi in the National Museum, and, as a matter of fact, hold that position to this day.

It was a great surprise to Mr. Lathrop to find that I had voyaged half way around the world without knowing what he called "the first principles of travelling." He tried to "round out my meager education," but I could not see any sense in dressing up and learning the endless rules of etiquette.

On the day before Christmas, we sighted Kotaraja. I had just finished a long, reassuring letter to my mother, telling her that, although I was so far away, I had never seen a serious accident of any kind. As I sealed the letter, a native steward climbed from a hatch close by. Suddenly, another steward gave him a kick and ran, and I heard a shout, "Amok! Amok! Amok!"

The two stewards grappled with each other on the stairway, one with a hatchet and the other with a knife. The cry rang out again, and the struggling pair fell headlong into Mr. Lathrop's cabin. The Captain and Mr. Lathrop reached the scene simultaneously and pulled the combatants out by the legs, as another steward appeared and wiped the blood off the baggage.

I opened my mother's letter and added a postscript.

Mr. Lathrop and I spent Christmas day in Kotaraja. We were not allowed to go outside the town, but were interested in seeing the block-houses and the barbed-wire entanglements constructed by the Dutch to assist in their warfare with the fierce, wild Achinese, those untamed, savage people of North Sumatra. This was one of the first occasions when barbed wire was thus used.

New Year's Eve found us off the little island of Penang. A Catholic priest was on board, a man of great refinement, who had a garden of his own and a fine collection of orchids. Mr. Lathrop and I talked with him well into the evening about tropical horticulture.

When the priest bade us good-night, I followed Mr. Lathrop to his cabin and he began to lay before me his ideas of what a botanist could do if he were given an opportunity to travel and collect the native vegetables, fruits, drug plants, grains and all the other types of useful plants as yet unknown in America. I cannot recall the conversation exactly, but I know that when the clock struck twelve and the new year of 1897 began, I had promised Mr. Lathrop that I would take up a study of the plants useful to man and, together with him, find a way to introduce their culture into America. It was a rather vague, ill-defined agreement, but it was a turning point. From that time on I began to pay attention to the economic plants around me, and to work with Mr. Lathrop on a scheme for a botanic garden which we thought should be started in the Hawaiian Islands.

Singapore in the nineties was the great emporium of the East, the place where you could have duck suits made to order for a bagatelle. Here the club life of the Orient was at its height, while in the harbor rocked craft from many lands. The picturesque sails of Chinese junks and catamarans flapped and bellied in the breeze; sampans and steamers passed in strange contrast.

The botanic garden was under the charge of Henry M. Ridley, an outstanding scientist. He received me with a cordiality which I shall never forget, and offered to show me the Brazilian rubber trees (*Hevea brasiliensis*) with which he was experimenting. He picked up a mustard tin from the table, saying,

"If you're not familiar with the latex of the rubber tree, perhaps you would like to have a specimen."

As we walked towards the grove, he complained of the lack of interest

Barbour Lathrop and David Fairchild off Kotaraja, Sumatra, Christmas, 1896. The cabin is equipped with a punkah which can be seen above the Author's shoulders.

The Wampi, a native of South China. It is related to the Citrus family. (Page 87)

84B

the British planters of the Malay States showed in this Brazilian rubber tree which, as I have said, has since become the rubber tree of the world. He had found what he felt certain was the correct way to tap the trees, and he believed that rubber-growing would be a success in that part of the world if only the planters could be persuaded to grow the Hevea.

Ridley drove his jackknife into the bark of one of the larger trees while I held the mustard tin to catch the brilliant white latex as it flowed from the cut. Soon the can was a quarter full. At my hotel, I watched the fermentation process which took place. Bubbles formed and the whole mass hardened, assuming the appearance of a piece of cheese. I lifted it out and manipulated it. I have it still, and it is probably one of the oldest pieces of rubber in the United States today.

Doctor Ridley is still alive, I believe. His discovery of the correct method of tapping the Brazilian rubber trees was the turning point in the cultivation of rubber on a plantation basis in the Orient. But neither he nor I had the faintest glimmering of the rôle those Brazilian trees were destined to play in the development of the world during the next quarter century. I often remember how avidly we discussed the possibility that, in some future era, rubber might become cheap enough to pave the streets of London and silence the clatter of horses' hoofs on the cobblestones. Little did we dream that rubber might roll on the streets as tires on motor vehicles, silencing the clatter of hoofs by superseding the horses.

Thirty years after my visit to Singapore, my colleagues in the Department of Agriculture and I had the pleasure of presenting Doctor Ridley with the Meyer medal in recognition of his great discovery.

In a few days Mr. Lathrop and I embarked on a little freight steamer, bound for Bangkok over the shallow waters of the Gulf of Siam which shimmered about us as smooth as glass. I have never seen the sea so still, and basking on its surface were hundreds of black and white striped sea-snakes. One realized only too well why it was necessary to put snake-guards on the hawsers and anchor chains to prevent these venomous reptiles from climbing into the port-holes.

It was hot when we reached Bangkok, steaming and sickeningly hot. The little hotel was chock-a-block full, and I was allotted the ballroom as a bedchamber. The proprietor was so proud of his newly installed electric-light system that he kept the lights burning day and night. Unfortunately the dazzling chandelier in my ballroom had no separate switch, and I was expected to sleep in this boiling-hot room in a glare equal to Broadway. When I asked the Chinese servant to turn the lights off he said, "No can do," and disappeared. After wooing slumber unsuccessfully for a torrid hour or so, I piled the tables, chairs and boxes into a tower and clambered upon them to screw out the bulbs. The room had been still enough when I crawled out from under the mosquito-netting, but as I turned out the last light, balancing on the perilous pyramid I had erected, I heard the rustling noise of something scuttling over the floor. It was a noise which I had not heard before, and I screwed the bulb back again to behold hundreds of the largest, blackest cockroaches I ever saw—thick, ugly brutes—running for cover under the baseboard of the ballroom. The place was alive with them.

Realizing that I could not spend the night perched on my pyramid like an uneasy acrobat, I turned off the light again and made a dive for the mosquito-netting. I tucked myself in, sincerely hoping that I was safe from the ugly beasts, to me the most loathsome of the insect tribe. In the days of sailing-vessels, when the sailors were in their bunks, they used to protect their fingers and toe-nails with leather coverings as the roaches particularly enjoyed gnawing these parts of their bodies. Since these insects harbor all kinds of intestinal parasites related to the nematodes infecting man, I think that my abhorrence of the cockroach is entirely warranted.

Bangkok has frequently been called the most fascinating city of the East. I shall never forget the klongs (canals), the swift-flowing Menam River, the floating houses and rafts. Above the city rise gorgeous temples, some crowned with gold, and others with roofs of blue and white mosaic composed of myriad fragments of china. In the bazaars, the short-haired, gayly dressed women, with their blackened teeth, add interest to the truly foreign scene.

As "distinguished American travellers," we at once met John Barrett, our young American Minister from Oregon. He was a rising diplomat who felt the dignity of his position and maintained it in spite of numerous difficulties with the Siamese. Just before we arrived, he had sent his secretary up the river to investigate a story told by a missionary named M. A. Cheek, and Barrett's secretary had been badly beaten by the Siamese soldiers.

Although Cheek had come to Siam as a missionary, he had left the mission field to engage in the teak lumber business. Assembling a large herd of elephants, he had begun to exploit the teak forests of upper Siam. The Siamese Government became jealous of his activities, dispersed his herd of elephants, and tried to discourage him in every way. This is not the place to go into the details of the Cheek case, which became famous, but the part Mr. Lathrop played in it has probably never been known.

The United States meant little in the Orient in those days. No American gunboat had ever crossed the bar of the Menam, and it was perhaps natural enough that they did not concern themselves with the protests of the young American Minister. Barrett understood the Siamese attitude, but he could not understand why he received no reply from our casual State Department when he cabled them that his secretary had been mal-treated by the Siamese. In one of his cables, Barrett suggested that it would help his prestige to receive a visit from the American gunboat which he knew was in the harbor of Singapore.

Mr. Lathrop shared Barrett's annoyance that these cablegrams should remain unanswered, and as he knew Melville Stone, President of the Associated Press, he decided to do something about the situation. So Mr. Lathrop and Barrett drew up an eighty-dollar cablegram describing the treatment which the American Minister to Siam was receiving. How they succeeded in getting the message through the Siamese cable office, I do not remember, but the headlines which appeared in the newspapers "back home," brought the gunboat to Bangkok in short order. When it dropped anchor off the Legation, the Cheek case was soon satisfactorily settled.

I remember a gay dinner party in Bangkok one night at the house of Doctor G. H. Hayes, an interesting American who was physician to the royal family. I still have my place-card with a cryptic line written on it by Doctor Hayes in answer to a scribbled question from me asking which was the finest fruit in Siam. Without a moment's hesitation he wrote, "It's the Wampi," and returned my card. I think of Doctor Hayes as I pluck the delicious fruits which ripen on the wampi tree in my place in Florida. It is near my daughter's studio which she has named the "Wamperi" in its honor.

The wampi (*Claucena Lansium*) belongs to the citrus family and bears fruits about the size of a large marble. They are yellow when ripe and have a flavor which reminds some people of a grape and some of a goose-berry. The juice makes a pleasant, cooling drink, and good jam is made from the fruits. The tree itself is symmetrical and bears pinnate leaves which exude a balsamic fragrance if crushed.

Doctor Hayes came upon Mr. Lathrop and me as we were drinking our tea one morning and asked us if the water had been boiled. Of course we did not know. The Chinaman had brought the tea and we had supinely accepted it.

"Well," said Doctor Hayes, "I have three cases of cholera on my hands today. It's everywhere about here, and I'd advise you to be damned sure that the Chinaman has boiled any water you drink, whether in tea or out of it."

His remark added to the indefinable sense of insecurity which constantly disturbed a traveller's peace of mind in the Orient in the nineties.

John Barrett secured invitations for us to attend all the court functions, and we saw the gorgeous pageant of the "Swinging," a semi-mystical performance designed to insure ample harvests of rice and other crops.

One evening we went in style to a wedding banquet at the home of the "spirit farmer," the licensed liquor dealer. John Barrett had his boatmen arrayed in colorful costume and rowed us up the Menam to the house.

We were distinguished guests, and were served after the members of the royal family who were present. During the thirteen-course dinner, every dish was strange to us except the rice. Each course was noiselessly placed on the table by a servant deferentially crawling on his knees. Not a person stood or walked erect while the Prince and his guests were at the table. At the close of the long meal, the wives appeared, and even those of royal birth all hitched themselves across the floor like a child who has not yet learned to creep. It was my first glimpse of that strange phenomenon known as court etiquette.

At the end of the evening's ceremony, each guest in turn was given a gorgeous, jewelled shell which a robed priest filled with sacred water. We were then requested to go into an adjoining room and pour this sacred water upon the bowed necks of the bride and groom who were kneeling on a dais, their throats connected by a delicate gold chain. By the time I reached the prostrate couple, they were drenched.

Our stay in Bangkok was short and crowded, but I did learn that back in the interior a special grapefruit was grown, the so-called "Nakon chaisri pomolo." I tried to make arrangements to get some, but it required another trip to Bangkok before I actually secured this grapefruit. The

The Lathrop–Fairchild Odyssey Begins

"King of Siam" orange had already been introduced into Florida by a missionary.

Before we left, we made a trip with Barrett up the river in his launch. One of his Chinese boys, climbing around the bow, lost his hold on the awning and fell overboard. Before the launch could turn, the stream swept him astern so fast that we feared he would drown. Fortunately he could swim, and he managed to catch hold of one of the houseboat rafts which are so characteristic of the river life. A Chinaman was washing clothes on a raft close to where our poor fellow clung half-exhausted. We shouted and waved our arms, trying to make the Chinaman pull the man out of the water, but he only glanced at the half-drowned man and went on scrubbing. This inhuman indifference was incomprehensible until I learned that, according to his philosophy, the Chinaman could not interfere with the will of the Almighty. If he saved the drowning man's life, all the fellow's actions, including any crimes which he might subsequently commit, would be on the washerman's head. Like many Oriental philosophies, this reasoning is rather difficult for a Westerner to understand.

I was loath to leave Bangkok and stood sorrowfully on deck as our small steamer dropped down to the mouth of the River Menam. Mr. Lathrop was chatting with the Captain, discussing the river and its peculiarities, but I did not pay much attention to their conversation. In the middle of the night I was awakened by the boat shaking as if it were going to pieces. I jumped up, rushed to the side and saw the water boiling around us. I concluded that the boat was going to the bottom and dashed into Mr. Lathrop's room shouting "Hurry, it's sinking."

He turned over lazily in his bunk and said, "What's sinking?"

"The boat," I said.

"Oh, pshaw! I don't believe it," he remarked.

Do what I would, I could not excite him in the least. Finally he said calmly,

"Go to bed, Algie" (one of his nicknames for me). "We're crossing the bar and the water is a bit low. The boat's all right." He then turned over and went peacefully back to sleep.

The Gulf of Siam, which had seemed a mill-pond coming up, was an awful body of water on our return trip to Singapore. The boat writhed and rolled as though it were going to turn inside out. For four days I lashed myself in my narrow bunk and wished myself home in Kansas where there was no water worth mentioning.

After a day or two in Singapore at the old Raffles Hotel, we took passage for Ceylon, but our stay there was barely long enough for a glimpse of the Peradeniya Garden, as Mr. Lathrop decided to take a steamer for Australia which was sailing shortly after our arrival. It was one of the ten-thousand-ton North German Lloyd boats then considered the most comfortable steamers afloat.

Perth, the capital of Western Australia, was but a shanty town in those days, containing nothing much worth seeing. It made little impression on us, but I should like to return there some time. My friend Morton Wheeler declared that in the country back of Perth there are more beautiful, flowering, desert shrubs suitable for introduction into our country than are to be found in almost any other place in the world.

Our next stop was Melbourne, where a German baron named Ferdinand von Mueller had lived for some years. Due to his enthusiasm for plants, the famous Melbourne Gardens had come into existence. Also, he was the author of an excellent book in which he gave brief descriptions of many useful plants which he believed suitable for cultivation there. It was called *Select Extra-Tropical Plants,* as the region of Australia in which he was interested was not within the tropics. This book has been a constant companion of my voyages. It is a reference book of utmost value.

Von Mueller's successor, W. R. Guilfoyle, showed us every possible courtesy. It was here that I first began to realize how poverty-stricken America was in collections of plants. Except for the Arnold Arboretum, we had no garden which could be compared with that of Melbourne.

One phase of the railway journey to Sydney came as a great surprise. We were turned out in the middle of the night to change cars when we crossed from Victoria into New South Wales, as the railway gauge in one state was different from that in the other! I understand that this condition remains the same today.

Sydney harbor is considered second only to Rio in beauty. The city itself was quite a metropolis and I was confused by the big hotels, shops and busy street traffic. Cities always do upset me anyway, and Sydney was particularly annoying, as it was more difficult to get out into the country than from some towns. However, I succeeded in getting down to Gippsland to see the giant eucalyptus trees. They were really enormous, but the

The Lathrop–Fairchild Odyssey Begins

huge buttresses extending far up their trunks made them appear less impressive than the Sequoias of California, although it is generally admitted that at least one of them was taller than the tallest recorded Sequoia.

As I stood gazing up at these forest giants, a lyre-bird passed close by me and disappeared in the brush. I had seen its picture in Wood's *Natural History* when I was a boy, but I had never believed that a bird with such a tail could actually exist.

I longed to see the kangaroos in the wild, and the wallabies, and particularly that strangest of all mammals, the duck-billed platypus which lays eggs and yet suckles its young. But Mr. Lathrop was not a leisurely traveller. I did, however, see the primitive Australian blacks, who impressed me rather pleasantly. There is something peculiarly attractive about men who find their way through the wild and have a fund of knowledge about the life that is hidden there.

I was also impressed by the immense tree-ferns with characteristic leaf scars such as those on the black, fossil fern trunks of the coal measures.

The Australian bush is rather terrifying. I suppose, if one lived in it, one would come to love it. But eucalyptus trees are so curiously alike that they become extremely monotonous, whether there be two, twenty, or two hundred. All the sickle-shaped leaves hang with their edges to the sun; great shreds of bark dangle from all the smooth, gray trunks and large branches. Unlike the trees in other forests, the eucalypti appear to grow regularly spaced so that each tree is the same distance from each of its neighbors. Whether for this reason or another, the Australian bush is known to confuse people so that they become lost in it with an ease that is quite surprising.

The Australian Department of Agriculture had recently been organized, and in Sydney I unexpectedly encountered a remarkable Yankee named Nathan Cobb, who was investigating wheat for them. Cobb and I later became intimate friends and he had a considerable influence on my life.

I have a vivid recollection of him in Sydney showing me a wooden stand which he had made. On it lay an eight-cornered dental tool with its point piercing the end of a wheat kernel. By turning the dental tool, he was able to take eight pictures of the wheat kernel, and could thus identify the varieties by the use of these careful photographs.

In those days not much was known about the structure of the wheat kernel, or why the flour of only wheat, rye, and the Abyssinian cereal, teff, would make bread. Cobb had teased the grains of starch from one of the cells in a grain of wheat, and had showed that the remainder was composed of gluten, without which, as every baker knows, dough will not hold the bubbles of carbonic acid gas produced by the yeast.

The channel between Australia and New Zealand resembles the English Channel in the roughness of its water, but unfortunately there is even more of it. As I have never been a particularly good sailor, I suffered acute discomfort until we reached Auckland.

Mr. Lathrop was by nature a reformer, and not only took up the cudgels for any one who was abused, but also attacked anything which he thought wrong in the management of hotels, ships, or men. I was under his observation much of the time and, of course, came in for my share of reforming, too. Somehow I could not seem to do anything quite to suit him. A constant irritation to him was the fact that I have no sense of direction, whereas he had a supersense as to where he was every minute. As I was continually losing my way, I was in perpetual disgrace. He did not hold with the modern theory that you should not discourage a person with criticism. He heaped it on!

I remember in Australia being sent back to the ticket office with some railroad tickets I had bought, because I had not thought to ask on which side of the train our seats were located. I felt utterly discouraged and disheartened and visioned a life of failure stretching before me. This certainly seemed to be Mr. Lathrop's picture of my future.

It was in this rather dreary frame of mind that I had embarked for New Zealand. After a rough night, I felt even more disconsolate at breakfast the next morning. Mr. Lathrop always talked at breakfast—at every meal in fact. He considered it an unthinkable situation for two people to sit and look at each other across the table without saying anything.

"People around us would think we were quarrelling," he explained.

This drab morning, I made monosyllabic responses to his pleasantries, hoping that he would soon run down. Mr. Lathrop endured this for a while, but at last exploded:

"My God, man! Can't you talk?" Throwing his napkin on the table, he strode from the room.

I can never gain credence for this story nowadays, as neither my friends nor my family will believe that there was ever a moment when I could not or did not talk!

The Lathrop–Fairchild Odyssey Begins

In Auckland I received a cablegram from the Department of Agriculture in Washington, signed by Bernhard E. Fernow, chief of the newly established Division of Forestry. He asked if I would accept a commission to collect the seeds of desert trees and shrubs in Australia for a desert arboretum which was contemplated in one of our southwestern States.

The cablegram had been much delayed and there remained less than two months before the authorization terminated. I was restive and anxious to be on my own, and was keen to accept this commission. Mr. Lathrop expressed numberless, unanswerable objections to the idea and I regretfully declined the appointment. However, this episode brought me into friendly relations with Doctor Fernow, as I shall have occasion to recount.

The evergreen forests of New Zealand have a curious silence and a character quite their own. Beneath the tall trees, one of the finest tree-ferns, the Prince of Wales Fern, covers great areas. The tallest specimens are barely waist high, but their foliage is so delicate that the sunshine slanting through the trees above them seems to penetrate through the leaves. In fact, their leaves are composed of but a single layer of cells, whereas most leaves are at least four or five cells in thickness. Even in conservatories, the Prince of Wales and other filmy ferns are often kept in special glass boxes to prevent the leaves from drying out.

The temperatures were cool in those mountain forests, cool but always moist. After the arid conditions in Australia, this part of New Zealand seemed particularly delightful. I wandered through the forest alone, seeing only a few Maoris, and came upon one who was frying grubs in his little shanty; great white grubs which he took from decayed and fallen tree trunks. The grubs looked appetizing, fried crisp in their own fat. I was very hungry and found them quite as palatable as fried oysters, although smaller. However, either the Maori had some special technique, or the grubs were a special kind. Years later my son Graham and I attempted to fry some large white grubs we extracted from a fallen tree on our place in Washington, but, instead of behaving like New Zealanders, the American grubs exploded. Much to my son's surprise, they scattered sawdust all over the stove!

It seems logical that grubs which feed on purely vegetable matter should be perfectly good to eat. Yet, like the candied grasshoppers sold by street vendors in Japan, grubs belong in the category of things which Western people do not consider table fare.

Much to my regret, there was not time to see the great forests of Kauri Pines, which were considered one of the wonders of the world. At one time, they covered vast areas of northern New Zealand.

This so-called Kauri Pine (*Agathis australis*), like the Araucarias, belongs to the broad-leaved conifers, and is easily the most stately of them all. Its great trunk is without buttresses, and grows a hundred feet from the ground before the huge branches of the dense crown begin.

The curious feature of the Kauri forests was the valuable liquid resin which in past ages had oozed from the injured or fallen trunks into the swampy lands where the trees grew. This gum formed masses, some larger than a man's head, of a hard, almost translucent resin much prized by the varnish makers. It was classed as one of the many kinds of Dammar of which the Orient is full. The fine varnishes and lacquers all contained this "Kauri Gum." To find these masses of resin, men wandered over the swamps with long, pointed iron rods, sounding for the buried lumps of gum. Like mining of any kind, it was a gambling profession. A depletion of the areas gradually took place, and a substitute was found in the Chinese Tung Oil, which, with special treatment, has been made to answer the same purpose.

Trees of the Australian Kauri Pine grew in the Peradeniya Gardens in Ceylon, and I thought them the handsomest trees in the garden. I was told that they gave promise of being valuable for forestry purposes and have expected to hear that the West Indies had been planted either with the Australian or the New Zealand species. Our own attempts to establish these trees in Florida have not been successful to date.

CHAPTER VIII

The Cannibal Isles

I WONDER IF the Cannibal Islands appeal to a boy's imagination today as they did back in the eighties. I doubt it, for, in those days, real cannibals were living, and cannibal feasts were not uncommon.

Because of the unfriendly attitude of the natives, few people had visited the Fiji Islands, but, in Sydney, we had met a charming man, a sugar planter by the name of Thomas Hughes, who spoke Fijian fluently. Mr. Lathrop arranged to charter an auxiliary steamer and cruise among the coral islands of the Fijis with Hughes as our guide and interpreter. It was about a seven-day run from Auckland to our trysting place, and there we found Hughes waiting for us with the shallow-draught vessel.

The trip contained a succession of dramatic moments, the first being our sight of a strange, full-rigged ship lying at anchor in the harbor of Suva. In the rigging flapped male laundry of every shape and color and, on shore, the little town swarmed with the owners. They had just landed and were Americans from all walks of life who had been induced to pool their money and sail to the South Sea Islands because their crazy leader told them of an Adamless island inhabited only by beautiful native women. The boat was filled with tools, seeds, and all the paraphernalia of civilization, and was well stocked with arms and ammunition.

Mr. Lathrop went at once to see the harassed Governor, who was pacing the floor in an agitated state of mind.

"In the first place," said the Governor, "there is no Adamless island in the South Seas. And in the second place, if there were one in this group, it would be British territory and I could not allow these crazy men to take possession. If they try to take over an island, I will have to send the native police from here in my gunboat to dislodge this mad expedition, and there might be considerable bloodshed."

95

The World Was My Garden

When we left Suva, the Governor was still trying the power of peaceful persuasion and, in the end, I believe that a few men settled in Suva, but most of the expedition went on to New Zealand, where they sold the ship and disbanded. Like many other dreams of romance, this one did not materialize.

The seas were shallow and there was always danger of our boat's running afoul of some coral-head. Hughes took us to a number of islands, but, although the natives were interesting, I was disappointed to find the islands much poorer in plant species than Java and Sumatra. There were coconut groves, of course, and graceful Seaforthia palms, but, on the whole, there was little of botanic interest.

We landed one morning on the island of one of the minor chiefs. He was a cannibal, of course, as they all were in those days. Mr. Hughes led us to the chief's house by a circuitous route, as it was a great breach of etiquette to go direct. Hughes was none too sure of the chief's temper, so we were particularly careful.

We finally came to a large, thatched building. The Fijian houses, built up from the ground on rocks, are most attractive. The steep pitch of their palm-thatched roofs, and their good proportions, are very pleasing.

Once inside, we found ourselves on finely woven mats of pandan leaves covering a layer of ferns so deep that it was difficult to walk. Our feet sank as though in fine sand. The chief, a tall, rather handsome man, was attired for our reception in a white dress shirt, a loincloth into which his shirt-tails were loosely tucked, and a black four-in-hand tie around his neck. Waving a long, fly-swatter switch made of the midribs of a coconut frond, he advanced slowly toward us.

Mr. Hughes explained to him that the President of the United States and his private secretary had come to call. The chief was highly honored and invited us to sit down with him on the floor. In a few minutes an attendant brought a great bowl of water which he set before us, and a basket containing chopped-up roots of the kava vine, *Piper methysticum*. Beside the bowl, the servant laid a bundle of coconut fibers and a carved coconut-shell, obviously a drinking cup.

I had read of kava-drinking in the South Sea Islands; that beautiful, scantily clad maidens prepare from the roots of this pepper-vine a drink which is said to be curiously intoxicating. However, my heart failed me when a bevy of half-naked old hags appeared and sat in a semicircle in front of us. From the basket, each took a piece of root, popped it into her mouth, and began to chew. One by one, these disappointing Harpies deposited their quids in the big bowl of water and walked out; the attendant then stirred the mess with his hands and dragged the coconut fibers through, partially straining the liquid. Dipping the carved coconut-shell into the bowl, he presented it first to his chief, who promptly drank it off.

Hughes's turn came next.

"I am sorry but you will have to drink it or offend the chief," he said, grinning at us.

What Mr. Lathrop murmured as he lifted the cup to his lips is not suitable for the printed page. His words were far from reassuring, but, somehow, when my turn came I managed to drink the bowlful obediently and, with an effort, kept it down. For days, however, I believed myself internally upset in consequence. Imagine my rage when I learned that neither Hughes nor Mr. Lathrop had really drunk the chief's kava at all.

The coiffure of the Fijians was extremely picturesque. Their thick, wiry hair stands out all over their heads and, although naturally black, it was usually bleached red with lime. Occasionally we saw a warrior smeared with freshly slaked lime while he was in the process of beautification. Originally of fine physique and clear brown skin, their appearance was often horrible due to the prevalence of skin diseases, trachoma and other affections of the eyes.

That evening, as the sun set behind the coconuts, a raft put off from shore. In the center was the chief, seated in a big armchair with one of his granddaughters beside him and, grouped about them, a retinue of warriors. As entertainment, Mr. Hughes had arranged to let them experience a phenomenon of nature which at that time was rare in the tropics. So, when they arrived, a piece of ice was presented to the chief. Shouting the Fijian word for "hot," the old man promptly dropped the ice, but eventually was persuaded to put it in his mouth. He spat it out immediately, but laughed and seemed to find it amusing. It was then passed down the line of warriors. Each one put the diminishing piece in his mouth, grinning as he took it out and passed it on.

Possibly the old chief was feeling a bit superior, too. Although we might know about this hard, wet stone which seemed so hot, we white men had never tasted delicious tidbits of a barbecued human being. I longed to discover how many feasts this peaceful-looking old man had participated in. With the advent of British rule and missionary propaganda, he had supposedly discarded his habit of feasting upon his fellow men.

After a quick trip up the Rewa River, we sailed over to Levuka, which is on the island of Ovalau. The consular official there was a German by birth who had brought an English wife out to this forlorn station in the Fijis. Mr. and Mrs. Herder immediately proposed that they take us to visit their friend, the King of the Cannibal Islands. The King lived on a little island called Mbau, which had been artificially created by bringing baskets of dirt and stones from the main island and dumping them into the shallow water over a coral reef.

Mr. and Mrs. Herder had a telephone running from the general store up to their dwelling house, and one day when the King came to the store for supplies, Herder had asked him if he would care to ring up Mrs. Herder. Mrs. Herder spoke Fijian fluently and answered the phone speaking in that language. The King dropped the receiver and rushed outside to look for her. He was greatly excited because while he could understand that Herder might speak in English or German to his wife up on the hill, he could not see how they had been able to play such pranks with his own language, which he knew so much better than they did.

We started off the next afternoon in Mr. Herder's antique launch. Launches were not as reliable as they are now, and this one stopped so repeatedly that we all became very nervous. As the afternoon wore on, storm clouds gathered in the sky and the wind began to raise waves on the shallow sea. We were on watch every moment for the coral-heads, but, when night came on, the expected happened and we ran hard on to one. It seemed that the trip might have a serious ending, when a light appeared in the distance and our guide declared that we had luckily struck on a reef within hearing of the King's little island.

Canoes came off and we were soon taken ashore and ushered to a guest house—a large, thatched affair with a single room. We put up a curtain, and Mr. and Mrs. Herder slept on one side of it and Mr. Lathrop and I on the other. The King sent his compliments and a pot of coffee. We had, of course, brought provisions with us, but we had not counted on the pigs. In the morning I was awakened by a

The Cannibal Isles

regiment of razor-backs rooting around among the supplies. I rushed to the rescue but could only salvage a part of our stores.

Although our breakfast was scanty, we were a merry party when we set off to pay our respects to the King of the Cannibal Islands. However, the King himself was rather an anticlimax. Instead of a ceremonial reception such as we had received a few days before, the King was playing cards when we came in and evidently had been drinking heavily. He merely turned, gave an unintelligible grunt or two, and went on with his game.

Even if the King did not live up to expectations, the island itself was interesting. We saw the houses of the King's harem and those of his special guardsmen. A stone post stood near one of the larger houses. We were told that, in the good old days, babies destined for the cannibal feasts had their brains dashed out against it.

Apparently the natives in the Fijis really liked the taste of human flesh. It is believed that at times when they had no captives, they killed and ate the babies of their own tribe. The South Sea Islands is the only place in the world where cannibalism advanced beyond a mystic, religious rite or sacrifice.

We were shown the special, three-tined, ceremonial fork used in cannibal feasts. It was made of a human arm bone. I secured from a photographer in the village of Levuka the only photograph of a human barbecue which I have ever seen (or heard of), and a description of the whole procedure. In fact, it was only a few years since cannibalism had been forbidden by the British.

In a large pit lined with smooth stones, a roaring fire was built. When the stones were red-hot, the body of the captive was laid upon them. Roots of the yam and taro, the two principal vegetables of the Fijians, were put in to be baked with the body and the pit was filled with banana leaves. An old judge, whom I met in Suva, told me that he had often overheard the campfire talk of warriors discussing those feasts. Most of them agreed that the thumb and palm of the hand were the choicest morsels.

Regretfully we left the Cannibal Isles and turned our faces toward civilization again. Our next port of call was Hawaii. Mr. Lathrop had been there many times, and had once acted as Hawaiian Consul in Japan;

so he found many friends on the pier to welcome him with Alohas and leis.

I doubt if there ever has been, or ever will be again, an existence more idyllic than that of the early missionaries and traders in the Hawaiian Islands. It would have been easy to forget home and ambition when basking there under the mangos and coconuts. In those days (1897), the Hawaiian Islands still had a queen to add a touch of pageantry to the natural beauty of the place.

An interesting and progressive group of Americans had already gathered in Hawaii; people such as the Athertons, Bishops, Castles, McFarlanes, Cookes and Wilders. Honolulu was gay with a constant round of entertainment.

In a tentative way, Mr. Lathrop and I had talked over plans for a botanical garden somewhere on the islands, and he introduced me to William T. Brigham, then the outstanding savant in Honolulu. Professor Brigham had created a museum, and had already amassed a great fund of information about the native peoples of the islands of the Pacific. I spent several evenings with him discussing our project. The only man on the island, he said, with ample means and a taste for plants was Sam Damon, who already had something of a garden on his own estate.

Mr. Lathrop knew Mr. Damon slightly, and took me to call upon him. The interview was exceedingly short. From something Mr. Damon said, Mr. Lathrop received the impression that he was only interested in a garden for himself where he could have the finest orchids and other specimen plants, and was not in the least interested in any distribution of plants to other people or in interesting others to start gardens. Mr. Lathrop promptly developed a violent prejudice against Sam Damon and left abruptly. I later came to know Damon's head gardener, McIntyre, quite well, and together we eventually carried out a plan which brought some superior mangos into the Islands from India.

One of the Spreckels, Claus I believe, was living in Hawaii, the largest island of the group. Mr. Lathrop knew him well, and sent me there to interview him.

Consequently, I landed one morning at Punaluu, half dead with seasickness and loss of sleep, after a trip on the uncomfortable little steamer which plied between the islands. I spent the morning flat on my back in the shadow of some cacti until I had sufficiently recovered to make my way to a coffee planter's house at Kona. It was a two-roomed shack in a devastated area of virgin forest which he was preparing for a coffee plantation. The man did not know much about coffee, but, like many American pioneers in the tropics, he thought that he would soon learn.

During this period, the Islands were the only tropical American outpost, the one place where Americans had gone in any number, bought land, and settled down to live. The sugar-growing craze had not yet begun, but the missionaries and their friends were casting about for means of making a livelihood. They and other Americans began to take an interest in the cultivation of tropical fruits in a way which the Dutch and the English had never done.

In the Hawaiian Islands, both avocados (commonly called alligator pears) and mangoes were beginning to attract attention. It seemed the moment to establish an experimental botanical garden, but Mr. Spreckels was not in the least interested. After interviewing him I felt much disappointed, for I should have liked nothing better than to settle down then and there and spend my life introducing plants into the Hawaiian Islands.

Obedient to my instructions to see the island, after my futile interview with Claus Spreckels, I spent a day in the crater of the neighboring volcano, Mauna Loa, which seemed far more impressive than either Vesuvius or the Gedeh in Java.

The terraced native plantings on the hillsides were charming with their taro patches. Taro is a crop ideally suited to the Islands, and both wet land and dry land varieties grow there. From the first, I enjoyed the flavor of taro tubers baked, mashed, boiled, or even fermented as in poi, although many people consider poi an acquired taste. We later introduced taro into Florida from the West Indies, where it is called dasheen. Dasheens deserve a place on the American menu. They are inexpensive, nutritious and have a pleasant flavor of chestnut. Dasheen chips are certainly superior to potato chips.

There were two particularly interesting trees in Honolulu which I believe are still standing. One was the oldest "Algarobillo" tree (*Prosopis dulcis*), a different tree entirely from the algaroba (*Ceratonia Siliqua*) of Spain, although the long pods of both are used as fodder for stock. The algarobillo had been introduced by the Catholic Fathers from the dry coastal regions of Peru, and it had spread throughout the Hawaiian

Islands. Its long, yellow pods, filled with a sweet, honey-like substance, had become a valuable food for cattle and hogs.

The other tree stood in the garden of Ah Fong, the wealthiest China-man of the Islands, whose daughters were accepted members of Honolulu society. It was a large specimen of the Chinese litchi, a relative of the ram-butan of Java. This delicious fruit has been slow in spreading to the Hawaiian Islands and America, and I believe that those of us who tried to introduce it were to blame, as we failed to understand its requirements. One of these is the fact that its root system, when young, apparently needs the presence of a microscopic fungus, which is embedded in the outer cortex of the rootlets and assists the plant in assimilation of foods in the soil.

I met the doctor who had charge of the leper colony at Molokai, and had a most interesting talk with him. I have always regretted that something interfered with my going to Molokai, for, as it turned out, I later had much to do with the expedition made by Joseph Rock into the forests of Burma to secure seeds of the Chaulmoogra tree. From the seeds of this tree, chaulmoogra oil and chaulmoogric acid are obtained, and the ethyl ester of this acid has proved quite efficacious in the treatment of leprosy.

Sailing from Hawaii is an emotional and beautiful experience. Our shoulders were laden with leis of fragrant frangipani blossoms, ginger, gardenias, and pikaki flowers. With abandon and feeling, a native orchestra played plaintive Hawaiian melodies which lingered long in our memory. As the gangplank dropped from the side of the ship, we and those ashore sang together Aloha Oe.

Next year Queen Liliuokalani came to Washington to make her last appeal to the President, who had determined to put the Hawaiian Islands under the American flag. I called upon her in a down-town hotel drawing-room. Her attendant departed on an errand and we were left alone. The Queen, a large, rather striking woman, well educated and interesting, was much depressed. She was returning in a few days to Honolulu and I think she realized that her visit had been a failure. To divert her mind, I spoke of the beautiful Hawaiian music. Going to the piano, she sang her favorite songs in a charming and affecting manner.

The Cannibal Isles

Regrets are futile, but I cannot but be sad that such a happy people, living in peace and plenty, should ever have been discovered by the white man and decimated by his diseases and civilization. I am glad that I saw a few of the quiet places of the world before the coming of automobiles and jazz.

CHAPTER IX

American Interlude

I HAD BEEN ABSENT from America almost four years. As the steamer entered Golden Gate, I was acutely conscious of that curious increase in the tempo of the life around me which I have experienced many times since. It is the rhythm of American life. A feverish, ambitious something which I have never felt when entering any other country.

When I had set off on my wanderings, I had resigned from the Department of Agriculture and, consequently, was worried now as to what I should do to find a position. I felt that Mr. Lathrop was at times a bit doubtful about his "investment," and I was eager to try my own wings.

While I was meditating in this manner, the ship docked and I forgot serious matters for the time being in my admiration of the American girls. I thought that I had never beheld so many beautiful and well-dressed girls in all my life as there were that day in San Francisco!

Mr. Lathrop took me to the Bohemian Club, of which he was a charter member, and showed me his portrait on the wall and many rare and interesting volumes which he had gathered during his travels. He introduced me to some delightful San Francisco people, and then bought me a ticket across the continent and let me go. I feared that I had failed him, and that I should perhaps never see him again. With this unwarranted conviction in my heart, I crossed the Great Divide and returned for the last time to my old home in Manhattan, Kansas.

There had been dramatic changes during the eight years since I left Kansas. My father and mother were living in the town instead of on the campus, their house having been destroyed by fire. Not only had the college president's house thus been reduced to ashes, but the college as an emotional entity was also in ruins.

American Interlude

A curious chapter of our national history had been enacted in Kansas—a chapter little known and fortunately of short duration, but of serious consequences in my father's life. A wave of "Populism" had swept the State. I am somewhat vague about this movement, except that it bore a certain resemblance to Bolshevism.

This wave of Populism soon engulfed the college. Politics of the most destructive type played a large part in the situation and soon controlled the Board of Regents. A man whom I shall call Johns was appointed teacher of Political Economy and his courses made obligatory. He was a cheap socialist and an unpleasant character. Other professors' appointments were terminated summarily, and my father resigned—refusing to participate in the dirty political row which developed with Johns as its center. The Populist Board of Regents then made Johns president of the college in father's place.

Father had been very popular with the students and there was a terrible uproar when they heard of Johns' appointment. They plotted a demonstration for the next Sunday when Johns would appear to conduct the chapel services for the first time.

Father heard of this plan, and decided that the college was more important than any individual, himself included. When the bell rang for chapel on Sunday, he entered with Johns, accompanied him to the pulpit, and remained for the service. Thus the demonstration died before it was born.

After a few unfortunate years, Johns disappeared. I last heard of him as a dealer in questionable Everglades lands.

In consequence of these events, it was a sad homecoming for me. My old professors were seeking positions elsewhere, and the entire atmosphere of the college had changed. My parents soon left for Berea, Kentucky, where they ended their days. They lie buried there on the hills overlooking the lovely valley of Kingdom Come, made famous by John Fox in his story of *The Little Shepherd of Kingdom Come*.

In August I reached Washington—without a job. The Secretary of Agriculture was James Wilson, whom I had known when he was Director of the Iowa Experiment Station. He was a Scotchman with a convincing way of using Biblical phrases and allusions. Best of all, he had the firm conviction that what agriculture needed most was more knowledge.

My old friends in the boarding house were glad to see me back and even became enthusiastic over the Javanese rijstaafel which our good-natured landlady helped me to prepare.

The idea of plant introduction as a government activity was germinating in other minds besides Mr. Lathrop's and mine. Secretary Wilson's first act after taking office had been to send N. E. Hansen to Russia in search of cold-resistant fruits and cereals for our great plains. Also, even before our student days in Germany, Swingle had read a scholarly paper before the horticulturists of Florida, giving a list of subtropical plants which should be introduced into that State.

It transpired that I had arrived in Washington at an opportune moment. Once more Swingle and I put our heads together. There existed, as a special service of the Department of Agriculture, a branch known as the Congressional Seed Distribution. It was spending several hundred thousand dollars a year, and each Congressman had a quota of several thousand packages of seed which he distributed to his constituents. The farming papers were full of jokes about the Congressional Seed Distribution and it was, of course, just a species of petty graft. Packets of pansy seed were reported to yield foxgloves, and petunias grew as hollyhocks. Also, many thoughtful folk felt that it was not a proper function of the government.

It seemed a logical proposition to Swingle and me that a portion of this seed fund should be diverted, and spent for the introduction of new and carefully selected crops. We therefore drew up tentative plans for a clause in the Appropriation Act which would set aside $20,000 for this purpose. I took the plans to Doctor A. C. True, head of the Office of Experiment Stations. Since the scheme was designed primarily to furnish the experiment stations with new seeds and plants, Doctor True was enthusiastic over it, and took me with him to present it to the Secretary.

Secretary Wilson was a tall, gaunt man with gray beard and deep-set eyes. He sat listening to us with his eyes half closed and, at intervals, made use of the near-by spittoon. In those days, even the majority of the Cabinet members and Justices of the Supreme Court chewed tobacco. When we finished talking, the Secretary seemed deep in thought. I waited breathlessly for his verdict.

"Yes," he said, "I think it's a good idea. Now, how can it be done, Doctor True?"

It was simple enough, according to Doctor True. We produced the clause which we wanted inserted into the Appropriation Bill, setting aside $20,000 of the Congressional Seed Fund for the establishment of a "Section of Foreign Plant Introduction."

Secretary Wilson pressed a button and called in the Disbursing Officer, Mr. Frank L. Evans. Mr. Evans agreed that the plan could easily be put through if the Secretary wanted it. But Mr. Evans advised a slight change in the title. He felt that the addition of the word "Seed" into the heading would pacify the grafters. So "Section of Foreign *Seed and* Plant Introduction" it became.

I argued against the lengthened name on the ground that it was a clumsy title, and said that seeds were parts of plants anyway. However, to my everlasting regret, the words "Seed and" were added as a matter of expediency and, for over thirty years, those utterly superfluous words have had to be handwritten, typewritten and printed, probably a hundred million times. The persistence of superfluous phrases in a large organization of any kind is a characteristic phenomenon, I suppose.

Secretary Wilson asked me to return to the Department of Agriculture and organize the new Section. But how was I to live in the meanwhile? The Appropriation Bill would not be passed before the following July, and this was October. Wilson was willing for me to start work at once if a salaried niche could be found for me. At this juncture, Doctor Fernow returned from the West and immediately hunted me up. He was still working on his scheme of a desert arboretum in the Southwest.

"I'll make room for you in the Forestry Division," he said, "and will find enough money to keep you going until the other appropriation comes in."

I set to work at once and wrote a bulletin on the aims and objects of plant introduction, which was published by the Division of Forestry with a note of explanation by Doctor Fernow, whose interest in the problems of American forestry was that of a thoroughly trained scientist.

In those days as now, no one could start an office without a stenographer and a typewriter. I did not know how to go about finding either one.

"There's Grace Cramer," Professor Thompson said, pointing to a girl in her teens with short skirts and her hair in a braid down her back. "She's a Kansas girl. Why don't you ask for her?"

My friends used to make fun of me for picking such a little girl. But

Grace entered into the spirit of the work with great enthusiasm and for many years carried the burden of correspondence and clerical management of what came to be known as the "Section of S. P. I." Her interest and quick adaptability soon made her a most important factor in the organization.

There was a superstition in the Department that any new Section had to have its beginning in the attic in order to amount to anything. So, although it meant climbing four flights of stairs, I was proud of starting our Section of S. P. I. under the rafters of the old brick building from which Doctor Galloway and his staff had recently graduated. However, I soon was moved to the second floor and then to the first. In fact, for over thirteen years, so mobile was the growing Department of Agriculture that we were moved bag and baggage in and out of leaky private houses, basements, store-rooms, and business blocks.

Almost the moment I was settled in our first location, tons of seeds and plants began to pour in upon me from Russia. Hansen felt that he had been sent out to collect, and he collected everything and collected it in quantity. It was all most embarrassing, as no provision whatever had been made to take care of his shipments and the situation soon became complicated, to say the least.

Secretary Wilson strode out with me one afternoon to find a room in which to handle this Russian deluge. He discovered an old carriage-house behind the red brick building.

"Don't let them crowd you, Fairchild," he said. But crowding one another was one of the chief occupations of Department employees.

My days, nights and holidays were spent at my desk. There was never half enough time for the correspondence, plans for exploration, and handling of Hansen's cold-resistant seeds and plants, which were a job in themselves. They had to be put up in packets and distributed to experimenters who had first to be discovered and then kept in touch with through endless correspondence. To add to the dilemma, we were afraid that spring would be upon us before the seeds were in the experimenters' hands.

But I was only one of many hard-working young men in the Capital. I used to walk home with Frank Hitchcock, who was building up the office of Foreign Markets, and talk of foreign countries and foreign-market possibilities. Knowing his indefatigable energy, I was not surprised that he became the "steam-roller" of the Republican party, card-cataloguing every politician, becoming chairman of the National Republican Committee, and Postmaster-General.

We had one rather comic interlude when we spent fantastic hours on a job unconnected with the Department. As I look back, it is difficult to understand how we took it as seriously as we did.

The Spanish-American War was on, and Frank Hitchcock became obsessed with the alarming idea that the Spanish fleet might sail up the Potomac any moment. Of course it was not his idea alone. There are always plenty of nervous politicians in Washington, particularly during a war. Anyway, Frank conceived the idea of mining the Potomac, and we spent jittery hours constructing mines and rowing around the river to place them. I well remember one uncomfortable occasion when I was connecting the wires to the bombs, knowing that a particularly clumsy fellow was at the battery-end of the wiring. I was sure that he would blow me sky-high any moment!

Recently an item appeared in a New York paper describing the finding of some old mines in the Potomac. The article said that the Navy Department believed that they had been there since Spanish war days, as there was a "tradition" that the river had been mined as protection against the Spanish fleet. It made me feel old, but important.

Years later, Frank Hitchcock flew in the first Wright airplane which was purchased by the Army in 1908, and was one of the first men to drop bombs from an airplane. I went out to see the demonstration with Thomas Nelson Page. The "bombs" were croquet balls and the target a white square chalked on the ground.

Through Mr. Lathrop, I made my social début in Washington that winter. His sister, Mrs. Thomas Nelson Page, had perhaps the most brilliant salon in the Capital. When I received an invitation to tea one day, I asked Frank Hitchcock what I should wear.

"If it is an afternoon affair before dark, just put on your Prince Albert, but if it is evening, you will have to wear full dress," he told me.

I decided that it was dark enough for full dress, and arrived thus attired. Unfortunately for my peace of mind, none of the foreign diplomats, authors and Washington élite agreed with me, and I was alone in my glory. In dreams, people occasionally find themselves in some crowded ballroom completely unclad, and I felt equally uncomfortable. Miss Belle Hagner put me in her eternal debt by assuring me that if I were going out to dinner it was perfectly all right to come dressed in a tail coat. This was an example of her charming tact and thoughtfulness, so much in evidence when she was private secretary to the First Lady of the Land during two administrations—perhaps the most popular secretary who ever handled the social affairs of the White House.

During that winter of 1897–98, Swingle and I had rooms together and fitted up a laboratory where I unlimbered my microscope, believing that after our strenuous days in the Department we would be able to do some quiet microscopical work at night. However, dust soon covered the microscope. Swingle was to go abroad in the spring and we began listing plants which should be introduced into America. This led to a system of explorers' notes for which we devised duplicating pads in order to make the indexing comparatively easy. We studied books in all the languages we could read, and made notes of every species of desirable plant on these cards in triplicate—one for the explorer in the field, one to be filed under the locality where the plant grew, and the other under the scientific name of the plant.

It may seem incredible, but card catalogues were in their infancy. In consequence, there was a freshness and excitement about developing a card catalogue system which I fear is lacking today. The famous Dewey system of library classification was beginning to attract attention. The first school for the training of librarians in this system had been opened in Albany by Mary Salome Cutler, who later married my brother Milton.

Due to my stay in Java, my thoughts constantly turned towards tropical plants. Florida was the only place where they could be grown in the United States, and I began to enquire about the conditions there.

Henry M. Flagler, whose hobby was the development of the eastern coast of Florida, had recently completed the Florida East Coast Railroad as far as Miami. There he erected one, of his yellow frame structures, "The Royal Palm Hotel." James Ingraham, general manager of the railroad, had interested Flagler in the creation of a laboratory to investigate diseases of tropical plants. In conjunction with the laboratory, it was hoped that there would be a small garden for tropical agriculture. Land for the garden had been offered by Mrs. Mary Brickell, who lived on the Miami River.

American Interlude

Therefore, in the winter of 1898, I was sent south, and reported to Mr. Ingraham in St. Augustine. I formed an immediate and immense liking for him, which never varied as long as he lived.

Those were days when the increasing market for grapefruit was making it a profitable crop in Florida. Some groves were netting over five hundred dollars an acre. The Florida "Crackers" really introduced the grapefruit to the American public, and their fondness for this fruit led to the planting of the first pomelo orchards. Because the fruit was borne in clusters, instead of growing singly like the orange, they called it *grape*fruit. The grapefruit areas of Texas, Arizona and California, and the newer ones in the West Indies, South Africa, and Rhodesia, have developed gradually with the growing popularity of this fruit.

The Ponce de Leon Hotel and its companion the Alcazar had recently been built in St. Augustine by young architects whom Flagler sent to Spain for a year's study. The buildings created quite a stir, as they were the first noteworthy examples of Spanish architecture in the eastern States. Mr. Ingraham took me through both hotels with great pride before we started south.

On our trip "down the line," we skirted the Everglades and he described a trip he had made across them in a dugout with Indian guides. Mr. Ingraham believed himself to be the first white man to cross this area. He had had to wade a considerable portion of the way where it was too shallow for the dugout, and of mosquitoes and water moccasins he had a good deal to say.

Palm Beach had already become a famous winter resort. The huge Royal Poinciana Hotel stretched its yellow length along the shore of Lake Worth, and the Breakers, another immense wooden hotel, faced the ocean. Plantings of young coconut-palms had been made, interspersed with a strange, fast-growing evergreen tree, the Casuarina (stupidly called Australian Pine).

I was anxious to meet any one who knew the existing plants and planting conditions of Florida, and I immediately became engrossed in an examination of everything growing around the hotels. William Fremdt, head gardener of the Royal Poinciana Hotel, was one of the few trained gardeners at that time in southern Florida. Because of his interest in tropical plants, some of our earliest introductions were sent to him, among them a Brazilian rubber tree, *Hevea brasiliensis*. This tree is alive today, having outlived the hotel. When I "called upon" the Hevea in 1935, carpenters were tearing down the great caravanserai, once famous as the longest frame building in the world.

I particularly wanted to see Elbridge Gale, my old professor of Horticulture at the Kansas College, who had settled in West Palm Beach. Professor Van Dieman, head of the Pomological Division of the Department, had entrusted to him some of the imported mango trees from India but, of that first introduction, I had heard that there was only one surviving, a tree of the variety known as "Mulgoba."

Professor Gale had called his simple home "Mangonia," and was as enthusiastic as ever over his plants. He showed me the Mulgoba mango tree, which had been killed to the ground by the freeze of 1895, but had sprouted again. At the time, it resembled a three-year-old, bushy fig tree. As this Mulgoba mango later produced the famous seedling known as the Haden, it is of interest in the history of Florida horticulture.

From Palm Beach, Mr. Ingraham took me to Miami, then the end of the Flagler system. We stayed in the rambling Royal Palm Hotel, a barn-like structure which was rather luxuriously equipped. A close friend of Mr. Flagler's told me that Flagler did not believe that the American public would ever care to go south of Palm Beach and, when plans for the projected city of Miami were shown him, he said that Miami would never amount to anything.

In 1898 I believe there were about a thousand people in the neighborhood of the Miami River, living here and there in the pine woods along sandy roads which led through the brush.

Probably the two most prominent property owners were Mrs. Tuttle and Mrs. Brickell. Mrs. Tuttle lived on the north side of the Miami River, whereas Mrs. Brickell, whose husband had recently died, lived on the south side. Mr. Flagler made an arrangement with Mrs. Tuttle which permitted his railroad to run through her property. Mrs. Brickell did not believe in his railroad, and fought Flagler from the start. She was still fighting him in 1898.

South across the Miami River, a little road led through the Brickell hammock along Biscayne Bay to Coconut Grove. On an acre of land between the road and the Bay, the small laboratory was being erected with the thousand dollars which Mr. Flagler had subscribed. H. J. Webber, a member of the Section of Vegetable Pathology, was in charge of the

The old road from Miami to Coconut Grove.

112A

Building the little laboratory on Brickell Avenue, Miami.

building operations and was, at the same time, clearing a six-acre tract of the hammock for the garden.

As I bicycled over the little bridge across the Miami River, I remembered that I was approaching the jungle trail which Mrs. Bryan Lathrop had described when we were in Java. After the East Indies, the Florida vegetation seemed a little disappointing in its lack of height, but it was dense enough to be called a jungle, and certainly was tropical in character. I was thrilled to find tropical territory in the United States, and a love for southern Florida was born in my heart and has grown with the passing years.

There is not much left of the great Brickell hammock now. Houses and backyards, garages and streets, have destroyed it. But, in the old days, this hammock was a place of great natural beauty.

I found the laboratory only half finished, and but a few plants in the garden—some guavas, a few citrus varieties, and a single Carob tree grown from seed which I had sent from the Mediterranean. To us young fellows, this six acres seemed quite marvellous. It was the first land available to us where we could plant anything we wanted.

Sad to say, even this little Miami garden was not to be ours permanently. There was a one-eyed clerk named Pennywit in the Disbursing Office in Washington who liked to feel important. With much ado, he claimed that neither Mr. Flagler's gift of an acre, nor the six-acre gift offered by Mrs. Brickell, could be accepted by the government. I believe that the law he quoted originated at a time when people feared that the Federal Government might encroach upon the States by acquiring important tracts of land.

"Either as a gift or purchase," said Pennywit, "it is against the law and cannot be done."

It is curious how much pleasure some minds derive from the discovery that a thing cannot be done, or is wrong. Pennywit used to audit our travelling accounts and pick flaws in them. As he had never travelled himself, he could not imagine why we did not have a receipt for every fifty cents expended. To our delight, Pennywit was sent on a trip and made so many slips that his own expense account was disallowed!

Due to Pennywit's quibbling, it was decided that we could only rent the land. This policy seems utterly incomprehensible, and eventually lost us the orginal six-acre garden.

The World Was My Garden

Of course in 1898 Miami Beach was just a sand key across Biscayne Bay. Its mangrove trees were visible in the distance and, with a boat, one might visit the alligator hole, a deep inlet completely surrounded by fantastic mangroves. Certainly there was nothing then to indicate that Miami and Miami Beach would become the greatest winter playground in the world. The place had, however, an indescribable charm and seemed a land far from conventionality and civilization. But the exciting thing to me was that here was a climate equable enough for my beloved tropical plants.

Full of enthusiasm, I returned to Washington only to find the Department absorbed in an entirely different problem. Both the great Southwest with its irrigation projects, and the fertile prairies of the Northwest were opening up. Settlers in these new regions were appealing almost daily to the Department of Agriculture for information as to what crops could be grown. Settlers in the Upper Mississippi Valley demanded hardier plants than were to be found in the eastern States. On the other hand, settlers in the Southwest needed drought-resistant plants, and definite knowledge of how to handle crops adapted to irrigation, which was a new system of agriculture in this country.

The real cowboy days were even then drawing to a close. The bitter fight between the cattle owners who wanted the great ranges left untouched and unfenced, and the farmers who wanted to grow crops on the ranges, had ended in the defeat of the ranchmen. As a result of the increasing agriculture, the farmers were appealing to the government for new crops for experimentation.

For my part, I was in love with Florida. The more I thought about it, the more convinced I became that a tide of immigration would some day flow south. A land in America where coconuts, pineapples, and mangos would grow must eventually attract settlers. Unfortunately, the Secretary of Agriculture was an Iowa man and felt little interest in Florida. He had never been there and, like all of his associates, considered it a backward country with a disagreeable summer climate. Moreover, the menu of an average American family consisted largely of ham and eggs, beefsteak and onions. Therefore, there was no encouragement for farmers to settle in a region which supposedly had neither soil rich enough to grow corn, nor a climate suited for the raising of cattle and hogs.

Still, I persisted in my desire to introduce tropical plants, maintaining that the southern portion of Florida would eventually need them for its development. Eventually, to my great joy, the Spanish War created a wave of interest in tropical agriculture when our soldiers returned from Cuba and Puerto Rico with tales of new and delicious fruits. Also, the problem of supplying our newly acquired possessions with the right crops became second only in importance to the development of the Northwest and Southwest.

Spring found me in Washington swamped by the great cases of nursery stock, and plants of all kinds, which arrived in increasing numbers. We had no facilities whatever for taking care of them, and things became more and more hectic. Also, my powers of organization were limited by the inexperience of the staff.

When the Appropriation Bill eventually passed, I would have at my disposal the perfectly unheard of sum of $20,000. So, plans for foreign explorations were formulated. Swingle had already sailed to make a collection of European table grapes, financed by the Office of Pomology, I believe.

Mark Carleton, our Kansas classmate, was working in the Department testing cereals to discover their resistance to rust. He had learned to read Russian and, in the course of his reading, had reached the conclusion that the Black Lands of Russia contained valuable varieties of wheat which we did not have. We decided that he should go to Russia and should start at once in order to visit the wheat fields before harvest time. I was as enthusiastic as Carleton about this expedition, and, just before he left, had him spend the night in my quarters to talk over his trip. The date of this conversation is fixed in my memory for, as we finished discussing his plans, we heard the newsboys crying "Extra Star!" The battle of Santiago was in progress, and I rushed out in my pajamas and returned with the paper announcing the destruction of Cervera's fleet. So the Potomac needed no further mining!

Paul de Kruif, in his book *The Hunger Fighters*, perhaps carried away by the drama of the situation, represents Carleton as having to fight for his opportunity to visit Russia. This was not the case. Carleton went out as a Special Agent of the little Section of S. P. I., with all possible coöperation on our part, and the reports he wrote of his collections were all published in the "Inventories" of our office. In later years, he frequently expressed gratitude for the start which I gave him when I authorized him to explore Russia. An Office of Cereal Investigation was formed soon after his return, and he was put in charge of it. This grew to be a large organization and is still in existence, but Carleton left the Department to work for the United Fruit Company and died in Central America many years later.

In 1898, when Carleton first went to Russia, scientists were predicting that by 1931 the increase in population would overtake the possible production of wheat as then grown. These prognostications spurred our desire to find drought-resistant wheat in order to extend cereal production into the arid regions of our country. Carleton had experimented with over a thousand varieties of wheat as well as oats, barley and rye, at the Department of Agriculture's trial plots near Garrett Park, Maryland. Therefore, even without his knowledge of Russian, he was splendidly qualified for this mission.

Carleton's famous "durum wheat" is a special class of early wheats which prefer dry conditions and are also very productive. In dry regions, durum wheats are resistant to wheat rust, smut and attacks of insects, but, curiously enough, in moist regions they are very subject to rust. The durum wheats are generally distinguished from the ordinary soft wheats by their unusually long, stiff awns or "beards" which are often dark in color. Also, their kernels, when held to the light, are almost translucent.

When the millers tried to convert durum wheat into flour, they ran into difficulties due to its hard kernel. At first they were unwilling to remodel their machines, and refused to pay a fair price for the new wheat until forced to do so by popular demand. The proteins in it are different from those in the soft wheats, and the superiority of durum wheats for making macaroni paste was soon acknowledged and had an effect on the millers. Eventually the advantage of mixing durum wheat flour with soft wheat flour came to be recognized, and gradually a new wheat culture emerged. We have forgotten how poor our bread was at the time of Carleton's trip to Russia but, in truth, we were eating an almost tasteless product, ignorant of the fact that most of Europe had a better flavored bread with far higher nutritive qualities than ours.

The spectacular success of the durum wheats focused public attention upon the value of plant introduction. It demonstrated the profits to the farmers which could result from the introduction of a new variety of one of the staple crops. Two years after the introduction of the Russian wheats, 60,000 bushels were produced, while only five years later 20,000,-000 bushels were grown.

While I was still struggling with the tons of material arriving from Russia, Mr. Lathrop appeared most unexpectedly. After a few words of greeting, he came directly to the point. He expressed himself as delighted by the Section of Seed and Plant Introduction, but proceeded to make me a proposition which quite staggered me.

"David," he said, "you're no more fit to build up a government office of plant introduction than I am to run a chicken farm—and I don't know a thing about chickens. You have no contacts with the rest of the world. How do you expect to secure your seeds and plants? You can't do that sort of thing by correspondence. If you try to run things from Washington, it will be a little pinchbeck affair. I am prepared to pay your expenses and take you with me on a quick reconnaissance of the world. When you get back you'll really know how to build up this office. Tell the Secretary of my proposal. It's true you've been around the world once, but you haven't seen South America, India, or South Africa."

After several sleepless nights, I decided that Mr. Lathrop was probably right. Undoubtedly some one else would be capable of running the office, and it would be a great mistake for me to refuse such an offer. On the other hand, I felt pretty sure of what Secretary Wilson would say about Mr. Lathrop's proposal. And he said it! I spent a most uncomfortable hour with him before he reluctantly consented for me to travel with Mr. Lathrop as a "Special Agent" of the Section of Seed and Plant Introduction.

It was a long time before Secretary Wilson forgave me for "running away" from my duties. Later, he recognized that my travels had opened up far greater opportunities for plant introduction than would have been possible with the limited funds of the Department, and he wrote me a letter complimenting me on the work I had done during those years of travel.

O. F. Cook had returned from Liberia and took charge of the Section of S. P. I. as my successor. To him the Section owes one of the important developments of the organization, namely, the origination of the printed "Inventory of Plants Introduced." This Inventory now contains the records of over a hundred thousand seeds or plants introduced into America through the Section since 1898.

Another creation of Doctor Cook's for which I am sincerely grateful is the title "Agricultural Explorer." To the general public, a title may not seem important, but, to any one who has travelled as I have under an appellation such as "Special Agent," a properly explanatory title means a good deal. I do not in the least resemble a spy, but the term "Special Agent" always had to be carefully explained in foreign countries and, even then, generally aroused suspicions.

It was five years before I again took the reins of the "S. P. I." into my hands once more, and settled down in charge there for twenty years.

CHAPTER X

The West Indies and South America

IN OCTOBER, 1898, I bade my surprised and envious friends good-bye and started West to join Mr. Lathrop in San Francisco. We were to sail from there for the Orient.

On our previous travels, before the creation of the Section of S. P. I., there had been no place where I could send plants. Now, however, I was not only to collect plants but also to establish contacts with recognized plantsmen. I had papers signed by the Secretary of Agriculture recommending me to consular officials and agricultural institutions in the various countries through which we expected to travel. These papers were wonderful creations, hand-printed on parchment, bedecked with ribbons, and emblazoned with the gold seal of the Department of Agriculture. Nothing the Department of State ever produced could compare with them, and they soon became affectionately known as "Dago Dazzlers."

By the time I reached San Francisco, something had happened to the boat on which Mr. Lathrop had intended sailing. There were few boats then and the next suitable steamer would not sail for a month or so. Mr. Lathrop heard of a boat from New Orleans which would take us to South America, so I headed east again almost immediately.

We stopped for several days in Santa Barbara and I met the noted Italian plant enthusiast, Doctor E. O. Franceschi, a man of striking appearance, with a long nose, sharp eyes, and alert manner. He had already imported many interesting trees and shrubs and was absorbed in promoting the horticulture of the region.

Santa Barbara in 1898 was but a simple, small town. Residents of the beautiful hillside villas today would not credit their eyes could they visualize the bare, sparsely settled roads where I drove with Doctor Franceschi.

119

The World Was My Garden

In his little nursery, Doctor Franceschi was growing many ornamentals and had already issued a catalogue remarkable for the number of new species which it contained. I recall that he was also much interested in an Italian vegetable called the zucchini, a variety of squash, the fruits of which are consumed when they are small, before the blossoms fall from them. This species has become established in the vegetable trade and is now common both in California and on the menus of New York hotels and restaurants. Franceschi was doing all this himself without governmental or other support, and was having a difficult time financially. I wrote to Washington and arranged that he be made a collaborator of our new Office and be supplied with any suitable plants which might come in.

We walked through some new estates, and Doctor Franceschi explained that he had charge of them and hoped to make a living supplying the owners with plants and advice about landscaping. Santa Barbara was so undeveloped that I considered him visionary and over-optimistic. However, he foresaw the future more clearly than I, and lived to see Santa Barbara become a great winter resort containing hundreds of beautiful villas like those on the Riviera.

The residents of Santa Barbara have preserved his old home as a tribute in recognition of his contributions to the beauty of the region. Certainly they are more than justified in doing him honor. Doctor Franceschi was a fine plantsman, and California owes much to his untiring interest and effort.

We arrived in New Orleans only to find that the steamer for Panama had been taken off. Utterly disgusted, Mr. Lathrop decided to sail from New York for Jamaica.

There were few passengers on the boat, but two of them were extraordinary personalities, Mr. and Mrs. Brooks Adams of Boston. Mr. Adams' brother, Henry Adams, wrote *The Education of Henry Adams,* a book found in every public library. Brooks Adams was writing a book on the trade routes of the world as influenced by wars. I met him three years later in Athens still pursuing his investigations.

Mr. Adams' personality became stamped on my memory because he annoyed Mr. Lathrop acutely every morning by inquiring of the Captain the latitude and longitude of the boat. Mr. Lathrop developed violent prejudices against people whom he met, and he particularly detested any one whom he considered a poseur.

The West Indies and South America

Kingston, Jamaica, was the first foreign port in which I began a serious study of the marketplaces, tasting the new fruits and vegetables, packing and shipping both seeds and cuttings of those which seemed desirable for introduction into the United States.

I was much impressed by the tropical yams, vines which are a species of *Dioscorea,* bearing aerial as well as subterranean tubers. These yams are entirely distinct from the forms of the sweet potato which grow in our southern States. Mr. Lathrop and I collected various types and had them cooked and served in numerous ways. We particularly liked a variety called the "yampee" which is a form of *D. trifida* producing small, flat tubers the size of a man's hand. An old friend of Mr. Lathrop's, a coffee planter in the mountains near Mandeville, served the yampee baked in the skin together with a black land crab which is found in the mountains of Jamaica, and I still think that the two dishes were among the most delicate I ever tasted.

On this visit to Jamaica I first encountered the chayote (*Sechium edule*), a delicious vegetable belonging to a monotypic genus (containing only one species) of the cucumber family. The chayote has a firmer, more agreeable texture than squash and an equally delicate flavor. Upon entering a chayote arbor, one sees hundreds of fruits hanging from the thick canopy of leaves like large green or white pears. When the small yellow flowers are fertilized, the fruits grow with astounding rapidity, maturing in only a few days.

Mr. Lathrop was enthusiastic about their flavor, and I shipped a quantity to Washington. After my return there, I spent much time trying to interest both my superiors in the Department and my "social" friends in this vegetable. When Mr. Houston was Secretary of Agriculture, Mrs. Fairchild and I gave a luncheon for him and served chayotes as our "pièce de résistance." The Secretary ate the chayotes with gusto, but later my triumph was turned to despair when I heard Mrs. Houston say that her husband never knew what he was eating!

Years later, my father-in-law, Alexander Graham Bell, became an ally in my struggle to overcome the resistance which the public always sets up against a new food. At one of his Wednesday evenings, he featured chayotes and this led to their being served at a National Geographic Society banquet. The guests at the banquet were most enthusiastic but, when they wished to secure some for themselves, we had to admit that chayotes were not on sale in any market. We could not persuade farmers to grow any quantity when there was no market, and we could not create a market when there were no chayotes!

The story of any introduction includes human-nature problems as well as horticultural difficulties. The chayote situation repeated itself over and over again. Many practical and delicious fruits and vegetables have been introduced, and successfully grown in our country, only to fall into oblivion because of the lack of funds to "push" the product into popularity.

Chayotes are still grown in Florida and other southern States but they are seldom seen in the northern markets of the country. In the New Orleans district, where they are grown occasionally, they are called mirlitons or "vegetable pears." They constitute an important part of the diet in Guatemala and are common in Madeira, North Africa, and the West Indies.

Chayotes have a single large seed embedded in the flesh. The entire fruit is planted and, as the "meat" decays, it provides nourishment for the young plant. In Georgia, the plants are protected through the winter by placing a box of straw over their crowns. This arrangement permits the plant to breathe, and yet gives some protection against low temperatures. A Mr. Pierpont had two vines which grew so fast in the good soil of his plantation near Savannah that in one season they covered arbors an acre in extent, and produced 1500 chayotes.

But to return to Jamaica.

Mr. Lathrop was a true gourmet. He had an extraordinary palate for flavors, and often amused himself by guessing the various ingredients in some complicated dish. Until I met him, my own experience with foods had been limited by my slender purse as well as by my years in Kansas, where French cooking was unknown.

Mr. Lathrop's approval of a new vegetable or fruit gave me great confidence in its good qualities. Everything we introduced would have to run a gauntlet of sarcastic criticism when it reached Washington, but, when Mr. Lathrop had said that a food was "good—delicious in fact," I was able to tell those who declared "the stuff wasn't fit to eat" that evidently they had uneducated palates. In other words, Mr. Lathrop's approval gave me courage and combative arguments without which I would often have been utterly discouraged.

An arbor of Chayotes, a delicately flavored vegetable popular in the South but little known in our northern States.

The Perulero variety of Chayote from Guatemala is the handsomest of them all.

Young Jamaican holding a West Indian Tropical Yam,

An old Jaboticaba tree growing in a private garden in Rio.

The purple fruits of the Jaboticaba grow on both trunk and branches of the tree. They are eaten like grapes and contain a delicious white pulp. (Page 134)

The West Indies and South America

In the nineties, Jamaica held a more important place in the esteem of Britishers than it does today. The decline of the planter class in the British West Indies was not as yet evident. Planters were making good money from coffee and sugar-cane. In the late nineties, a new era of sugar-cane cultivation began in which hundreds of new varieties were originated from seed. Until then, sugar-cane had been grown only from cuttings of the stems and most planters did not know that the cane ever produced seeds.

At the Botanic Gardens at Hope and Castleton were two remarkable plantsmen, Doctor William Fawcett and Mr. William Harris, who gave me both advice and propagating materials, including the Laguna grape, and a double variety of poinsettia.

My collections were ordinarily confined to "useful" plants, as I would have been severely criticised had I wasted Department funds on mere ornamentals. However, my belief in the future of South Florida made me occasionally include a few species which I carefully designated as "dooryard plants."

I was annoyed by the shortness of our stay in Jamaica; I felt that I had barely begun to study its resources. But Mr. Lathrop said that ours was not an expedition to exhaust the plant resources of any one island; if I wanted to see the world, I could not spend months in one place.

After one day at the island of Granada, we arrived in Barbados. Doctor J. B. Harrison, the chemist of the sugar-cane experiment station, was an enthusiastic investigator who received me most cordially. He was still thrilled by his recent discovery that sugar-cane produces seedlings, and told me that he had first noticed, growing between the rows of sugarcane, green blades which everybody considered to be common grass. His scientific curiosity prompted him to plant some of these blades where he could study them. To his delight, they proved to be sugar-canes—the first seedling sugar-canes observed by anybody in the West Indies. Today, probably most of the sugar-cane has come from selected seedlings, often from seeds of deliberately cross-pollinated varieties.

For some time I had been in correspondence with Mr. J. H. Hart, director of the Botanical Garden at Port of Spain, Trinidad. I found him absorbed in problems connected with the fermenting of the cacao bean. The development of the cacao (cocoa) plantations on the Gold Coast of

123

West Africa had not yet occurred, and the finest cacao in the world came from Trinidad and the coast of Venezuela.

Mr. Hart shared my interest in tropical plants and, with characteristic generosity, sent us a collection of mangos which included the Gordon, the Peters No. 1, and the Père Louis. These mangos were, I believe, the first to be introduced through our Section of S. P. I. Although they have been superseded by better varieties from the East Indies, I still grow the Gordon, a variety named after Sir Arthur Gordon, a former governor of Trinidad.

From Trinidad we went to La Guaira, the main port of Venezuela, and made the magnificent trip over the coastal range of the Andes to Caracas, the capital.

I was immediately attracted by an ingredient which we noticed in many of the Venezuelan soups. In appearance it resembled slices of potato floating in the soup, but the delicate flavor was quite individual, partaking slightly of artichoke. This food staple was known as "Apio," the Spanish word for celery. It differs from any European vegetable and, as far as I know, is not grown extensively anywhere but in Venezuela and Colombia. Botanically it is *Arracacia xanthorrhiza*.

This plant is associated in my mind with Mr. Lathrop's severe illness in Caracas because I returned one afternoon from photographing a field of Arracacia and found him in bed with a tropical fever. I was badly frightened, for I was totally inexperienced in medical matters.

Thanks to his stubborn constitution, he was able in a week or so to sail on a French steamer for Panama.

A pleasant young Englishman by the name of Hoffman was sailing too and was of great assistance in getting Mr. Lathrop from Caracas to the boat. We were sorry to part from Hoffman when he left the steamer at Barranquilla to go up to Bogotá, a slow journey of uncertain length by steamer and muleback. As he departed, he waved to Mr. Lathrop lying in a deck-chair, and said jokingly,

"I'll meet you at the Empire," meaning the famous London music hall.

Eight months later we arrived in London after journeying down the West Coast of South America and across the Andes.

"Let's go to the theatre," said Mr. Lathrop as soon as we had dined; "how about the Empire?" I laughed and said, "Maybe we'll see Hoffman." Mr. Lathrop had forgotten Hoffman's parting remark, and scoffed at the idea of our ever seeing him again. Between the acts we wandered out into the foyer. Believe it or not, there stood Hoffman! He had arrived in London the day before from Bogotá!

Our trip to Panama was most uncomfortable. Mr. Lathrop hated French boats anyway, and was unable to eat anything on the menu. Worse yet, the delicacies which we had stored in the ice-box had a strange way of disappearing. At Maracaibo, I went ashore to buy some milk for him, only to discover that there was absolutely none to be had. Nobody drank milk; it was only used for café au lait. I hired a horse and wagon and drove out into the country to see if I could not get a bottle of milk from some of the planters. But it was a feast day, and the planters were in town. It was maddening! I saw cows; I knew how to milk; but my Spanish was so bad that I could not secure the necessary permission from any one. Eventually, by paying five dollars a bottle, I procured two beer bottles full of milk, only to have one of these also disappear from the ice-chest.

At one time Mr. Lathrop had practically existed on milk and had to carry bottles of it with him wherever he went. In those days stomach ulcers were not well understood, and no method of their removal or cure had been discovered. Frequently during the course of our travels, Mr. Lathrop suffered intensely. It would be unjust to emphasize his irascibility without telling of this ailment which annoyed him for over thirty years until, when he was seventy years old, he was operated upon by the Mayos at Rochester, Minnesota. I imagine that his disposition would have been quite different had he not been plagued by this stomach trouble which probably also contributed to his extreme restlessness.

I had heard so much about Panama and Chagres fever that I viewed with apprehension the wooden shacks and low, malarial coast line of Colon. In 1899, the French attempt to build the Canal had collapsed, but as yet the yellow fever mosquito had not been suspected of carrying the disease. The old French hospitals were still standing and, in the open space beneath them, on the ground lay great piles of iron crosses imported from France, each with a serial number on it ready to mark the grave of some patient. In fact, all patients were measured for coffins immediately they were admitted to the hospital. Practically none ever emerged alive.

The World Was My Garden

The Sisters of St. Vincent, who were in charge of the hospitals, kept potted plants on the window-sills to make the wards more cheerful. Under the pots were saucers and, as the plants were watered daily, water always stood in the saucers. In consequence, mosquitoes bred there, and fed on the yellow fever patients, thus becoming infected themselves with the parasite. In turn they infected nearby patients who might otherwise have escaped. The heavy toll of life turned the French project into a ghastly failure, although it was led by the great engineer, Ferdinand de Lesseps, who had built the Suez Canal.

These facts were not yet appreciated as I stood on the deck and watched the boat come into the primitive harbor of Colon. Even without knowledge of the French disaster, the dangers of a stay in Panama would have been present in my mind, for I had heard of the many deaths among those "Forty-niners" who had crossed Panama on their way to California. I knew families whose sons lay buried beside the Chagres Trail, victims of what was then called "Chagres River fever." There was a mystery about this fever which engendered a haunting fear. However, after landing we found that the European residents in Colon and Panama City seemed to be quite oblivious of any danger and, in consequence, we ourselves soon forgot about it.

The short railroad journey across the Isthmus is a fascinating experience for a naturalist. The dense, tropical vegetation of Panama has a charm lacking in high mountain scenery in the tropics. Even today, after the depressing destruction which man has wrought along the Canal, this ride across the Isthmus of Panama must rank as one of the most interesting short trips in the world.

In his weak physical condition, Mr. Lathrop was not inclined to tarry long in Panama. The few days which we spent there I employed collecting seeds. A banker named Gerardo Lewis showed me a pretty little orange he was growing which he thought had been introduced in Panama by some Chinese gardener. I sent seeds of it to the experimental garden near Miami, where they grew and fruited. Many seedlings were distributed from there and the plant is now much prized as an ornamental. It became known locally as the Panama Orange, but Swingle discovered that this name was a misnomer, as the fruit is the true Calamondin (*Citrus mitis*).

One of these so-called Panama orange trees, loaded with its small, acid fruits of vivid orange, stood amid the débris left by a hurricane which wrecked my place in November, 1935. Strangely enough, the wind was apparently unable to blow the little round fruits from their slender, willow-like twigs.

A cattle boat, redolent with the fragrance of manure, took us down the west coast of South America. At various ports, we anchored in the open roadstead to unload the poor animals which constituted our cargo. They were yanked up by the horns with block and tackle and handled with less consideration than sacks of coffee or sugar.

My attention was attracted one morning by a jar on the dining-saloon table marked "Miel de Palma," palm honey. It proved a delicious addition to griddle cakes and one of the most delicate syrups I had ever tasted. It was made from the sap of a large, native palm (*Jubaea spectabilis*), groves of which were found along the dry coasts of Peru and Chile. In order to procure the sap, the trees must be cut down. As a result, these magnificent palms were rapidly being destroyed by the syrup makers. It is curious that of the many varieties of palm sugars, most of which have distinctive and delicate flavors, none has found its way into America in any noticeable quantity. I should think that they could easily compete with maple syrup.

Mr. Lathrop had often spoken of a place called Paita on the coast of South America where he said that I would not find a single thing to collect, because there was not a plant to be found there. When we sailed from Panama, he said,

"Now, Algie, we're going to stop at Paita, but it is one place where you won't want to linger, I assure you." (Algie was a nickname he had invented for me. I was rarely called anything but Algie or Fairy.)

One morning he hammered on my cabin door at daybreak:

"Get up, Algie. We're off Paita, but something has happened to it."

I hurried on deck, and there before me spread a land which appeared as verdant as the coast of Ireland. When we reached shore the miracle was explained. The first rain in eight years had fallen in Paita! Seeds scattered by the wind over the desert landscape had retained their vitality in that arid climate, strange as it may seem. A few days after the rain, some oxalis and other flowering annuals began to appear and, when we arrived, there were blades of grass two or three inches high. Each blade was perhaps a foot or two away from its neighbor but, seen from the boat, they had effectively tinted the landscape bright green. The Paita houses were built of dried mud, roofs and all. During the phenomenal visitation of a torrential rainfall, the houses had dissolved, much to the discomfiture of the inhabitants.

However, Mr. Lathrop was right. It took only a few hours to collect specimens of the flora of Paita to send back to the National Herbarium. I think there were but eight or ten species in all.

Among the seeds which I sought and found during our South American trip was that of a primitive cotton growing along the banks of a short river called the Piura. Often, for years at a time, nothing but a dry river-bed extended from the foot-hills of the Andes to the Pacific. At intervals, conditions in the Andes produced a flood of water in the Piura River. After the flood, the natives planted the cotton along the wet river bottom. I felt sure that this primitive tree-cotton would interest American cotton growers, although they would not care to imitate the peculiar conditions of its cultivation.

When we reached Lima, I found the markets more to my taste than bull fights, although Mr. Lathrop recommended both. The stands were piled with unusual fruits and vegetables, some of which I had read about, and others entirely new to me. Every waking moment I was busy either cleaning seeds or finding out as much as I could about them preparatory to sending them to Washington.

Many of the seeds which I collected were of plants growing in the high altitudes, and these proved unsuited for cultivation in the United States. The history of some of my other introductions again involved the question of American taste. For instance, neither the delicious flavor of a golden-yellow potato with deep-set eyes, nor quinoa, the famous cereal of the Incas, has appealed to our public. Yet a Scotchman told me that he considered a porridge of quinoa better than the finest oatmeal. When young, the inflorescence of quinoa is also eaten as a vegetable, and resembles broccoli.

One introduction which I made from Peru in 1899 led to the establishment of a new type of alfalfa in southern California. I knew that the alfalfa then being cultivated in California came from seed brought in from Chile by the Catholic Fathers. In Lima, I heard that there was a Peruvian variety called "Omas" which had proven superior to the Chilean alfalfa. It was longer-lived, could be cut more frequently, and was better for irrigated lands. I therefore secured a quantity of seed and sent it home.

Twenty years later, when travelling through southern California, I saw in a local newspaper a reference to the alfalfa grown in that section as "Hairy Peruvian." I investigated and was much elated to find that my introduction had developed into a real farm crop.

Mr. Lathrop told me that during a former trip to Peru he had visited the high Andes and had been much impressed by the way the natives carried heavy loads up the mountainsides, seemingly without exhaustion. He found that they all chewed the leaves of a plant called Coca (*Erythroxylon Coca*) which they used either fresh or dried, mixed with a small quantity of ashes. To the use of this leaf, they attributed their ability to carry loads for long distances in high altitudes. Since no one could tell him its chemical content, Mr. Lathrop sent a quantity of the leaves to the Academy of Sciences in San Francisco for analysis, with a description of its use. Much to Mr. Lathrop's disgust, nothing was ever done in the matter. He was particularly annoyed later when a German chemist isolated the useful drug cocaine from coca leaves, a drug which soon became the best known alleviative of human suffering.

It was a bitter disappointment to Mr. Lathrop that he had not had the satisfaction of bringing cocaine to public notice. Whenever I gather the red berries from my coca bush, I am reminded of Mr. Lathrop's first introduction.

Just as I was beginning to learn something about Peruvian possibilities, Mr. Lathrop announced that unless we wished to spend the rest of our lives in Peru we must be moving on. We therefore took steamer for Valparaiso, the port of Santiago, capital of Chile.

Professor Charles S. Sargent of the Arnold Arboretum had told me that the best nurseryman in South America lived in Santiago and, indeed, Señor Izquierdo, a Chilean senator, proved to be one of the most intelligent and congenial plantsmen I ever met. For years he and I carried on a lively correspondence which was to the mutual advantage of both Chile and the United States, for it resulted in the exchange of many plants. Our Plant Inventories contain records of all sorts of Chilean fruit and vegetables which I imported for trial in the irrigated regions of California because of the similarity of the Chilean and California climates.

Fruiting cacti were just beginning to attract the attention of settlers in the Southwest, and I collected every promising species I saw. These eventually found their way into the collections which we established in California, about which I shall have more to say.

In Santiago I also discovered a hardy variety of Avocado (alligator pear) reputed to have stood a temperature of twenty-three degrees and to have been uninjured by snows. This seemed to me a valuable find for California, as we already knew that the major portion of the irrigated areas of California were subject to frost. It was a small, black-fruited avocado, at times a little stringy. I secured all the fruits that I could, amounting to about a bushel, packed them in boxes, and forwarded them to Cook, who was then in charge of the Office.

I wondered for years what became of this hardy Chilean avocado, the very first avocado, in fact, which was introduced by the Section of S. P. I.

One day when walking through the plantation of C. P. Taft, one of the pioneers in the California avocado industry, I saw some small purple fruits on one of his trees. They seemed familiar and I asked from where they had come.

"That's the first avocado I ever planted," said Mr. Taft. "The seed was sent to me in 1899 from the Department of Agriculture. They are fibrous, but I keep the variety because it gave me my first helpful experience with the avocado."

The reason I had been unable to trace this importation was that the system of recording distributions of seeds and plants to experimenters had not then been perfected. Files are now maintained by means of which the distribution of any imported variety of seed can be ascertained in a few moments' time. Few recipients of experimental plants and seeds from the Section of S. P. I. during the past thirty-seven years realize that a complete record of the shipment is kept. I have frequently confounded experimenters who were complaining that they received few plants from the Government, by sending them a long list of those they had had. Also, on innumerable occasions these records have enabled us to determine the species or variety of a plant growing on the property of some early experimenter after the property had changed hands. In fact, the whole country has been treated as a vast arboretum, the records of which have been kept in Washington.

During that visit to Chile I secured the Algarobillo (*Caesalpinia brevifolia*) which furnishes one of the best tanning materials, and the Soapbark tree (*Quillaja Saponaria*). Many thousand tons of the bark of this tree were shipped from Chile to Europe for the cleansing of wool. Yet neither of these importations resulted in anything of value in the United States. Unfortunately there seems to be no relation between the obvious usefulness of an introduction and its probability of success.

We left Santiago on April 16 much excited—or at least I was—by the idea that we were to cross the Andes on muleback. The railroad had not been finished and it was necessary to arrange our baggage so that it could be carried by mule train over the high pass to Las Cuevas in the Argentine.

We had heard disturbing rumors about the thefts committed on travellers crossing the Andes, and as an extra precaution Mr. Lathrop packed a small suitcase containing most of his valuables in the center of one of the big trunks. This valise was the only thing stolen from our baggage. The thieves, although located somewhere along the route, must have been in league with the hotel people or they could hardly have known of a suitcase being hidden in the trunk. Anyway, I well remember Mr. Lathrop's fury at the occurrence.

The enormous height of the Andes, their incredibly steep, rocky slopes, and their aridity, made an indelible impression upon me as my mule plodded up the narrow trail. Great condors soared overhead and I stared at them fascinated, little thinking that great man-made birds would one day fly over these mighty peaks. When we had almost reached the top, my mule slipped on the ice and suddenly fell to his knees. There was a horrid moment of suspense while he struggled to save himself and me from plunging down a thousand-foot abyss. But my luck held and I lived to tell the tale.

After an uncomfortable night in the hostelry at Las Cuevas, we boarded a train for the long, dreary journey across the great open plains of Argentina. Twelve days after leaving Santiago we were in beautiful Buenos Aires, enjoying its great club houses, fine parks and gay boulevards.

The Director of the Botanic Gardens gave me, among other things, some seed of the "Bella Sombra" (*Phytolacca dioica*), a rapidly growing shade tree, one of the finest in the country, and also an interesting roadside tree, *Tipuana Tipu,* brought from the Chaco. Both of these trees are now popular in California and Florida. I also procured the seed of *Carica quercifolia,* a hardy relative of the papaya, which has grown well in California.

I made a special trip to La Plata to meet one of the most interesting men in that part of the world, an Italian botanist named Doctor Carlos Spegazzini, who had travelled extensively in the Gran Chaco. Among other things, he had brought back a cactus (Opuntia) which, being spineless, furnished an excellent fodder for the cattle of the dry plains of northern Argentina. In writing to the Department, I recommended that this cactus be carefully tested in our Southwest, and consequently they sent one of the plants to Luther Burbank at Santa Rosa, California. I reported that, according to Doctor Spegazzini, many cattle belonging to the Indians of the Chaco lived almost entirely upon this spineless cactus during the summer months.

I mention sending this material to the Department in 1899 because it was much later when Burbank announced his development of a spineless cactus which he encouraged the public to believe would solve the problem of cattle fodder for desert regions. The joints of his "spineless cactus" were shipped far and wide to enthusiastic believers, but, so far as I know, the existence of spineless forms in other parts of the world previous to his own selective work was never admitted by Burbank.

As with many apparently perfect things in this life, there was a "catch" to the cactus as a panacea for deserts. David Griffiths made extensive researches into its usefulness, and discovered that, like all other plants, the cactus will not actually *grow* without water. It can *remain alive* without water much longer than most plants, but it makes practically no growth. In its South American habitat, there was a short rainy season which made all the difference in the world.

Having made these interesting contacts in the Argentine, we moved on to Rio de Janeiro, where Mr. Lathrop's cousin, Mr. Charles Page Bryan, was Minister to Brazil. He insisted that we stay with him at the Legation in Petropolis, where all the foreigners lived. A cog railway from the Bay of Rio ran up through the clouds to this heavenly spot in the mountains. Our Legation had a business office in Rio, and Mr. Bryan spent about four hours each day going down the cog railroad and returning at night to Petropolis.

The West Indies and South America

This commuting to Rio prevented my spending more than a few hours a day in the Botanical Gardens but there was no alternative. The risks entailed by spending a night in Rio were not to be taken lightly by one who had never had yellow fever. Europeans had discovered that by living in Petropolis they escaped the fever which was then considered endemic in Rio, but they must leave Rio before sunset and be safely in the mountains by nightfall.

It is difficult to give an adequate description of the mental attitude of the Europeans in Rio in those days towards the mysterious, usually fatal, disease of yellow fever. I met just one person who felt that there might be a connection between yellow fever and the mosquito. While crossing the Bay one afternoon on the ferry-boat, I chatted with a young German who had a theory that people not immune to the disease escaped the fever by leaving Rio before sunset because the mosquitoes did not appear in numbers until five o'clock in the afternoon. He had not heard of Theobold Smith's discovery of the tick as a carrier of Texas cattle fever, but was searching for a species of bacteria. I never heard of him again, and have even lost his name from my notes.

Since those days of fear a great change has come about. As I write these lines, I have no feeling of apprehension for my son Graham who is now in Central Brazil studying some still unknown factors in the yellow fever situation, with his headquarters in one of the Rockefeller Laboratories there. As there is now a practical method for immunizing people against the fever, Graham went with a reasonable feeling of safety to live in a region which had a particularly unsavory reputation in the past.

The magnificent avenue of Brazilian Royal Palms (*Oreodoxia oleracea* or, as Cook has more correctly named them, *Roystonea oleracea*) impressed me by its beauty, but I am ashamed to say that I failed to collect seeds of this palm, which is taller and more stately than the Cuban Royal. Being a palm which did not produce anything good to eat, it did not rank as a so-called economic plant. A third of a century has passed and half a million people have settled in Florida, where this splendid Brazilian Royal Palm should grow. That we have not yet brought it into common cultivation represents an oversight for which I have no explanation.

Mr. Bryan arranged a trip for us to São Paulo. I was not enthusiastic, as it was a ministerial affair which, although interesting, would hardly

133

permit the collection of much plant material. But, in São Paulo, an American, Orville A. Derby, put me in immediate touch with the plantsmen of the district and after I returned home sent me material which I particularly wanted.

São Paulo, lying at a considerable altitude above the sea, was the home of the coffee exporters who had their business in Santos, the great coffee exporting port of Brazil. Yellow fever was so prevalent in Santos that no one cared to live in the town. Instead, they travelled back and forth, spending five or six hours a day on the train.

Then, as today, São Paulo was the center of the greatest coffee-growing region in the world. The owner of the principal newspaper in Rio had invited our party to visit his large plantation and spend the night. He was a delightful host, as so many Portuguese gentlemen of my acquaintance have proved to be.

He had preserved a small forest of magnificent Jequitiba trees (*Couratari legalis*) which were among the most beautiful tree giants I have ever seen. Their light gray, columnar trunks rose eighty feet or more in the air without a side branch, and the glorious crowns, composed of branches often three feet in diameter, gave a more stately appearance than the large Eucalyptus of South Australia or even the giant redwoods of California.

The weather was dry and a peculiar red dust blew from the coffee lands and stuck to our clothes like red talcum powder. We were rosy as a sunset when we reached our host's country house. As was to be expected, at dinner the conversation was largely political. When there is world peace, there is a good coffee market and internal calm in Brazil. The United States was then the chief market for Brazilian coffee. As the Spanish-American conflict was just over, war and possible eventualities in case of more war were much discussed—a theme of never-ending interest to diplomats.

São Paulo is in a climatic region similar in many respects to that of Florida, and would have yielded much in the way of interesting plants, I am sure, had I been given time to explore for them. The Jaboticaba (*Myrciaria cauliflora*), for example, was one of them. Sixteen years after my visit, an expedition from the Section of S. P. I. imported this tree from Brazil. During the years it has been growing in Florida, two freezes and two hurricanes have demonstrated that it is adapted to conditions there.

Its resistance to cold and the toughness of its wood entitle it to a serious trial in the southern half of the State.

Our first attempt to depart from Rio ended disgracefully from Mr. Lathrop's point of view. We missed the boat, a thing quite inexcusable in his mind. It was the only time during our travels together that we failed to catch either a boat or a train. Fortunately for me, it was not my fault! Mr. Lathrop's cousin, the Minister, had entrusted the management of our baggage to the Legation staff, and they failed to get the luggage down from Petropolis in time. Even the not inconsiderable powers of persuasion of both Mr. Lathrop and the American Minister could not induce the German captain to wait one extra moment for that baggage.

For my part, I was delighted, for it gave me a little longer in Brazil, and I hastened back to the Botanical Garden, leaving Mr. Lathrop to vent his wrath on the luckless staff.

CHAPTER XI

Cotton in Egypt

W<small>E FINALLY SAILED</small> from Rio the last of May, on a Royal Mail steamer bound for Southampton. Mr. Lathrop wanted to spend some months in England before going to Egypt in the fall. I doubted whether there were any useful plants in Great Britain which had not already been introduced into America but enjoyed the idea of England as a dramatic change, horticulturally speaking, from the jungles of Brazil.

At Kew Gardens, where I presented my credentials to Sir William Thistleton Dyer, the Director, I was overawed by the efficient organization of the Gardens. It was amazing to find that Sir Thistleton handled the extensive correspondence without the use of typewriters. With brief handwritten notes he apparently accomplished more than many of my American friends achieved with the aid of several stenographers.

At that time there were few botanical gardens in America; the Arnold Arboretum at Boston founded in 1872; the Shaw Botanic Garden at St. Louis founded in 1859, and a garden recently established in New York City in 1894. As I was not collecting for a botanic garden, but was searching for economic plants, Kew, although fascinating, was of little help to me just then, and I turned to the marketplaces.

The morning after our amazing meeting with Hoffman at the Empire, I discovered a bean about which I knew nothing although it was a favorite vegetable in Great Britain—the Windsor broad bean (*Vicia Faba*). It seemed curious that so popular a vegetable was not grown extensively in the United States. I therefore shipped home numerous varieties.

Like most people, I did not understand the essential differences between the climates of Great Britain and the United States. The spring season in America is apt to be short and the weather in May frequently becomes almost tropical in character. This is in great contrast to the long, cool springs of England, Ireland, and the northern part of Europe. In this European type of climate, the broad bean luxuriates, while with us it is unhappy, if it grows at all, and is easy prey for the black bean Aphis which attacks it.

I still hope that there may be mountain regions in America where the broad bean can be grown successfully, but the attempts made in the nineties all came to naught. It is a great pity, because the young beans, as a purée, make a delicious dish.

The study of the broad bean later led me into an investigation of the horse bean, a small-seeded variety. The horse bean forms an important summer forage crop in Belgium and a winter forage crop in the whole Mediterranean region where it is grown during the essentially frostless winter.

Mr. Lathrop's attack of fever in Caracas had left him in rather bad shape, and he determined to take the cure at Carlsbad. I had never been in that part of Europe, and welcomed the opportunity to see something of Bohemian agriculture.

The first interesting information I gathered was that the American practice of letting horse-radish grow for years, until it develops tough, fibrous roots to grate when needed, is not the best method.

At the little town of Teschen in Bohemia, I tasted "Maliner Kren," a delicate and delicious horse-radish. With the Austrian method of cultivation, only the young shoots were eaten. The technique of horse-radish cultivation was so interesting that I wrote a bulletin about it which was published by the Section of S. P. I.

I later visited a horse-radish grower in Camden, New Jersey, who had received a cutting of this "Maliner Kren" and had made quite a success of its cultivation.

After Mr. Lathrop had finished his cure, we crossed the Alps into Italy. He insisted on going to Venice although I felt sure that there could be no plants in a town with canals for streets and a horse kept in the Botanical Garden as a curiosity to show the children. Mr. Lathrop seemed confident that I would find something even in Venice, and I did discover an enormous flat squash which was unknown in America. I also sent home some of the sweet peppers which have since become common in the United States. Broccoli was also a common vegetable among the Italians, although unheard of here at that time.

Venice enchanted me. I wandered its narrow ways for hours with Mr. Lathrop's delightful friend, F. Hopkinson Smith, a noted builder of light-houses, writer of best-sellers, and landscape artist. It was a rare privilege to see Venice with a man who so appreciated and understood its beauty.

Hopkinson Smith showed me a rusty old anchor, half buried in the ground, and told me that a Venetian, after acquiring a fortune in America, returned home and purchased a shipyard close to this picturesque antique. The man started in to clean up the place, and ordered the removal of the anchor. Immediately he was visited by a delegation of city fathers who informed him that the anchor could not be touched. It was one of the landmarks of Venice and there were laws preventing any alteration which might detract from the picturesque appearance of the city.

I have often thought of the wisdom of the Venetians and wished that a like intelligence could be transferred to city authorities in my own country who placidly allow the removal of historic landmarks and even encourage their destruction.

In the fisherman's quarter, of which he was particularly fond, Hopkinson Smith smiled and said:

"I suppose that I have made enough money from my paintings of this part of Venice to buy most of it."

As the monks of San Lazaro were said to take an interest in horticulture, I took a gondola there and knocked at the monastery door. Father Giacomo Issaverdens, an Armenian monk with a long white beard, showed me about. At York Harbor the previous summer I had met William Dean Howells and his delightful family, and while reading Howells' fascinating book on Venice had noticed the name of this monk. The Father was delighted that I knew the Howells and gave me a fragment of a painting by Tiepolo to take to Miss Howells. We had a long conversation under a lovely old peach tree in the monastery grounds and, among other helpful suggestions, Father Giacomo told me of a seedless grape which he thought I could find by going to Padua.

When I reported this to Mr. Lathrop, he was strongly in favor of my going to Padua and I arranged to take the morning train.

Cotton in Egypt

My little room in Marco Polo's old palace was so low that I could barely stand erect in it, but it had a window on the Grand Canal. That night I was awakened by the most heavenly music I have ever heard. I leaned from the window and looked across the Canal towards the famous old church of Santa Maria della Salute. In the moonlight I saw the graceful silhouette of a man standing in a gondola playing a flute. To my delight, the gondola turned into the side canal almost directly under my window. The musician was dressed in black from head to foot and was alone except for the gondoliers. How softly the gondola passed! All too soon the music died away. But those moments still live in memory as an exquisitely perfect experience. In the morning I learned that the mysterious musician was an Italian nobleman who often wandered through the canals on moonlight nights, playing his flute.

At Padua, I went directly to the monastery to present my letter from Father Giacomo. The Father Superior sent me to a nurseryman in a little village not far away, and there, over an arbor, grew the seedless grape of which Father Giacomo had told me. The nurseryman claimed that a Roman gentleman, at one time majordomo to the Pope, had sent several cuttings from vines in the Vatican garden. I was not inclined to believe this story, but thought that the grape had probably been introduced from the region of Smyrna by some Armenian monk in the near-by monastery.

The vine was covered with grapes, some bunches sixteen inches long, and the individual grapes were rose-colored, seedless, and of an excellent flavor.

The idea of seedless fruits was new and popular in the nineties. Furthermore, Burbank had aroused a great deal of interest by his writings discussing the possibilities of producing even a stoneless plum. Upon my return to Venice, I packed the cuttings which I had secured, feeling that my pessimism with regard to plant hunting in Venice had after all been quite unwarranted.

Later we secured a considerable shipment of this "Sultania Rosea Seedless Raisin Grape" from the nurseryman. Quite extensive plantings were made in California, where its color attracted a good deal of attention. A pale green variety, also seedless, which had been introduced from Smyrna, is known today as the Thompson Seedless grape and is extensively sold on the American markets. I understand that it is a better shipper than the "Sultania Rosea of Saonara," but the Sultania has been used by Doctor

The World Was My Garden

A. B. Stout of the New York Botanic Garden to produce an American seedless variety for the grape-growing area of the Atlantic States.

A strip of incredibly fertile land on either side of a great river running through a desert—that is Egypt. Mr. Lathrop had promised me a long stay there, as I had as yet seen nothing of the agricultural section of the country. The ancient method of basin irrigation still prevailed, with agricultural practices dating back thousands of years, but from this narrow strip of fertile land have come food and material to create and maintain a civilization six thousand years old. It was my purpose to study the living food plants of the people and their methods of growing them.

Except for the rice terraces of Java, I had never seen an irrigated field and therefore made a careful study of this agricultural practice. Much which I reported to my colleagues in Washington about the Egyptian methods of irrigation was as new to them as to me. Although the valleys of both the Rio Grande and Colorado River were somewhat comparable to the Nile valley in richness, almost nothing was known about irrigation by the promoters selling farm lands along the banks of our southwestern rivers. Therefore a study of the food crops upon which the Egyptians had lived during the centuries was of vital importance.

There was sesame, which I had known only as a password for Ali Baba. The word now became the name of *Sesamum indicum,* an important oil-producing plant grown extensively in the Nile silt. The seeds are sold in America under the name of bene. Growing with the sesame was a species of lettuce cultivated for its oily seeds. In adjoining fields grew garbanzo (*Cicer arietinum*), a leguminous annual producing seeds which are roasted and eaten like peanuts by the farming classes, or boiled as an ingredient for their soups. (This garbanzo I later discovered to be one of the principal food crops of Spain, Palestine and other countries. It is even used as a substitute for coffee.) Then there were the Egyptian varieties of okra, red pepper, vegetable marrow, pumpkin, cucumber and peanuts. They were all interesting, and I sent home seeds for trial in our new irrigated regions. Great quantities of onions were being shipped down from the islands of the Upper Nile, and exported to England. I procured seeds of these, too, and they proved valuable and have been grown extensively in Texas.

When the Empress Eugénie visited Egypt to celebrate the opening of

Market boats in the Grand Canal, Venice.
Above: Father Giacomo Issaverdens.

The enormous Venetian Flat Squash.

In the early morning, endless caravans laden with Berseem plodded along the long avenue under the Lebbeck trees planted in honor of Empress Eugénie's visit to Cairo for the opening of the Suez Canal.

The first Egyptian cotton in Arizona was grown by Dr. A. J. Chandler from seed the Author sent from Cairo in 1899. The photograph shows Dr. Chandler at the right.

the Suez Canal, the Khedive built a five-mile avenue from the principal bridge across the Nile to the Great Pyramids. Beside this boulevard were planted two rows of lebbek trees, which cast a dense and restful shade. The dignity of five straight miles of leafy green stretching across the glaring sunlight of the fields was most impressive. I sent seeds to Washington and wrote an article describing this avenue. Today hundreds of lebbek trees are growing happily in the Florida sunshine.

The lebbek (*Albizzia Lebbek*) is commonly known as the "Woman's Tongue Tree" because of the clatter which its dry pods make when they are ripe. It has found the climate so congenial in southern Florida that it has "gone native" there, and many homes are shaded by descendants of the seeds sent from Egypt so many years ago.

During a visit to Egypt in 1926, it was distressing to find the great avenue of trees had been cut down because they proved to be a host plant of an insect destructive to the cotton, the important money crop of Egypt. Before our Civil War, the cotton fields of our southern States furnished most of the cotton used by the manufacturers of Europe. The war checked this export for five years, and gave a tremendous impetus to cotton-growing in Egypt.

Mr. George Foaden, an English agriculturist, was secretary of the Khedival Agricultural Society and gave me valuable information about the Egyptian cotton situation. The so-called Egyptian varieties were then believed to have originated from the Brown Peruvian or Piura cotton, which I mentioned when describing our visit to Peru. While the fibers of this Piura cotton were shorter than the ordinary cotton of our southern States, and were deep tan in color, the crosses of it with cottons already growing in Egypt had long, silky, nearly white fibers. This resulting type of cotton was popular with the cotton spinners, and considerable imports of Egyptian cotton were made annually into the United States. A new variety, Jannovitch, had recently appeared as a sport in Egypt. Its seed was selling for twenty dollars a bushel. It was thought to be a hybrid between the Sea Island cotton of Georgia and descendants of the Peruvian cotton introduced into Egypt in the sixties. Christian Stamm, an old German horticulturist, gave me a small sample of this Jannovitch cotton with samples of other well-known varieties to send home.

As Mr. Lathrop and I were drinking our coffee that evening on the verandah of Shepheard's Hotel, and watching caravans of camels and donkeys passing on their way to and from the desert, I described the cotton situation and told him of the samples I was sending to Washington. After a moment, he said quite simply,

"Yes, Fairy, but if you think it is such a good thing, why don't you send more of it?"

Knowing the crowded condition of affairs in the Department of Agriculture, I hesitated, for I had a harrowing memory of our embarrassment when Professor Hansen's immense shipments poured in from Russia two years before. However, Mr. Lathrop persuaded me to send two bushels of each variety, which was fortunate as, unbeknownst to me, my old friend Webber had begun his cotton-breeding experiments. Some of the seed was sent on to Arizona and California and from it grew the first cotton of that section, drawing attention to the possibility of Egyptian cotton culture in the Southwest.

Three years later, I photographed Doctor A. J. Chandler standing beside his cotton plants grown in Arizona from the Egyptian seed. During the next quarter of a century, T. H. Kearney, O. F. Cook, Swingle, and Webber all spent much time in the improvement of the Egyptian cotton, and Chandler, Arizona, is now the center of a large cotton-growing district.

Although it was winter, there were patches of what appeared to be clover, and the cattle browsing in these fields seemed to be enjoying a delectable meal. The plant proved to be *Trifolium alexandrinum,* the berseem of the Egyptians, an entirely distinct clover from any I knew. It was the most profitable forage crop of Egypt, perfect for irrigated lands and ready for cutting fifty days after planting. Berseem had never been introduced into America, although it seemed ideal for the irrigated lands of the Southwest. Alas, the supposedly Nile-like valleys of the Colorado and Rio Grande proved too cold for this clover in winter; in summer it would not grow there at all. Since then it has been successfully introduced into Tunis and the Po valley of Italy.

This stay in Egypt was so completely filled with studies of the irrigated crops of the Nile that I had no time for the date palm, but, later, both Swingle and I spent much time introducing date growing as an industry for Arizona and southern California.

CHAPTER XII

Across the Java Sea

Mr. Lathrop decided to visit some of the less known islands of the Java Sea, and also New Guinea, which was practically unexplored. So we sailed from Port Saïd for Ceylon, where we caught a "P. and O." boat on December 10, 1899, reaching Java on the first day of the year 1900.

My friends in Buitenzorg were amazed to see me again so soon, and to find me more interested in plants than in termites. I must admit that I cast longing glances at the laboratory in the Botanical Garden, and had it not been for Mr. Lathrop, would probably have taken time out to work there again.

The Netherlands India Packet Boat Company had recently inaugurated a service through the eastern portion of the great Malayan Archipelago. This immense group of islands extends for a distance equal to that from New York to San Francisco. The boats were small, two thousand tons or so, and were manned with Malay crews and Dutch officers. Both our captain and his first mate were over six feet tall, and were splendid seamen. Mr. Lathrop and I were the only white passengers during most of the voyage, and our trip resembled a yachting cruise in the Java Sea.

Soerabaja, the principal port of eastern Java, was a small town but, even then, gave promise of becoming the important emporium for this vast Archipelago. Its harbor was filled with an amazing variety of native sailboats from all the near-by islands—many of the sails decorated with beautiful, though simple, designs in color.

Our first stop was at Bali, and our second at Lombok, two of the smaller islands lying east of Java. At the time of our visit, the relations between the Dutch and the inhabitants were not entirely friendly, and during the short stay of the steamer we were requested not to stray far from the port. The natives were armed and it was not considered safe to go into the interior without a guard. Bali is supposed to have the most interesting civilization of any island in the Archipelago. My only impression was of the comeliness of the Balinese women and the fact that all of the men were armed. The native market yielded little in the way of interesting plants, and altogether our days on these two islands were a disappointment.

I had with me Wallace's *Malay Archipelago,* as great a book in many respects as Charles Darwin's *Voyage of the Beagle.* That these two friends each wrote a really great book is not as astonishing as the fact that independently they each evolved the theory of "natural selection." I wish that more young people today would read these masterpieces instead of being satisfied with superficial books of travel.

It was disappointing to find that our boat would make only one stop at Celebes. Except for the fact that it had proved a great collecting field for Wallace, this island was little known geographically.

Our stop at Makassar, then a little one-street town along the water's edge, was also disappointing. I longed to see the interior of the island, but such a trip meant an expedition with suitable equipment and required weeks of strenuous jungle travel. The markets in the town, contrary to my expectations, yielded little except varieties of rice, Lima beans, and special sorts of peanuts. There were none of the marvellous tropical fruits which I had seen in Java.

From Makassar we crossed the Banda Sea to the island of Amboina. The weather was ideal, as usual in this part of the world; deep blue water as smooth as glass, the surface broken only by the flying fishes and rollicking porpoises. We enjoyed watching the native passengers—Chinese, Amboinese, Javanese, Sundanese, Madurese, and a few Arab traders. The days passed all too quickly. I became very friendly with the first mate, a pleasant young Hollander named Dirk. He was about my own age, and had come East determined to spend his life in this part of the world.

A half-caste Amboinese woman had come on board at Makassar with her two pretty young daughters, bound for Amboina. Dirk told me that the woman was a granddaughter of Rumphius, the famous naturalist, whose name is as well known to zoölogists as is that of Linnæus to botanists and horticulturists. Rumphius was a Hanoverian who came to the

Indies as a youth and lived and died in the island of Amboina. He had a passionate love of nature and a genius for descriptive writing.

Some years before Linnæus, Rumphius had described hundreds of species of plants and animals which he found around him, or which were sent to him from other parts of the Archipelago. Finally the voluminous manuscript, the result of many years' labor, written painstakingly in longhand, was ready to go to Holland to be printed. During the voyage, the Dutch vessel was attacked by the French and went to the bottom carrying the precious manuscript. Rumphius was old and practically blind, but he set to work with the aid of one of his daughters, and laboriously rewrote his book.

Not wishing to risk another disaster due to the uncertainties of ocean travel in those days, the government in Batavia had Rumphius' new manuscript copied by hand, a task which consumed several years. Tragically, when the book finally reached Holland, it met with a cool reception from the stupid officials to whom it was submitted and was not published for fifty years. Consequently, it did not appear until after Linnæus had published his *Species Plantarum*. Thus Linnæus received the credit for first accurately describing many plants which Rumphius had really described earlier. Rumphius, however, never used the binomial system.

I soon realized that Dirk was fascinated by the prettier of Rumphius' great-granddaughters and, when we reached Amboina, he asked for my assistance in his courtship. I was non-plussed for, after all, Rumphius' blood was pretty well diluted, and a half-caste is but a half-caste. Dirk said that he would be left in charge of unloading the cargo, but later in the evening, he wanted to take the two girls for a moonlight ride in the launch with me as companion and chaperon.

I had met the girls' father, a Dutch schoolteacher, when he greeted his family at the dock, but I was disturbed by Dirk's request that I go to their home to deliver his invitation. My command of Dutch was hardly adequate for such an interview, but I marched up to the house and explained matters as best I could. I must have appeared harmless, for the father consented and, when Dirk had finished his work, we chugged around the harbor in the launch, after which I escorted the girls safely home.

I was not conscious of wrong-doing until Mr. Lathrop returned from dining ashore with the Captain, and pointed out in no uncertain terms that I should not encourage the young Hollander's attentions to a half-caste girl. On the other hand, it was Dirk's contention that if he married a Dutch wife, she would not be happy in Java, and the climate would not agree with her. He claimed that it would be much better to marry a girl accustomed to the tropics where he expected to spend his life. Dirk did marry her, as I learned years afterwards, but was soon transferred to Holland. His Amboinese wife lost her head amid the freedom of European life and behaved so badly that he had to divorce her. I am sure that there is a real moral to this tale somewhere.

Try as I might, I seemed unable to find any one on the island of Amboina who would collect plants for me. But the day before we sailed I discovered a plantsman in a most unexpected manner. I had set out to visit the grave of Rumphius, and was walking down a country road. To my amazement, a completely naked white girl ran directly across the road in front of me and disappeared into a native house. There were few white people on the island, and certainly one hardly expected to encounter them in the nude.

A man then appeared at the door of the house and spoke to me in French. He was an intelligent person, a collector of insects who also had a good knowledge of the botanical names of many of the plants on the island. He said that he had tried to gain a livelihood by collecting specimens for European museums, but at the time he seemed pretty close to starvation. We were glad to help him, and I carried on a correspondence with him for some time. Whether the girl was wife or daughter I do not know. I never saw her again.

It is only a short run from Amboina to Banda, which is generally considered to be the loveliest of all the islands in the Archipelago. Banda also had historical importance for, like Amboina, it was one of the "Spice Islands" for the control of which the Dutch and Portuguese had fought so determinedly. From these islands, shiploads of nutmegs and cloves were brought in sailing vessels around the Cape of Good Hope to Amsterdam, where they were sold at fantastic prices. Spices were much in demand in the old days to add flavor to the heavy meat dishes which made up the greater part of the menus of the wealthy.

It was early morning when our ship slipped through the narrow passageway into the beautiful lagoon. We came to anchor at a little pier within sight of the nutmeg plantations and low, white houses of the Perkineers, the nutmeg planters.

Across the Java Sea

I have been in many out-of-the-way places, but I think that Banda seemed more out of the world than any other spot I have ever visited. I wandered for hours with one of the Perkineers through his plantation. There are few fruit trees more beautiful than nutmeg trees with their glossy leaves and pear-shaped, straw-colored fruits. As the fruits ripen, they crack open and show the brilliant crimson mace which covers the seed or nutmeg with a thin, waxy covering. The vivid color of the fruit and deep green foliage make the trees among the most dramatic and colorful of the tropical plant world.

Banda contained almost the only nutmeg plantations in the entire world. Attempts to cultivate the trees in other regions had not as yet proved successful. Since then, the island of Grenada in the West Indies has become quite a rival.

The Perkineer took me through his plantation to the shore of the lagoon, and showed me where tons of nutmegs had been burned so that there should be no surplus in the markets of Europe. This was my first experience with controlled agriculture. I was puzzled then as to its being a good policy, and I am puzzled still. It is questionable whether in the long run it was advantageous to maintain the price of nutmegs at so high a figure that sly Yankees sold wooden nutmegs in New England and the British established nutmeg growing in the West Indies.

Off the eastern end of Ceram, the captain stopped for a few hours to let us explore an even more beautiful atoll than those which we had seen in the Pacific when we visited the Fiji Islands. A beach of pure white sand shaded by graceful coconut palms surrounded the calm waters of the lovely lagoon. At the edge of the water, picturesque thatched cottages of bamboo stood on piles to raise them above the waves of any storm which might sweep across the island. Wandering among the houses were naked children and women in batik dresses of Javanese design. The atmosphere of the place was completely tranquil and care-free. It is small wonder that the thought of life on an atoll in the tropic seas often lures the adventurous youth of the world.

Curiously enough, life on a small island appears in one's dreams as something peculiarly delightful, but in actuality proves strangely disappointing. A small island is apt to become merely a prison. The fact that one cannot escape from it at will produces an increasing feeling of oppression.

The World Was My Garden

The island of Ceram had been so little explored that, when we anchored off a little clearing called Boela, our ship was a tremendous event. With much chatter and avid curiosity, the naked inhabitants paddled excitedly around the vessel in their catamarans.

The water was so shallow that we had to be carried ashore on the backs of the natives. The one on whose shoulders I was perched was infected with a horrid skin disease, caused by a ringworm fungus, singularly loathsome in appearance. Although the disease is neither dangerous nor particularly painful, I sincerely wished that I had secured a healthier-looking "horse."

As I was deposited on the beach, I saw hundreds of jumping-fish in the mud around me. They were little fellows, about six inches long, and hopped over the beach so fast that I could not catch them. They were quite amazing, even climbing up the sloping trunks of the mangrove trees, where they sat gazing about from their perches. These climbing fish (*Periophthalmus*) are all well known to zoölogists, but I had never heard of them and could scarcely believe my eyes.

In fact, the beach presented a most animated scene. Thousands of little shells were travelling this way and that across the beach. I stooped to examine them and found each one inhabited by a tiny hermit-crab. Suddenly a dog barked and I beheld before me a small canine, completely hairless except for a yellowish topknot rising from its head. As though these were not enough wonders for one day, on the trail to the village strutted an amazing pair of chickens covered with frizzled plumage of bright indigo blue.

This combination of novelties quite paralyzed me with astonishment. I was convinced that I was seeing sights which no other naturalist had ever beheld. It was disappointing upon my return to Washington to be shown pictures of the climbing fish, and to find hermit-crabs so abundant on the coast of Florida they may be gathered by the bushel. Also, I understand that hairless Mexican dogs can be seen in almost any dog show. Later that same day, I had discovered that a young American oil driller on the island had dyed the chickens by throwing them into a vat of indigo. Still, even such disillusionment cannot extinguish the memory of those first extraordinary moments in Ceram!

Penetrating the jungle to some distance from the clearing, I photographed some of the enormous, buttressed tree-trunks. The pictures were

Barbour Lathrop beneath the towering palms on an atoll near Ceram.

148A

Watching the white man's magic on an island in the Banda Sea.

Fruits of the Chinese Litchi.

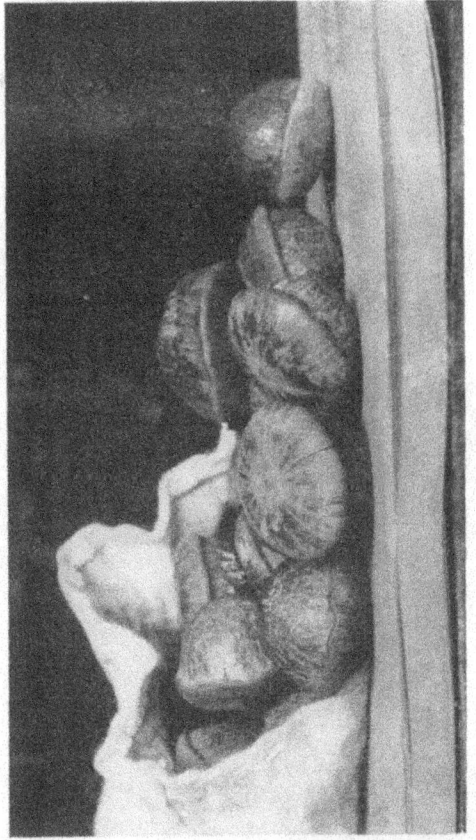

Blossom and fruit (*right*) of the Chinese edible Acorn. (Page 153)

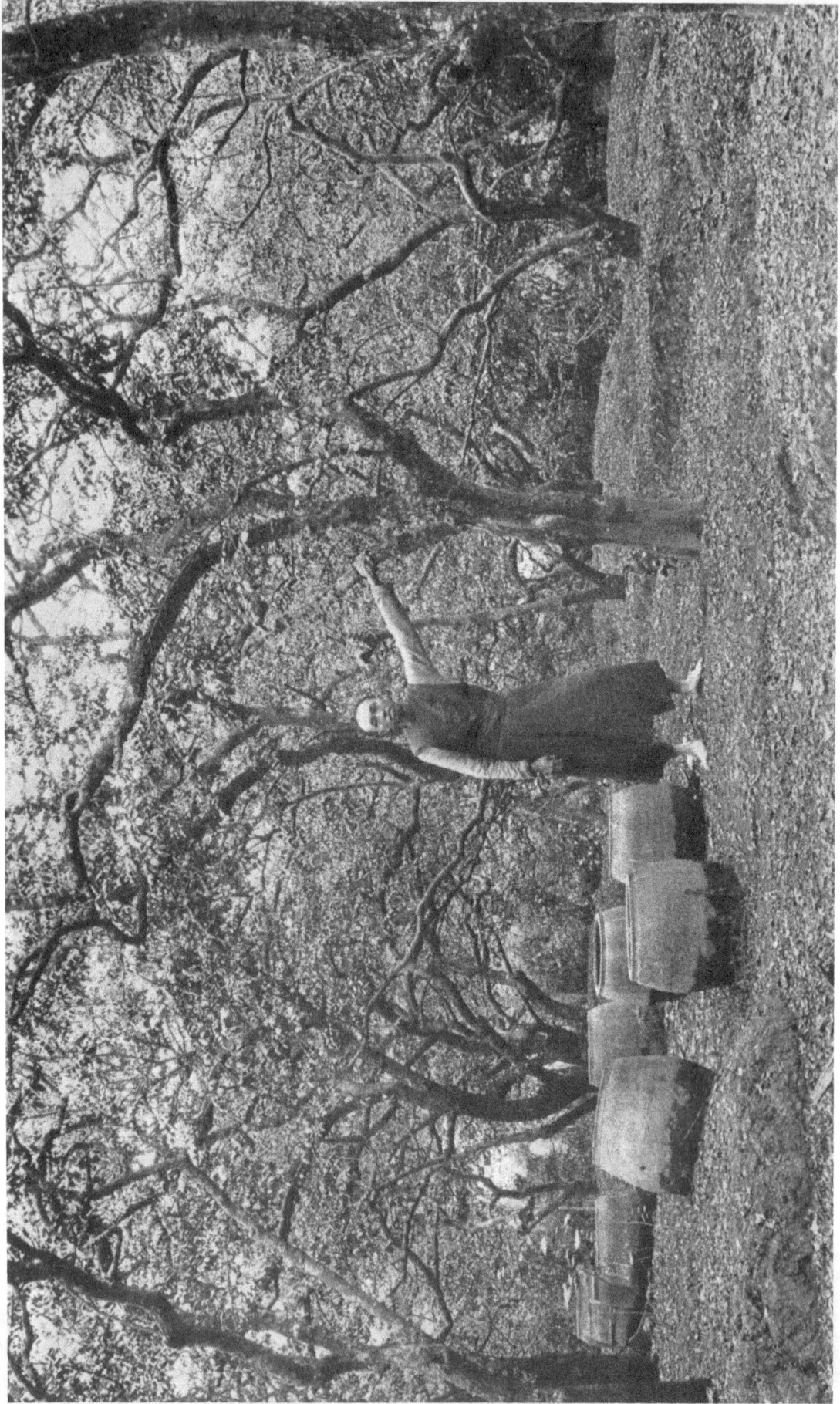

The earthen pots seen in an orchard of Carambolas near Canton were filled with the universal fertilizer of the Chinese and smelled to heaven. The Carambola, which plays an important part in the cuisine of the Chinese, grows and fruits well in South Florida. (Pages 155–156)

quite interesting and were later published in the *Literary Digest*. I also had a happy time collecting the beautiful butterflies which flitted through the rays of sunlight pouring into the dense jungle.

We had stopped at Ceram to deliver some iron pipes to the young American and, as we steamed away, we waved good-bye to him standing forlornly in the little clearing.

From now until the boat returned to Java, our trip resembled an enchanted cruise. During the next twenty days we made eighteen stops at islands varying from a few square miles in size to the immense terrain of New Guinea. At each place we met different types of natives who had grown up on their respective islands knowing nothing of the world beyond the waves breaking on their white beaches. These people spoke different dialects, or different languages, came of different racial stocks, and had their own individual civilization. These islands of the Malay Archipelago form a group so unique that I have always longed to return there. I would have been extremely unhappy when we sailed away had I realized I should never see them again.

A war was on when we arrived at Toeal, one of the smaller of the Kei islands, and the warriors were a startling sight with their faces streaked with white paint. I have forgotten the occasion of the trouble, but the inhabitants of a near-by island, which was in full view, were also armed to the teeth and ready to kill any inhabitant of Toeal who ventured across the narrow strait between the islands. The fierce hatreds between the inhabitants of the different islands were not deep-seated, and there were long periods of peace between them.

Some of the warriors conducted us to the chief's house to show us the great wealth of their chieftain. I could not understand their obvious pride until our captain explained that a pile of useless old cannon stored under the chief's house was considered a possession of great value. The fact that the chief lacked powder to shoot them with made no difference. They were cannon just the same, and wonderful in their eyes.

I was unable to go ashore at several of the islands, for I came down with an attack of malaria. Our servant, Pandok, insisted on rubbing me with cajeput oil—the first time I had ever heard the word. I little dreamed that the oil came from the leaves of a tree (*Melaleuca leucadendron*) which I would later help John Gifford to disseminate throughout South Florida and which would naturalize itself there until today thousands of cajeputs are growing wild in addition to the thousands planted in gardens as ornamentals.

When the tall, good-looking natives of Letti came on board our ship they had an especial interest for us. It was said that many of them were descendants of a Dutch officer who had been stationed on the little island previous to the Napoleonic wars. During the wars he had been completely forgotten, but later, when Holland regained possession of the Dutch East Indies, an official of Batavia discovered a record of this man's having been sent to Letti. A ship was therefore despatched to the tiny island, but the official had long since passed away. However, he had left a large number of descendants as proof of his residence there.

When we arrived at Dobo, in the Aru Islands, a steamer of the Packet Boat Company was in the harbor, having just returned from the coast of New Guinea. Every one was much excited, as members of her crew had been attacked by Papuans. The vessel had anchored off the New Guinea coast while the engineer and the boatswain had been rowed ashore, taking with them some trinkets to barter with the natives. Finding the natives apparently friendly, the engineer decided to cross a narrow spit of land, and called to his friend to row around and meet him in the cove on the far side of a grove of trees. He no sooner disappeared into the grove, than a shower of arrows struck the rowboat. One of them penetrated the boatswain's chest and knocked him unconscious. The Malay sailors rowing the boat were terrified, and rowed away from the shore as rapidly as possible, leaving the engineer to his fate.

I had a long talk with the boatswain, who showed me the wound in his chest and a black and blue spot on his back at the point to which the Papuan arrow had penetrated. As this was the day after the affair, it had evidently not been a poisonous arrow, for he was still alive and later completely recovered.

It seemed terrible to have left the engineer in the hands of the savages, but no relief expedition could have been organized in time to save him. As our next stop was to be on the coast of New Guinea, this introduction to its dangers, while exciting, made it clear that a plant-collecting expedition into the interior would require not only special equipment but a military escort.

The approach to the great island of New Guinea in the early morning was magnificent. The bay was dotted with little islands of graceful palms rising from white coral beaches. In the background, mountains rose tier after tier into the blue haze of the tropical morning. I studied these great mountain ranges and realized anew that we were approaching one of the largest and most romantic islands in the world, almost of size to be a continent; an island which no white man had ever crossed; a land where everything was primitive and savage.

At Fakfak we found a half-caste official and a few soldiers. Our Captain, who had frequently landed in New Guinea and knew the temper of the savages, gave us instructions regarding our behavior towards them.

"Smile at them," he said. "They have a very real sense of humor. Don't be serious with them; it makes them suspicious."

The Dutch had not constructed a proper wharf at Fakfak, so the steamer drew up to a pier made of wobbly stakes stuck in the water, covered with crooked poles over which it was difficult to walk. The path led directly to the house of one of the chiefs. As we approached, the Captain said to me quietly,

"We'd better avoid this chap; he's a surly brute."

We walked inland for some distance, followed by a crowd of Papuans who watched every movement and were curious about everything we did. They were sufficiently accustomed to white men to be bored with the beads and mirrors we brought. This was my first experience with definitely unfriendly savages. While I enjoyed the novelty of looking into the faces of primitive men, the fact that I could not speak a word of their language, and knew that they were distinctly hostile, gave me a decidedly unpleasant feeling. I think we were all relieved when we sailed safely away again.

The boat made another stop in New Guinea at Sekar, where the Dutch had a number of soldiers and more of an organization. The head man of the village presented me with an immense shaddock, one of the first I ever saw. I sent the seeds to Washington, with the collection of economic plants which I made on the various islands.

It seems amazing that a vast territory like New Guinea could be considered to belong to a European nation which had no more extensive holdings than small, futile settlements along its coast. Some years later at The Hague I met Paul Hubrecht, the first man to cross New Guinea from north to south. It was most interesting to hear his account of his explorations. During his weeks of wandering he had developed great respect for these savage peoples and their customs.

We spent a day at Celaru on Timor, the only island in the Archipelago which is not entirely controlled by Holland. Of this island the Portuguese still own over half.

As I was walking through the street of one of the native villages, my attention was attracted by strains of weird music. Following the sound, I came upon a native sitting on his verandah strumming the strangest instrument imaginable. It was made of a piece of bamboo, bits of wire, and the immense leaf of a fan palm. The wires were strung around the bamboo and stretched taut over a circular bridge. The dry palm leaf was bent around the contraption to form a sounding-board. It was one of those ingenious affairs which usually lie silent in some museum. This one met the same fate, for we bought it and sent it to the museum in San Francisco.

I am often reminded of Timor by a species of Eucalyptus (*Eucalyptus alba*) from there which has made about the best growth of any eucalyptus trees which we have introduced into southern Florida. Its success may be due to the fact that this species comes from a moister region than the eucalyptus areas of Australia. It also seems well adapted to Caribbean conditions.

We passed by rail through Java, stopping only long enough to visit the remarkable ruins of the temple of Boroboedoer, which rivals Angkor Wat in grandeur. From Batavia we caught a steamer for Singapore on February 24, 1900, and took a "British India" steamer from there to Manila.

The Philippine Islands had recently become American territory, and all sorts of plans for their pacification and development were in the air. We wanted to have a look at our new territory to see whether the introduction of new plants might contribute to its future wealth and well-being.

Two or three men-of-war lay in Manila Bay when we landed, but the thoughts of Governor-General Otis were reaching out beyond war toward peace. He gave us a cordial reception, discussed with us the possibility of coffee-growing in the islands, and gave us a permit to visit any of the islands at will. When we showed our letter to Lieutenant Ahern and our other acquaintances, they laughed heartily.

"That's all very well," they said, "but how are you going to get through

the cordon of soldiers which surrounds Manila? The islands aren't paci-
fied yet by a long shot. You cannot get outside the confines of Manila
without a guard, and no European is allowed on the streets after dark."

Those ten days in Manila gave us a glimpse of the "Days of the Empire." Everything was so militarized that I even found a cavalry officer in
charge of the little Botanic Garden. He may have known his manual of
arms, but he did not know one plant from another. I had a struggle to
keep him from "cleaning up" the place and running a highway through
it, but finally managed to show him that it would mean the destruction
of the most valuable specimens.

Lieutenant Ahern was of different caliber and was much interested in
the forestry system of Java. As I had learned a good deal about it during
my stay there, he spent much time in the hotel with us discussing the
possibilities of establishing a similar forestry system for the Philippines.
Ahern lived for many years in the Philippines, and eventually did in-
augurate a system there which was a model of its kind.

Of course I visited the markets in search of mangoes. The Philippine
mangoes were famous throughout the Orient, and I was anxious to study
them on their native heath. Almost all of the mango varieties which I
had tasted in the Dutch East Indies seemed much too fibrous. On the
outskirts of Manila I found a large Carabao mango, the finest of the
Philippine varieties, and collected seeds and cuttings.

We were much disappointed that, due to the unsettled conditions, it
was impossible to do any work in the Philippines, and we sailed for
Hongkong on March 16.

Hongkong was the gateway of southern China, and I found it both
bewildering and fascinating. There were a discipline and an orderliness
about the Chinese life in strong contrast to the careless, happy-go-lucky
existence of the Malays. Best of all, the markets were filled with all sorts
of new foods which arrived in sampans from the mainland across the bay.

On my first visit to the market, I noticed some strange-looking acorns
about the size of a hickory-nut, with a shell quite as difficult to crack.
Within, each shell contained a brilliant white kernel as sweet and delicate
as that of a filbert. They were the edible acorns of *Quercus cornea,* and
I sent a quantity of them to Washington in my first shipment of plant
material from China. They were said to come from groves somewhere
up the West River which were only accessible by boat.

These acorns grew, and the oaks were distributed in our southern States. Many of the trees were not cared for, but one sent to Mr. W. P. Wilson at Langdon, Mississippi, was well tended and grew to a large evergreen tree. When it produced its first fruits, I visited and photographed it.

Today I imagine there are only three or four specimens established in the United States, two of which are in the Plant Introduction Garden in Savannah, Georgia. Recent attempts to procure more seed have been unsuccessful, the report being that the acorns have completely disappeared from the markets of Canton and Hongkong. Until fertile seeds can be produced by crossing the flowers of the individual trees here, little progress will be made in popularizing this evergreen oak, which is not only beautiful but also bears delicious, edible acorns.

Mr. Lathrop had visited Canton several times before, and therefore stayed in Hongkong and sent me up the river by myself to see that crowded metropolis. Possibly the very fact that I wandered alone through the narrow streets made all the greater impression on me. I shall never forget the canals, sampans, truck gardens, pagodas, tiny orchards, and thousands upon thousands of hanging shop-signs decorated with Chinese characters.

Probably one's most overwhelming sensation is of the unbelievable congestion of the mass of human beings. Yet it is no more true of Canton than of other Chinese cities. It is characteristic, more or less, of the entire Celestial Empire, an empire possessing not only art and culture but also remarkable agricultural practices dating back at least forty centuries.

The first morning after I arrived, I crossed the bridge from the Shameen, an island where all the foreigners lived, and halted in amazement in front of a Chinese butcher shop. The butcher was busily engaged in chopping up the entrails of a sheep with a large knife, which descended regularly with unerring aim, cutting off a narrow ring of intestine with every stroke. These rings he gathered into a large bowl. Then he carefully scraped the remaining fragments into a small bowl and preserved them for poorer customers. It was a new idea to me that sheep's entrails could be eaten; and it brought home to me the realization that no possible food was wasted in China. But it took me some time to learn that whereas in material nothing is squandered, yet in human effort there is endless waste. Even in the utter lack of decent roadways, China seemed one of the most wasteful countries in the world. They had never deemed it worth while to build thoroughfares which would lessen the back-breaking labor of the human beings condemned to drag their little carts for weary miles. In China there has always been a feeling of limitless time, and an overabundance of creatures of labor therein.

The streets were lined on either side with shops displaying endless wares, all strange to me. Before I had elbowed my way far through the crowded throroughfare, I was conscious of an overpowering stench which steadily increased. I soon realized that I was being crowded by a coolie bearing a bamboo pole over his shoulder at each end of which hung an earthenware pot. During the moments necessary for him to pass me I felt that I was actually smelling the entire sewage system of China. Upon the Chinese about me it seemed to make no impression, being a common, every-day occurrence. The more I investigated this sewage system, the more amazed I was. Upon it depended the fertility of the truck gardens which supplied the immense, crowded city of Canton with food. The desire to preserve every particle of material available as plant food is firmly fixed in the minds of the Chinese people.

Professor F. H. King, Chief of the Division of Soil Management of the Department of Agriculture, wrote a book entitled *Farmers of Forty Centuries.** It is by far the most illuminating book ever published on the agriculture of China. In it Professor King shows that the soils of China have been cultivated so long that their valuable plant food, phosphorus, would long since have disappeared had the Chinese agriculturists not inaugurated some method to return that precious element to the soil. Without phosphorus, plants cannot grow. Still, it would seem that had the Chinese been chemists, they might have developed other and less obnoxious methods of sewage disposal, and recovered the necessary phosphorus by chemical means instead of continuing with an ancient method which condemns many people to lives of disgusting drudgery. I could not but admire a discipline which prevents the wastage of even bits of leaves and the droppings of the silkworms, but I wondered whether the maintenance of such a discipline would raise the cultural level of a people.

Pursued by these thoughts and the fragrance of their cause, I made my

Farmers of Forty Centuries, by F. H. King; edited by J. P. Bruce; Harcourt, Brace and Company; New York.

way to the famous water-clock of Canton, believed to be one of the oldest devices for keeping time. I was told a story from the Chinese classics of a young man who devised a different and better clock. His friends took him to the Emperor, thinking that he would be much pleased with the young inventor.

"Destroy it, my son," the Emperor said. "We have too many new things already."

Accompanied by a young, English-speaking Chinaman to guide me through the incredible maze of narrow streets, I fared forth from Canton into the surrounding country. Everywhere there were mounds scattered through the vegetable gardens and small orchards of fruit trees. My companion explained that they were graves scattered over the countryside because people are buried on their own land, not in regular cemeteries as in Western countries.

When I lived beside the cemetery in New Brunswick, New Jersey, I had pondered upon the difficulty of shifting a cemetery, no matter how valuable the land became. It now gave me a strange feeling to look across the level plain where hundreds of busy gardeners were working and to see as the most characteristic feature of the landscape these thousands of burial mounds, protected by the "feng-shui," spirits of the departed. The Chinese farmers could not move these graves and could not destroy them, for, if they did, ill luck would certainly pursue their farming operations. The feng-shui is a very real thing, closely related to another conservative factor of Chinese life, ancestor worship.

We visited some guava and carambola plantations. I knew the guava, but I had never seen an orchard of the carambola (*Averrhoa Carambola*). I was much impressed by the apparent productivity of the trees and sampled the curious, juicy fruits which the Chinese use in many ways. Fish cooked with a carambola sauce is especially delicious. I have since seen the carambola in the West Indies, and it is occasionally grown in South Florida where it fruits abundantly.

As we walked along between the tiny fields, we frequently came to depressions of considerable size full of water, and I learned my first lesson about the water agriculture of this remarkable people. Among the strange ingredients which composed the chow mein, I had noticed disks of a crisp, white vegetable of sweet, delicate flavor. These were slices of the water chestnut, a tuber produced by *Eleocharis dulcis,* a species of rush.

Across the Java Sea

The shallow ponds of a considerable extent, planted thickly with this water chestnut, suggested a possible culture for the numerous fresh-water lakes of Florida. I shipped home a quantity of tubers, and the plant has grown and produced tubers in Florida. But it is necessary for some one to wade into the mud to plant the crop, and this has prevented the water chestnut from being extensively grown with us.

The rice culture of the Chinese, like that of the Malays, involved back-breaking work in the paddy-fields, where all day long men and women bend, up to their knees in mud, as they set out each individual rice plant. Still, I knew that rice was being grown in the Carolinas, and I collected all the different kinds of rices I could find.

The development of rice culture in Louisiana and later in California interested me tremendously on my return home. I had felt that rice was one agricultural crop in which America could never compete with the Orient. But I reckoned without consideration of American ingeniousness. In the United States, they evolved a method of drying out the land and planting the seed by machinery. The land is then flooded again. In fact, the development of rice culture in California has resulted in the production of more rice than is consumed in the United States, and we have a profitable export trade to Japan.

During this brief visit of mine to China, I first got in touch with Augustine Henry, a remarkable Irishman living in Szechuan, one of the interior provinces of China. An acquaintance of Mr. Lathrop had given us a vivid account of the many interesting plants to be found there, and I wrote to Augustine Henry to inquire about them. He replied by sending me a splendid pamphlet which he had written for the mission-aries, giving them instructions about collecting plant specimens. At the end of his letter, he answered my query as to how we could procure seeds and plants from the interior of China by giving me the following advice: "Don't waste money on postage—send a man."

This word of wisdom made a deep impression upon me and had a great influence on my policy when I returned to the Department of Agriculture. Largely because of this advice, I inaugurated an explora-tion of that vast country.

I returned to Hongkong with the feeling that I had been living in a dream. Surely only a nightmare could fill my brain with such fantastic people, practices, and customs. Everything which I had experienced during my days in Canton seemed utterly unbelievable.

A small steamer was leaving Hongkong for Siam on March 29, and Mr. Lathrop suggested our returning to see what we could find in Bangkok. I was particularly delighted, as I hoped that this time I could secure the Siamese seedless pomelo, or grapefruit, which I had heard of when we were there in 1896.

We found Bangkok hot as ever, its temples beautiful as before, and its canals still animated with quaint craft. The European community welcomed us again with courtesy and hospitality and we met many interesting people. Among them was Doctor McFarland, the American dentist, as interested in plants as in dentistry. I made arrangements with him to prepare grafted plants of the seedless pomelo, which was grown some distance from Bangkok. I also made collections of the Siamese rices and, for the first time, became acquainted with the Bael fruit. I was bitterly disappointed in not being able to go up the river into the jungle vegetation of the interior. It was now the last of March. Steamships in those days were not powerful enough to buck the monsoon which began with the month of May; so it was necessary for us to leave that part of the world before the monsoon broke.

We had made the acquaintance of a charming English couple, Mr. and Mrs. Farnham, and were glad to find them and their child on board when we sailed for Singapore. The boat was crowded with a motley collection of natives, and a dirty lot they were, too. On the third day out we began to hear rumors of illness. The little English girl was taken sick, and the mother's place at the table was vacant, too.

The next day Mr. Lathrop, who was occupying the second officer's cabin, said to me,

"Do you know, I believe the baby has cholera. I'm going to give them my cabin. Don't say anything about it."

The child became worse. Its cries made conversation impossible at the dining table.

When we came into the harbor of Singapore all the passengers were much concerned about the possibility of being quarantined on the boat. It was rumored that there were several cases of smallpox on board as well as cholera.

In our travels, it was always Mr. Lathrop's custom to be the first man off the boat in any harbor. He invariably had the best accommodations and expected special privileges. This time he decided that he would go off with the doctor and make arrangements at the customhouse. I was to bring the baggage off with me later. As Mr. Lathrop left, he said:

"Be sure you see Mr. and Mrs. Farnham and find out about the child, and whether we can do anything for them."

Shortly after Mr. Lathrop started for shore, the little girl died. I went to the Captain and asked him if I could do anything for the Farnhams. I insisted that I must give them Mr. Lathrop's message, but the Captain told me that I should not disturb them; that there was nothing we could do. So I collected our baggage and went ashore.

Mr. Lathrop was waiting for me on the pier, and his first words were, "What about the baby? Did you see the Farnhams?"

When he found that I had not seen them he was perfectly furious.

"You let the Captain discourage you from seeing them? You didn't give them my message? Call that launch there. We'll go right back to the boat."

Without stopping to attend to our baggage, we returned to the boat, Mr. Lathrop lecturing me all the way about what he termed "the decencies of life." There we found that every passenger, including some old friends of the Farnhams, had deserted the boat like rats from a sinking ship. Mrs. Farnham, prostrated with the anguish of the situation, was waiting helplessly in Mr. Lathrop's cabin for a coffin in which to lay her child.

In a few minutes, Mr. Lathrop and I were back in the launch and off to the shore again. We found a coffin and returned with it to the boat, waiting while the heart-broken mother laid the little child in it. We then brought them ashore. They had acquaintances with whom they were going to stay, but their friends did not come to meet them and Mr. Lathrop found them a carriage. With the little coffin on the seat beside them, they drove away. We never saw them again.

This ghastly story which so well illustrates Mr. Lathrop's warm heart also gives an idea of Singapore in those days. Tropical diseases were abundant and deaths frequent. Most people were too much concerned about their own safety to help their friends if they were taken ill. Furthermore, cholera was a disease so deadly that it was hardly to be expected that your best friend would risk his life by taking care of you.

The World Was My Garden

We spent a week in Singapore and I saw Doctor Ridley again and visited a mangosteen plantation. We then hurried on to Colombo, as we wanted to make a longish stay in Ceylon before sailing on the last North German Lloyd boat of the season.

We reached Ceylon on the 19th of April and went up to Kandy, which was probably the greatest tourist center in the tropics. I spent my time to great advantage in the marvellous Botanic Garden of Peradeniya, which had become a center of research under the able directorship of Doctor J. C. Willis.

As I had never visited the garden of Hakgala, we then went up to Newara Eliya (pronounced "Nuralia"), the mountain resort of Ceylon. My expectation of spending hours in the garden was doomed to disappointment. The day after our arrival I came down with a severe attack of typhoid fever. Typhoid was another disease which lay in wait for the unwary traveller in those days. Although a good deal was known about its treatment, inoculation to immunize travellers was still unheard of.

Luckily for me, Doctor Valentine Duke, a Scotch doctor, took a genuine interest in my case. When the manager of the hotel realized that there was a case of typhoid in the house, he insisted that I be moved at once. The doctor and Mr. Lathrop insisted that I must not be moved, and guaranteed that every precaution would be taken to quarantine the wing of the hotel in which I was staying. Mr. Lathrop had seen the little hospital to which I would have to go, and was far from satisfied with it. When the manager persisted, Mr. Lathrop produced his revolver and laid it on his bureau, announcing that if they attempted to move me, he would shoot. Mr. Lathrop then sent an urgent telegram to the manager of the hotel in Kandy, of which our hotel was an annex, and he came up and settled the affair.

Doctor Duke prescribed carbolic acid in hot milk, which must have been a proper treatment, for I pulled through.

The last good steamer of the season was due to leave Colombo May 14. Mr. Lathrop was not anxious to spend the summer in Ceylon, and became very nervous as the time approached.

Although I had not as yet been able to sit up, the good doctor assured us that I would be well enough to travel, and Mr. Lathrop went down to Colombo to secure cabins on the *Prinz Heinrich*.

On May 13, Doctor Duke put me on the train and sent with me a

Singhalese nurse who had never been on a railway train in his life. The excitement of the trip impressed him more than his responsibility to his patient, and I had a lonely journey down the mountainside. The boy had his head out of the window most of the time and I could not make him hear when I wanted anything. The only way I could attract his attention was by throwing a shoe at him. Nevertheless, I felt stronger with every mile of the journey.

The moon was rising over the hills when Mr. Lathrop met me at the station. I thought that I had never seen anything so beautiful, but to Mr. Lathrop it did not appear different from any other moon which he had seen.

The newness of even commonplace things after a severe illness is compensation, of a sort, for the suffering and discomfort of the disease. It is an indictment of our poor use of our imaginations when we are well, that we lose the thrill of living, and allow the charm of little things to escape us.

My convalescence on the *Prinz Heinrich* was ideal. I saw none of the passengers on the twenty-five day trip to Southampton, but the doctor was a most amusing person. He had been stationed in German New Guinea, where the Germans had begun one of their first attempts at colonization.

The doctor described a typical bureaucrat who had been sent out from Berlin to take charge of the colony. When he had arrived, he was enraged to find that the Governor's abode was merely a thatched house like all the rest, and that there were no streets—only pathways leading off into the jungle. His first act was to erect a sign, which he had brought with him from Berlin, at the entrance to this house. It said, "Eintritt Verboten" in large letters. He had also brought with him the usual questionnaires, and my friend the doctor was present when he interrogated the inhabitants of the village. The natives appeared one by one before him while he asked each his name, the date of his birth and, to the doctor's especial amusement, what his religion was. I have laughed over this story many times, but fear it is no more characteristic of German functionaries than those of any nation.

CHAPTER XIII

From Finland to Dalmatia

M
R. LATHROP was pretty well worn out when we reached London and was ready for another cure at Carlsbad. He was a rather crusty person when drinking the saline waters, and I was not keen to go there with him again. So he started me off on a trip to Sweden and Finland about the middle of June, with instructions to join him at Bremen in August.

We parted in Hamburg and I took the train for Sweden. In the dining-car, an unusually interesting woman of middle age and her pretty daughter were at the table next to me. At the end of the meal I offered a German gold piece in payment and asked the rate of exchange. The waiter stated the value of the gold piece in Swedish money.

Suddenly the lady spoke to the waiter. "It is a shame to cheat the gentleman," she said. "You know that his money is worth more than you have said. If you will not give him the proper exchange, I will."

I bowed and thanked her, and she explained that she hated to see travellers to her country imposed upon. This courtesy was delightfully unexpected, and led to a long conversation which lasted until I got off the train to go to Göteborg. Madame Soederstrom gave me her card and told me to let her know when I arrived in Stockholm.

To leave the mountains of Ceylon in May and find oneself in Sweden in June was a dramatic experience. The northern summer was upon the land and the gardens were brilliant with annuals.

Doctor Nilsson and I had corresponded ever since we met at the World's Fair in Chicago, and he invited me to stay with him at Svälo{\ddot{o}}f. I was the first American to visit the Plant Breeding Institute which he had created for the General Swedish Seed Company, and I was most grateful for this glimpse of the results of plant selection applied to cereals. The process consisted in saving the seed of unusually productive and desirable individual wheat and barley plants, and raising their progeny. From the progeny of these selected individuals, Nilsson developed pedigreed strains which were peculiarly uniform and also heavy yielding. Today, pedigreed strains of cereals are playing a dramatic rôle in the evolution of the wheat and barley culture of the world.

As early as 1890, Willett M. Hays in Minnesota had reached the conclusion that there might be "Shakespeares" among plants. He began selecting individual wheat plants at harvest-time, testing his selections by his "centgener behaviour" method. However, the "Princess" and "Svanhals" barleys, which Doctor Nilsson sent to Washington at my request in the spring of 1901, were among the first pedigreed cereal varieties to be planted in America. Up to that time, a field of pedigreed grain was unheard of.

When I arrived at the hotel in Stockholm, I was puzzled by the deference shown me. Nothing about my appearance had ever before earned deferential behavior from hotel managers. Soon, however, I discovered the cause. Madame Soederstrom, wife of Senator Soederstrom, had left a message inviting me to spend a week-end at their country house on one of the islands in the Baltic.

It was a delightful week-end. The Senator was much interested in agriculture and gave me helpful letters of introduction to his friends in other parts of Sweden. The two young ladies of the family both spoke English and German, and enchanted me by their knowledge of the flora of their country. As the daylight lingered almost throughout the night, we spent our evenings on the water, talking of many things. They assured me that all the girls of their acquaintance knew the scientific names of common Swedish plants as well as they did themselves. Naturally, I was tremendously impressed; with the exception of a few professional botanists, I did not know even one American girl who could give the Latin names of half a dozen plants.

During this pleasant visit, I was introduced to a variety of interesting Swedish foods. The many dishes which compose the Swedish hors-d'œuvres, Smörgåsbord, are so fantastically elaborate that the first evening I took it for granted that they constituted the entire meal. Consequently, I ate so many of the cheeses, pickles, caviar, and other delicacies, that little appetite remained for the formal dinner which followed. Accompanying the meal were a varied assortment of wines and liqueurs which included a most insidious drink, Swedish punch. I soon became seriously alarmed about my ability to play my part as appreciative guest and, combined with my personal embarrassment was an increasing astonishment at the capabilities of a Russian count who was my fellow house-guest.

One feature of the menu, the Swedish spies-brod, or oatmeal cake, captivated me so completely that I visited the factory where it was made. The round, flat, thin, crisp wafers of spies-brod, each with a hole in the center, were piled in great stacks as they came from the presses. I had a hundred of them shipped in sealed tins to Washington where they proved to be as new and interesting as my collection of macaroni from Italy. Although Chinese eating places were appearing in the larger cities in America, Swedish restaurants were as yet practically unknown.

In Stockholm, the Botanic Garden was ablaze with arctic plants. The rocks were covered with magnificent clumps of sedum and half an acre was planted close with columbines.

A species of Hedysarum was of interest, as it is a high Alpine fodder plant which grows above the timber line and is suited to mountain climates. Also, my fancy was particularly taken by a western Asiatic species of verbascum (*Verbascum speciosum*). Its immense flower-spikes reminded me of the mullein of my childhood (*Verbascum Thapsus*) but it had many branches which remain covered with blossoms for more than a month. With its seeds, I sent a warning that this verbascum might prove a weed but, as it is easily rooted out, we need not be too nervous. Sufficient time has not yet elapsed for us to know whether some Asiatic verbascums may become weeds in this country and brighten the roadways with their golden yellow.

It would have been pleasant to spend the entire summer in Sweden but my itinerary included a trip to Finland.

On the map, the Baltic appears like any other sea, but, when one voyages on its waters, islands are so numerous and close together that the boat might be progressing along a winding stream.

Doctor Gösta Grotenfelt, Director of the Agricultural Institute of Mustiala, had thoughtfully arranged for me to see and learn much of Finland during my stay and to obtain some real knowledge of its agriculture. It is a land of lakes and swamps, and the utilization of these marshy acres has well demonstrated the intelligence and industry of the Finns.

Finland grows the European cranberry, or "Foxberry" as it is sometimes called, which occupies the place on the northern European menu that our larger fruited cranberry does in America. Its smaller fruits have an even more aromatic flavor and are generally much preferred by Europeans to the American species. As far as I know, our attempts to cultivate the European cranberry have not been particularly successful.

Doctor Grotenfelt was carrying on numerous experiments with a North Finnish variety of oat and a four-rowed Lapland barley, the latter growing even thirty kilometers north of the arctic circle.

One happy result of this trip was that I saw a native turnip which had been in cultivation among the peasants from time immemorial. They planted it each year immediately after burning-over the fields. I sent seeds of this turnip to Washington, and it has become popular in Alaska, where it was widely distributed by Professor Georgesson. It proved resistant to a root worm, a serious turnip pest. Mrs. Georgesson has written me much about this vegetable as grown in Alaska, claiming that it is as sweet as an apple.

The experiment station for breeding cold-resistant cereals was on the west coast of Sweden, at Luleå, about seventy miles from the arctic circle. I went to Uleåborg on the north coast of the Gulf of Bothnia and made my way across the Gulf of Torneå and so to Luleå by train, crossing the Finnish-Swedish frontier at Torneå. It was the farthest north I had ever been. For some reason, the snappy weather had not stimulated the energy or enterprise of the inhabitants, and I was not in the least impressed by the efficiency of the Laplanders. In fact, many of them seemed completely indolent. Certainly they appeared far less agreeable than the inhabitants of the tropical zone from which I had recently come.

The experiments to produce hardier and better-yielding barleys for Lapland were most interesting and ingenious. The seedlings were kept at a low temperature and the growing season materially shortened. Those individual plants which were able to mature grain in this artificially shortened season were selected as seed parents for grain to be grown in the far North where the period between spring and fall was so short that ordinary varieties failed to fruit. Since then, much has been discovered about the effects of chilling grains to hasten the occurrence of flowering, a process called "vernalization."

As per schedule, I met Mr. Lathrop in Bremen on the 15th of August, and we sailed from there for New York.

The first train found me speeding back to Washington. During the two years which had elapsed, the little Section of S. P. I. had seen many changes. Doctor Cook had been transferred to Puerto Rico and in his place Jared G. Smith was struggling manfully with the problem of five explorers in the field and a rapidly growing correspondence about the plant material that was being sent in. Swingle was exploring in Algeria and Asia Minor; Carleton had brought back his Durum wheats from Russia; Doctor S. A. Knapp was collecting short-kernelled rices in Japan; G. D. Brill had been sent on a collecting trip up the Yangtze Valley; and I had been sending plants from the Arctic Circle and the Equator indiscriminately.

Mr. Lathrop had been irritated by some Government regulations, and had not only assumed the payment of my travelling expenses but had paid me a small salary as well. This had relieved the slender finances of the Section of all cost except the freight bills for the plants which we sent in.

Secretary Wilson received me cordially enough, but explained that he did not know of any place in the Department that was vacant. I thus found myself for the second time virtually out of a job. The plant material which I had sent in and the letters and reports which I had contributed had, however, made an impression on the Secretary. After reflection, he decided to send me back to Europe to make a study of the European hops and barleys.

The Secretary of Agriculture was in a curious position. The brewers wanted better hops and barleys, but, at the same time, the W. C. T. U. was becoming a political power.

In order to accomplish anything that year, it was imperative to reach Bavaria in time for the hop harvest; so I had barely time to say hello and good-bye to my Washington friends before sailing for Antwerp.

Until I arrived in Bavaria, I had no idea how extensive were the exportations of Bavarian and Bohemian hops into America. Even less had I realized that the growers would object to my securing root cuttings with which to improve the quality of American-grown hops. However, I soon found that I was looked upon with suspicion and dislike.

The beer brewed in America had a strong, bitter flavor imparted to it by the larger and stronger flavored hops grown in the United States, and was not comparable in quality to that of Munich and Pilsen. Also, the brewmasters of the great Milwaukee and St. Louis firms were handicapped by the fact that the American barleys contained a higher percentage of protein than the barleys used for making malt in Bohemia and Bavaria.

My research entailed a careful study of both hop and barley varieties. During these investigations, I met many interesting technical experts, including the brewers' chemists and barley experts of southern Germany.

The most famous barley in Europe was the Hanna barley of Moravia— a "two-rowed" barley distinct from the ordinary six-rowed barleys then commonly grown in America. When I interviewed the manager of the Hofbrau Haus in Munich, he confided to me that he was having trouble with his directors because he insisted on using this Hanna barley from Moravia instead of buying the Bavarian barleys. Inasmuch as Hofbrau Haus was a state institution, the farmers naturally objected to his buying imported barley.

It was very difficult to determine the reasons for the superiority of the heavy Munich beers and the light beers of Pilsen; there were too many factors involved. The quality of the barley and hops was only a part of the story. The introduction of Hanna barley into America was the starting-point in the long series of investigations to which Doctor H. V. Harlan has devoted his life.

After my return from Europe, subsequent researches took me to the principal breweries in America. To my surprise, I discovered that even the best brewmasters were unfamiliar with the botanical characteristics of the different species of barley, the two-rowed and the six-rowed. Although they might know them apart in a superficial way, they had not studied their characters sufficiently to know that the six-rowed barleys contained two kinds of kernels, the straight and the crooked, and the two-rowed barleys were composed of only one kind of kernel.

The most famous hop in Europe—in fact in the world—was the Saaz seedless hop. The individual strobiles, as the fruits are called, looked like diminutive, paper pine-cones. Under each scale of the cone are innumerable yellow, glandular hairs filled with the bitter substance called lupulin. The quality of this lupulin determines the flavor of the beer. The hops then grown in America were frequently three or four times as large as the Saaz hop and were responsible for the over-strong flavor of our beer.

I visited the town of Saaz and had no difficulty in securing cuttings. I also spent some time in German towns where special varieties were grown, such as Spalt and Wolnzach. Eventually, I heard of a new hop which was making a name for itself. It had originated in the little town of Polepp in Bohemia and was known as the red Semsch hop. Although it had been discovered as a sport as far back as 1853, it had taken a great many years to become popular.

After finishing my investigations in Bavaria, I landed bag and baggage in Polepp. As the train left me standing by the tiny station, I decided that probably no such luggage had ever before been seen in the little town.

The proprietor of the "Gasthaus," with a long, drooping pipe hanging from his mouth, dragged my Gladstone bag to one of the two guest rooms in the inn. He told me that his beer hall was the gathering place of the hop growers, and added the discouraging information that they had formed a sort of union for the purpose of preventing the spread of the red Semsch hop.

How could I hope to convert them to my philosophy of a free exchange of plant varieties between the different nations of the world? It seemed that the best thing was to settle down in this little hotel and make friends with the growers. Possibly they might then be willing to let me have some cuttings.

The red Semsch hop was the first hop with a definite history of origin. As far as any one knew, all the others had been grown for centuries. This one originated in the garden of Wenzel Semsch. He was a keen observer and noticed that this particular hop-vine was very productive and ripened its "cones" even earlier than the famous Saaz variety. He propagated from it and, owing to its heavy yield, it became in time the hop of the whole region.

I soon became acquainted with the growers as they spent their evenings chatting over a glass of beer in the hotel. They offered to take me to the house in which Wenzel Semsch was born. The discoverer's son received me cordially and allowed me to photograph a handsome, silver pitcher which had been presented to his father shortly before his death.

I told the hop growers that it seemed unfortunate that some more permanent record had not been made of the discovery of this wonderful

Herr Wirth, genial pro-
prietor of the little
"Gasthaus" at Polepp,
Bohemia.

The Carob is a favorite
shade tree of Mediter-
ranean lands. The illus-
tration shows an un-
usual variety which
has both stamens and
pistil. (See page 198)

Old olive groves remind one of paintings by Corot.

The best Corinth seedless grapes grow on the island of Zante. Every year each vine is girdled near the ground to make its flowers set their fruits better. In California, where the Corinth grape is now growing, the same practice is followed. (Page 175)

At right, an old vine, showing on its stem the marks of girdling.

This scene in a dimly lighted oil mill on the Grecian Island of Zante seems a glimpse of the very beginning of agriculture. Olive Oil appears to have been used for food, light, and anointing even by the ancient Egyptians.

168D

hop. I suggested that a tablet be placed on Wenzel Semsch's house, and offered a generous contribution towards the expense. The town's folk were delighted with the idea and immediately undertook to have a tablet made. Later I received formal notification of the unveiling of this memorial which I had made possible. Eventually, one dark, drizzly day one of the prominent growers came to see me and asked me if I did not want some cuttings of the Semsch hop. I said that I certainly did, and told him why. He said:

"Well, there are some members of the Society who won't approve, so I cannot do this openly, but I will ship you one hundred cuttings to a station down the line."

Thus it was that the Semsch hop found its way into America. The hops were grown and tested in the United States, but before the American growers became educated to its value and possibilities, Prohibition descended upon the breweries, and the hop fields of America were ploughed up.

When I had left Washington with an authorization to investigate European hops and barleys, I had been told that further work would probably be required of me which would necessitate my going to Egypt. So from Bavaria I travelled to Trieste and awaited instructions. Due to some delay in Washington, my authorization was slow in coming and I spent nearly a month in this Adriatic port.

During this interlude, I made a trip to a region north of Trieste where I heard that Indian corn was extensively grown. There was little resemblance between the corn fields of the Karst and those of Kansas; also the agricultural methods used were most primitive. For the most part, the corn was hung up to dry by braiding the dry husks together—a very decorative method but not particularly scientific.

The man to whom I had a letter of introduction was not a corn enthusiast, and insisted that people who ate corn were subject to a special disease. As Exhibit "A" he produced a man whose hands were certainly diseased, and I took a photograph of him. It was the first case of pellagra which I had ever seen and of course I did not recognize it as such. At that time it would have been impossible to convince any Kansas farmer that a man could starve to death on corn. The so-called deficiency diseases, such as beriberi and pellagra, were not even suspected.

The World Was My Garden

My next visit was to a Biological Station on the coast of Istria south of Trieste, at a little town called Rovigno. Istria was noted for its filberts, or hazelnuts, and the Director assisted me in securing the different varieties grown in that region. Aside from a few plantations in the State of Washington, commercial growing of hazelnuts in America had not been particularly successful because of certain leaf diseases which were presumably of American origin. It seemed possible that the Istrian species (*Corylus tubulosa*) might be resistant to these diseases. However, none of the Istrian varieties has fruited well in California, although *C. tubulosa* has proved to be an excellent pollen-producing variety and is still kept by filbert growers in Washington and Oregon for this purpose.

Not far from Trieste, on the shore, stands the picturesque palace of Miramar, a spot haunted by memories of Maximilian and Carlotta. Here the curtain rose on the tragic drama in which they played the leading rôles.

I wandered through the deserted gardens with the head gardener. There were many American trees, among them California Sequoias. The Emperor had been very fond of plants, and many of the trees had grown from seeds which Maximilian sent from Mexico. In one of the last letters he penned, he inquired about some of his American plants.

On entering the palace, I noticed that the entresol was curiously lighted from above, and the gardener explained that the circular plate-glass over my head was really the floor of an aquarium, which had been filled with goldfish, creating a unique effect.

Because of his passion for the sea, Maximilian's bedroom was built to imitate the quarters of a ship. On the walls were enormous pictures of Queen Victoria when she was a small girl. Many of them were signed by her and they assumed especial significance, for only that morning the newspapers had announced her death at Osborne. I left Miramar weighted down with melancholy.

To occupy my evenings, I used to drink coffee with a group of boulevardiers at the principal café in Trieste, and listen to their discussions of politics and scandal. Most of them were so poor that each cup of coffee had to suffice for two. A few were well-to-do, and I could not understand why they were content to live such utterly aimless lives. Before I left, I wagered one of these that I could cram more romance into a week of travel than he would have in a year of gossip in the little café and making love to the actresses at the theatre.

It was February when my authorization from the Department of Agriculture finally arrived and I immediately took a small steamer down the Dalmatian coast, glad to be on my way again.

It proved to be an exciting trip. When we were but a few hours out of Trieste, we ran into one of those terrific storms which sweep down the Adriatic. The boat was driven out of its course and violently tossed about. In the afternoon we were so near to a collision with another vessel that its bowsprit swept our aft deck. I had to jump aside to escape being knocked over. The little cabin of the ship was filled with passengers in various stages of collapse from the rough sea. As I stepped inside to find shelter from the storm, a very pretty young woman swooned and fell into my arms. I laid her on the sofa and worked over her until she regained consciousness. She was profuse in her expressions of gratitude, and proved to be an Austrian countess on her way to Egypt.

That evening I left the boat at the little island of Lissa, which seemed a very out-of-the-way part of the world. To my amazement, another American was sitting at the little café on the Plaza. He had just returned from Cambodia, and had pretty thoroughly explored Indo-China and that quarter of the globe. I spent several fascinated hours listening to his stories.

Before I went to bed, I wrote to my boulevardier to tell him that I had had more excitement during the past twenty-four hours than he could have during a lifetime in Trieste.

This trip down the Dalmatian coast in February, 1901, although physically uncomfortable because of the cold, afforded me a glimpse of a region altogether picturesque, replete with the crumbling relics of an ancient culture. People, architecture, and vegetation were all interesting.

One of the most typical trees of the Mediterranean basin is the carob (*Ceratonia Siliqua*). When ripe, its great brown pods are filled with a pulp as sweet as honey and of a peculiar flavor. The tree produces these pods in such enormous quantities that they form the chief horse fodder of the region. Also, the great branches, covered with dark, glossy green leaves, make the carob the favorite shade tree. Remove the carob (or "carrubo" as it is called by the Italians) from the hillsides of southern

Italy and Sicily, and you rob the landscape of one of its most beautiful and characteristic features.

I determined to make a collection of the different varieties of carob, and made my first shipment from Lissa. It was there that I first noticed the perseverance with which the carob establishes itself, climbing the rocky hillsides with an irresistible phalanx of tiny seedlings.

Numerous varieties of European grapes were already established in California, but the Department believed that there were others which might prove superior in various respects. So, after collecting the local grape varieties in Lissa, I spent a few days on the neighboring island of Lesina making a collection of both the wine and table grapes on that charming island.

At Spalato, an intelligent chemist, proprietor of the only drug-store in the place, solemnly promised to send me cuttings of the Maraschino cherry from the town of Zara, but as soon as I departed he forgot both his promise and me. It was not until twenty-five years later that I again visited the region and procured this interesting cherry from which is made the famous Maraschino liqueur. The true Maraschino cherry is not the cherry which eludes one at the bottom of a cocktail glass. The "cocktail" cherry is a much larger cherry, flavored with Maraschino, and artificially colored. The cherry from which Maraschino is made is the marasca, a sour cherry of Dalmatia.

After a few hours at the ancient and picturesque town of Ragusa, we entered the beautiful harbor of Cattaro, which is an inlet of the Adriatic divided into three lovely lakes. Rising from the innermost lake, Mt. Lovhcen presides over the old town with its cathedral built in the ninth century. Cattaro has been Roman, Byzantine, Serbian, Bosnian, and Austrian. Today its name has become Kotor, and it is a seaport of Jugoslavia.

The day we arrived, Cattaro was celebrating the birthday of its patron saint. The streets were gay with Dalmatian costumes. Huge men, over six feet tall, wandered about in scarlet waistcoats covered with elaborate gold embroidery. On their heads perched the tiniest possible red caps, some of them hardly larger than a silver dollar. These were held in place by elastics under the men's chins in the manner of little girls' hats during my youth. A troop of soldiers paraded by, armed with ancient flintlock muskets inlaid with mother-of-pearl. Priests paced the streets in processions bearing the relics of the saint. Women and children kept dodging back and forth under the canopy sheltering the holy relics, hoping thus to acquire some benefit.

Eventually, the gaily costumed soldiers lined up before the church with their flintlock guns aimed at the sky. I suddenly realized that I was about to hear a volley fired by flintlocks like those used in the American Revolution. When the triggers were pulled, the resulting noise was most disappointing—resembling nothing more than a bunch of fire-crackers popping one after another. Some of the magnificent guns would not go off at all!

Cattaro must enjoy a most salubrious climate, for I found an unusually large variety of the English walnut with nuts of twice the ordinary size; a giant olive with fruits an inch in diameter; and a mammoth lemon from Mesopotamia. I sent cuttings from all of these to Washington with seeds of the cylindrical cypresses to try in California.

A plant which interested me almost more than any in the district was a curious perennial cabbage known as "Capuzzo." This vegetable—apparently a form of kale—forms one of the principal foods of the Dalmatians, and is grown especially in the regions of Ragusa and Cattaro. In certain portions of Dalmatia it is stored, and its leaves are blanched and used for salads. They are quite showy when served with lettuce.

A coastal steamer took me to Corfu, perhaps the most beautiful island in the whole Adriatic region. Although Corfu is a Greek possession, the educated people all spoke French.

In my album are many photographs of an immense lemon, a fruit weighing two and a quarter pounds. It was named the "Colla Giant" for a horticulturist of the island, Mr. Antonio Colla. Mr. Colla told me that these large lemons were in much demand among the Jews, who used them in the ceremony of the Passover and were willing to pay quite fabulous prices for perfect fruits. I collected seeds, and Swingle grew the Colla Giant for years in his citrus collection but, as far as I know, it has never been cultivated exclusively for Jewish consumption.

The German Kaiser had built a palace in the mountains of Corfu, and frequently stayed there for short periods. The palace was closed, but I saw the grounds, which were well maintained but contained too many marble and bronze statues to suit my taste.

The costumes of the people of Corfu were striking, and the head-dresses of the women were truly amazing. All the women seemed to possess masses of hair, enormous braids wound around and around their heads. When I learned that these braids are heirlooms, passed down from generation to generation, I felt differently about them!

A storm came up just as I regretfully stowed my voluminous baggage in a rowboat for the trip out to the steamer which was to take me to Greece. The wind increased and, by the time I reached the boat, the night was pitch-dark and the waves were running high. No lantern was hanging at the gangway and I had the greatest difficulty in getting on board. In a furious state of mind, I dragged my luggage into the main saloon, where I found some of the crew. As I directed the lazy stewards to take the bags to my cabin, I used some rather strong words. Turning, I observed a tall, aristocratic gentleman smiling at me. I apologized for my language, but he said heartily,

"I think you were perfectly justified. I am sure that I would have been even more violent than you."

Later I overheard the passengers whispering something about some person of importance being on board and learned that my friendly acquaintance was Crown Prince Constantine. He was most unpopular with his countrymen at this time because of his part in the disastrous campaign against the Turks in 1897. In the Balkan War of 1912 he regained their respect and was twice king of Greece until his final abdication in 1922.

I was first off the boat as we pulled into the harbor, for I was curious to know how the people of Greece would receive the Prince, and wanted to photograph him as he landed. Not a hat was raised nor a hand clapped as he came ashore. My photograph was given to him by one of the ladies of the court whom I met later in Athens, the only record of his return to Greece.

There were several definite crops which I had come to investigate, one being the seedless grape from which the "Corinths" or "Zante currants" were produced. Every American housewife was familiar with the Zante currant. In fact no wedding cake would be authentic without these dried Greek currants, which are not really currants at all but the dried fruit of a small, black, seedless grape.

Normally, one would expect Zante currants to come from the island of Zante, off the coast of Greece, but, instead, the best ones were grown in the mountains of the Peloponnesus near a little town called Panariti, back of Xylokastron.

A reliable grower whose home was at Panariti offered to procure cuttings from that region and seemed to have no objection to selling them for export. I therefore commissioned him to secure several hundred and bring them down on donkeyback to Patras. While waiting for them, I made a trip to Zante. The boat was unspeakably dirty, but Zante proved to be a beautiful island; I spent nearly a week there, and met a number of well-informed Greeks with whom I corresponded for years.

It was February, and the grapevines were bare. But a striking feature of the method of cultivation was more apparent than if they had been in leaf. On every vine were scars produced by a regular system of girdling which had apparently been practiced since prehistoric times. The growers had a special tool with which they removed a ring of bark from around the stem. It was their conviction that this girdling was necessary for the production of the largest and most delicately flavored fruits.

There were no drying trays in any of the vineyards, and I was curious to know how the fresh fruits were converted into raisins. When they showed me the drying fields, they explained that a mixture of cow dung was spread on the bare ground during the harvesting season. This rapidly dried into an absorbent pulp. The pickers then spread the fresh fruit on this dried cow manure, and the hot rays of the summer sun completed the process of drying the grapes. This method had been in use for so many centuries that the question of its unattractive or unsanitary character did not seem to occur to any one. Revolting as it is, I doubt that any one has ever contracted a disease from eating Zante currants dried on cow manure.

In baskets of twine on the walls of the Zante houses, hung large, green cantaloupes which had been plucked before the first frost in autumn and allowed to ripen in cool places. The luscious, green flesh separated easily from the rind and was as inviting as a mound of pistachio ice cream. Through correspondence, the Department had begun the introduction of these winter melons into California. The so-called Persian melons, which sell at fancy prices in the Eastern markets, have been bred from them. In fact, one can now buy as fine winter melons on the American market as are to be found anywhere in the Mediterranean region.

The World Was My Garden

Upon my return to Patras, I became interested in the history of the "Corinth Bank," an arrangement or agreement for controlling the quantities of currants exported from Greece. It had the same purpose as that used by the planters of Banda when they destroyed their nutmegs on the shore of the lagoon.

The history of the Greek currant, like the history of so many cultures, had in it the tragedy of over-production. Curiously enough, one of the causes of that tragedy could be traced to America. An American insect parasite devastated the French vineyards by attacking the roots of the vines and killing them. So widespread did it become, and so many acres were destroyed by this Phylloxera (as it is called) that the French wine merchants were in despair. They did not have enough grapes to keep their wine-presses busy. The parasite had not yet reached Greece, and the French discovered that the Zante currants would make good wine. Consequently, they imported immense quantities of the currants from Greece, creating a grape boom in Corinth. Soon every available hillside was covered with the seedless Corinth vine.

This condition of affairs continued for many years until the French rehabilitated their own vineyards by grafting their wine-grape varieties on American roots which were not susceptible to the Phylloxera. The French vineyards recovered so rapidly that the demand for Greek Corinths immediately ceased and the Greek peasants were left with vineyards on their hands and no market for their product.

To meet this situation, the Greek government established the Corinth Bank and a system of quotas by which each peasant was allowed to sell only a portion of his crop for export. The remainder of the grapes were converted into alcoholic spirits. In this way, the export price of Corinths was maintained, and the farmers were tided over the bitter period after the collapse of the boom.

It seems possible that the amount of money required to run this Corinth Bank could have been better employed in encouraging the growers to plant something besides the seedless grape. When I visited Greece many years later I observed that the quarantine regulations which had been established against Phylloxera were still so stringent that even new varieties of potatoes were not allowed to come into the country, and the work of plant introduction was practically at a standstill.

From Patras I went to Athens, where it was my good fortune to meet

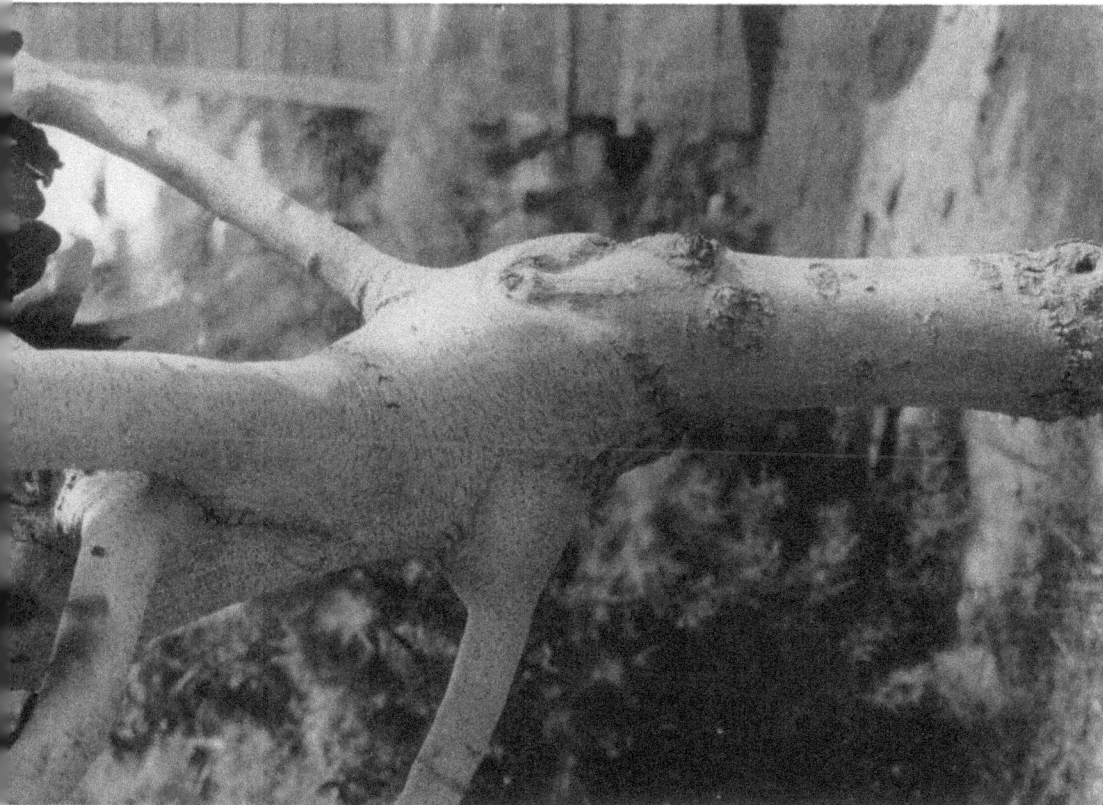

This Pistache stock was two years old when the bud was put in. Note the arrested development, demonstrating the effect of an uncongenial graft.

Pistache, Sfax variety, bearing an abundance of fruit.

On a steamer in the Ægean, a Greek asleep with his kombológia, or "conversation beads," in his hand.

a German botanist, Doctor Th. de Heldreich, who was familiar with both the cultivated and wild plants of Greece, many of which he had described botanically for the first time. When we visited the markets, he called my attention to some tiny green beans no larger than grains of rice, which he declared to be delicious when cooked either alone or with rice in the national Greek dish called pilaff. He recognized the beans as coming from India and thought that they were probably a variety of gram (*Phaseolus mungo*). I had collected seeds of this vegetable in Celebes, where it is used in soups, but it had not made any great impression upon me. The seeds which I now purchased in Athens were sent to experimenters in our southern States and were grown here and there, but it was not until the Chinese introduced them into America, as Mung beans in the form of bean sprouts in their chop sueys, that they were recognized as a valuable vegetable.

Professor de Heldreich had written an interesting book on the useful plants of Greece (*Die Nutzpflanzen Griechenlands*). When he discovered my interest in all food plants, he introduced me to chalvá—a very ancient form of sweetmeat made from seeds of the sesame. The oil from sesame seeds is also used extensively in the adulteration of olive oil.

Pistachio ice cream had appeared in America by this time, and Swingle had received some scions of *Pistacia vera* from Syria. He wrote asking me to keep my eyes open for other varieties when I was in Greece.

At the Agricultural Station in Athens, I was offered some three-year-old trees which had been grafted on a related species called the Terebinth. (The pistache does not come true from seed.) They were rather large specimens and I feared there would be complaints about the freight bill in Washington. However, I packed them in crates and shipped them from the Piræus with my Corinth cuttings. It turned out that I need not have worried, for they were the first budded pistachio trees to reach America, and contributed substantially to the interest which was being aroused in this, one of the most delicate of all table nuts.

Professor de Heldreich said that I would find seedless lemons on the island of Poros; walnuts with shells almost as thin as paper on the island of Naxos; and the famous Valonia oak on the island of Crete. The idea of a seedless lemon appealed the most, and I chose Poros as my next destination.

The passengers on the little boat were all Greeks, but one had lived in California and quoted endless statistics to prove that a man could live cheaper and better in California than in Greece. As we stood chatting on deck, three of the passengers were toying with strings of beads like rosaries, counting the beads in a rather aimless way. I asked the California Greek what they were doing.

"Have you never seen the kombológia of the Greeks?" he said in surprise. (He pronounced the "g" as "y.") "I think the English call them 'conversation beads.' They give occupation for one's hands to take the place of cigarettes, and are much cheaper. Many of my friends use them who cannot afford to buy even the cheap Greek cigarettes. The best quality of beads are made from a species of seaweed which grows in the Mediterranean and forms large, round, almost black stems."

The kombológia has an interesting history. The Greeks borrowed it from the Turks, with whom the beads possibly had a religious or ritualistic origin. In Greece they were never used as prayer beads, but degenerated into a device to offset nervousness. As this nervousness manifests itself most while listening to others' talk, the "rosaries" came to be called "conversation beads." Kombológion (the singular of the noun) means literally "beads-talk." Doctor Robert H. Fife, of Columbia University, tells me that he has seen the Athenians play with them with lightning-like rapidity while engaged in conversation or listening to a lecture.

In our country, fantastic sums of money are spent acquiring a drug habit in the form of cigarettes—often only smoked as employment for unoccupied hands during conversation. It does seem that, in this case, the Greeks had more than a word for it!

Later, in Washington, a friend of Sir Esmé Howard told me that the Ambassador never smoked but, instead, kept a curious kind of rosary in his pocket which he frequently fingered at conferences during the day. Sir Esmé was at one time British consul in Crete, and he probably acquired the habit while stationed there.

In the Poros orchards, I cut open dozens of fruit, but could not convince myself that there were really seedless lemons on the island. Many had only one seed, and occasionally some had none, but other fruits on the same tree might have a good many seeds. The lack of cross-pollination had much to do with the seedlessness. Empty-handed I returned to Piræus and sailed for Alexandria by way of Crete.

The Director of Agriculture in Crete was planting seedlings of the

Valonia oak, using Albanian women as laborers. He could not speak French or English and seemed suspicious of me and secretive about the acorns of this valuable oak. The acorn-cups furnish one of the best tanning materials in the world, used in Russia and elsewhere in the Levant in making morocco leather. Had it not been for the extensive development of chemical tanning, the Valonia oak and other of the tannin-producing plants might have been successful industries with us notwithstanding the length of time required to grow the trees.

CHAPTER XIV

Land of the Pharaohs

WHEN I REACHED Alexandria, happy at the thought of experiencing spring in the land of the Pharaohs, I found a cablegram announcing the sudden death of my father. It was a tremendous shock and I was crushed with grief and loneliness. For some time the faces at the table d'hôte—Greeks, French, Italians and Arabs—seemed repulsive to me, and I could not bring myself even to speak to the Greek sitting next to me. Finally I made up my mind that it was foolish to isolate myself, and began a conversation in French. His name was Pantelides and he owned the street-car line on the tiny Greek island of Chios; this was his third trip to Alexandria. During his first visit his daughter had died; during his second, one of his sons had died; and now he had just received word of the death of his wife. Needless to say, my sorrow paled beside his.

We had long conversations, and Pantelides told me of a "seedless" pomegranate growing in the island of Chios. He promised to send me plants of it when he returned home, and was true to his word. It is not completely seedless, but contains seeds with such tender shells that they can easily be eaten. Unfortunately, the color of this pomegranate prejudices people against it and it has not been widely planted.

Mr. Pantelides and I corresponded for many years but eventually the letters ceased. In 1930, during a cruise among the Ægean islands on the research yacht *Utowana,* we landed at Chios and I saw a door-plate with "Pantelides" upon it. When I knocked, a middle-aged man came to the door and I introduced myself. "Fairchild!" he said, repeating the name slowly several times. Then he went to his desk and returned with the calling card which I had given his father twenty-nine years before.

Land of the Pharaohs

"My father has been dead for some time," he said, "but I have read all your letters to him, and have always wondered what kind of a man you were."

But to return to 1901!

After meeting Mr. Pantelides I felt less lonely. Anyhow, I had no alternative but to push on. Life is nothing but pushing on, after all. Then, too, the gorgeous Egyptian sunshine, and fascinating opportunities for finding something new, all helped to take me out of myself.

Botanizing for plants in the field or forests is completely different from searching for cultivated plants in a region which has been civilized for thousands of years. In Egypt, I was not looking for possible herbarium specimens, but was studying the crops of the country.

B. Nathan & Co. of Alexandria, a firm having connections in the Sudan, had been recommended to me, and assisted me in collecting seeds of the cultivated plants growing there, particularly seeds of the grain sorghums. In this collection of sorghums was a variety with the Arabic name of "Feterita," which proved peculiarly well adapted to irrigated portions of southern California and Arizona. Its propagation and utilization as a grain sorghum contributed substantially to the development of those regions, and it is now grown on over 300,000 acres in Texas as well. In fact, this chance introduction has become a crop of annual value running into millions, and its Arabic name "feterita" is the name by which it is commonly known.

Aside from a few plants which I had introduced during my first visit to Egypt, the whole field of Egyptian agricultural plants lay before me.

First of all came the dates. Nothing was then known in America with regard to the dates of Egypt. Swingle had sent date varieties from Algeria, including the *Deglet noor,* which had the reputation of being the finest date in the world; and the American market was supplied around Thanksgiving time with the so-called "Fard" dates of the Persian Gulf region, which arrived in great, solid blocks and were sold by the pound. They were probably the stickiest thing that a housewife could buy at the grocery store. Dates in packages had not been invented.

Mounting a donkey, I started for the date-growing region of the Nile delta, where I found myself among big plantations differing from any which I had seen before. There is a stiffness and rigidity about a date palm entirely different from the coconut. The leaflets are sharply pointed and folded with the edges upwards, whereas the leaflets of other tropical palms, as a rule, are folded with the edges downward. A date palm leaf catches moisture, while the coconut leaf is designed to shed it.

In the first date orchard which I visited, the owner was hand-pollinating his trees. One of the men was sitting on the ground tearing apart the long, slender stems of a male flower-cluster. The plume-like cluster, resembling carved ivory, has side branches or stems zigzagging at short, regular intervals from the main stem. In each angle rests a flower bud which drops its yellow pollen in a tiny cloud. The Arab was deftly tying two or three together. High up in one of the near-by palms, held to the trunk by a strap around his body, another Arab was fastening these male clusters securely in the center of the young flower-clusters of a female palm.

The date palm, unlike the majority of other palms, produces its male and female flowers on different trees, and I was witnessing an agricultural practice dating back far beyond the dawn of recorded history. The Assyrians practised cross-fertilization for a thousand years or so before European man realized that plants, as well as other organisms, have sex. They had discovered that, if they wanted dates, they must transfer the pollen from the male date palms to the female trees, and that it was not necessary to have more than a few male palms in an orchard of females. Ancient pictographs give conclusive evidence of this fact. One, dating over eight hundred years before Christ, portrays a divinity dusting pollen from a male flower-cluster onto the flowers of a female palm. One of the oldest tablets in the world is a Babylonian fragment which is supposed to have been made three thousand years before Christ. On it is a crude picture of a date palm. Date palms are believed to have been grown eight thousand years ago in Babylonia, and to be the palm mentioned in the Bible in connection with Palm Sunday.

My immediate concern was the problem of securing a living collection of the palms whose Arabic names and descriptions I was busily writing in my notebook. Date growers never plant seeds; it is necessary to use suckers (young offshoots or sprouts) which come up around the young palms and can be cut off and planted. With the assistance of a friendly sheik, who taught me much about the date palms of the delta region, I shipped home suckers of six varieties. There were others which should also be sent, but my authorization was too small to finance the shipment. The American consular agent in Alexandria had no funds available, but sent me to an Alexandrian Greek, Emanuel Zervudachi, reputed to be the wealthiest man in Egypt, who employed many date growers. The Consul thought that Zervudachi might be willing to finance the shipment and wait for reimbursement from Washington.

Zervudachi was very cordial, introduced me to his wife, said to be the most beautiful woman in Egypt, and took me to a model farm where he was experimenting with Swiss cattle and water buffaloes and also had a large plantation of young date palms.

He agreed to finance the shipment of dates, and promised to keep the weight of the shipment as low as possible in order to minimize the freight bill. Zervudachi did land the plants in Washington successfully, but some of the suckers were enormous. When they reached the Arizona date garden, which was in charge of Professor R. A. Forbes, Forbes jokingly called them "Fairchild's Pillars of Karnak," and I was criticised for not having procured smaller suckers.

I mention this shipment not only because it contained a date variety which proved valuable, but because when I returned to Egypt I was shocked to learn that Mr. Zervudachi had acquired his great estate illegally. His methods of oppressing the fellaheen, or Egyptian peasants, had been investigated by the British courts and he had been thrown into prison, where I believe he died.

Among the interesting men in Alexandria was Lang Anderson, an English irrigation engineer, who was engaged in reclaiming a large tract of the Nile Delta. This land had been badly impregnated with salt ever since Napoleon flooded it during his campaign in Egypt, and it had remained unfit for cultivation since that time. Anderson had perfected a system of double canals at different levels to wash out the salt from the soil.

There had been a great deal of trouble with salt and alkali in the newly irrigated sections of our southwestern States, so I sent home detailed information about Anderson's work. This proved so interesting to our agricultural engineers that T. H. Kearney and Thomas H. Means were sent over to Egypt to investigate.

Anderson was working under the direction of the Khedive, and was amused to find that his reclamation project was solving a curious administrative problem as well as curing the salty condition. The Egyptian

Government maintained a salt monopoly, and levied a tax on all salt coming into the country. Consequently, the price of salt was high and a certain amount of smuggling went on. Donkey drivers, bringing produce to the cities in great baskets on their donkeys' backs, would take a route passing near the edge of a great salt lake in this delta land, and would gather quantities of salt as they passed by. Police were stationed along the route with instructions to confiscate the produce of any driver found guilty of smuggling. These policemen found it an easy matter to slip a handful of salt into the produce baskets and then confiscate the produce.

The altercations between the donkey drivers and policemen became more and more annoying to the Khedive. Finally, as a simple solution of the problem, Mr. Anderson appeared with the suggestion that he wash the salt back into the sea and at the same time reclaim the land.

My investigations of Bavarian and Bohemian barley had given me a particular interest in this crop, and I heard that a variety of barley called "Mariut," grown by the Bedouins in the delta region of Egypt, was being exported in considerable quantities to England and Scotland. This barley brought a higher price than ordinary barley. It was planted on the desert in the winter season and, if spring rains occurred, good crops resulted. Occasionally as much as nine inches of rain occur in this desert coastal region, and the profits from the Mariut barley were excellent, particularly as it could be delivered in England as early as May. I sent home a shipment and it proved so satisfactory that Harlan later made a special journey to the region to study conditions under which it was grown and make selections of seed.

Carleton had given so much attention to the question of the rust of cereals, that I was interested to find in irrigated regions of Egypt varieties of wheat which seemed particularly resistant to the disease. I therefore spent much time in the durum wheat fields selecting outstanding specimens for trial at home.

By this time the macaroni manufacturers in America had at last become interested in durum wheats and had begun to improve the quality of their product which, previous to 1900, was inferior, to say the least.

In the public gardens of Cairo a curious tree was dangling long sausage-like fruits on very long, rubbery stems. It was a "Sausage" tree, *Kigelia pinnata,* and I immediately begged a sausage from the gardener to send home. One of the seedlings raised from this seed was planted beside the

In our arid Southwest, Feterita, a sorghum from the Sudan, is used as a catch crop. It produces a good yield in seasons of drought, and its resistance to the smut disease has made it valuable for breeding purposes.

An Egyptian Fellah drawing water for his little patch of sorghum.

The curious, dangling fruits of a Sausage Tree growing beside a filling station at Coconut Grove, Florida, attract thousands of motorists every year and bring revenue to its owner. It grew from a seed secured in Egypt by the Author.

184B

The Author beneath a 30-year-old Tree of Life (*Ficus Sycamorus*) in the Plant Introduction Garden in Miami. From its wood, mummy cases with their portrait masks
were fashioned by the ancient Egyptians, and from it, 5000 years ago, was carved the
famous Sheikh al-balad statue which stands in the Museum in Cairo.

184c

These amazing Pigeon Cotes on Mr. Bayerlé's estate near Cairo housed 10,000 pigeons. They were made of earthenware jars set so that their mouths turned inward. The towers were hollow and the pigeon droppings gathered at their base. He asked no questions as to whose grain his pigeons ate (page 189).

road from Coconut Grove to Cutler, Florida. During the thirty-odd years which have passed, the tree has grown and prospered while, beside it, the traffic on the road has evolved from horse and buggy to Rolls-Royce and Greyhound Bus.

Today this tree hangs its enormous "sausages" over a busy highway and literally stops the traffic. Only rarely does a car go by without being slowed or stopped by the amazing spectacle. The present owner of the tree, in a sense, depends upon it for his living, as he sells gas and food to motorists. Moving-pictures of the tree have been exhibited all over the country, and I often think how amazed the parent tree in Cairo would be to see its offspring entertaining winter tourists in Florida.

Hearing that olives were grown in the oasis of the Fayum, I started off keen to see both the olives and oasis.

The olives proved disappointing, but I spent unforgettable days among the marvels of Luxor, Karnak, and Thebes. The great pylons, decorated with the cartouches of kings who had reigned thousands of years ago, impressed me beyond words. The desire for preservation after death was such an obsession in the minds of Egyptian rulers that they commenced digging their tombs in the Valley of the Kings as soon as they began to reign. As long as they lived, they dug deeper and deeper into the hillsides. When they died they were buried in immense stone sarcophagi at the ends of long tunnels in the rock; tunnels decorated in color with panels illustrating the events of their lifetimes. Poor fellows! How little knowledge of human nature they must have had not to realize that all their plans would be frustrated. Even the cartouches on the great columns were chiselled off by their successors, who in turn engraved their own names deeper into the rock.

Some of the frescoes in the tombs were so vivid that they seemed to have been painted only yesterday. One or two portrayed grapevines loaded with fruit, reminding me that the grape is a plant of great antiquity. Our American "Concord" grape was a seedling of the northern Fox grape discovered by Ephraim W. Bull of Concord in 1849. This was a strictly American species, *Vitis Labrusca,* but the "European grape," *Vitis vinifera,* is probably a native of the Caspian or the Caucasus and came into culture before the dawn of history. These Egyptian frescoes were the oldest recordings I had ever seen and interested me tremendously.

The World Was My Garden

At Thebes, beside the main avenue leading up to the palace of Queen Hatshepsut of the eighteenth dynasty (1570 B.C.), were the remains of grapevines said to have been planted by Queen Hatshepsut herself. In the bone-dry atmosphere of Egypt, stumps such as these could remain almost indefinitely without decaying.

This queen was evidently a plant lover. On one of the walls of the temple was a bas-relief depicting men loading some trees in tubs onto the deck of a sailboat, one of the "dahabeahs" of the Nile. My curiosity about this queen of three thousand years ago led me to do some research, and I discovered that the carving represented the introduction into Egypt by Queen Hatshepsut of the incense tree from the land of Punt. She had evidently sent an expedition down the coast of Africa to bring the incense trees for the embellishment of her palace.

A warm feeling of understanding surged through me for this woman who, like myself, appreciated the value and romance of plant introduction. Here on the walls of her palace in Thebes, she had commanded a bas-relief to be cut commemorating her importation of a new tree into her domain. It was quite thrilling, for, as far as I know, there are not a half dozen memorials commemorating the introduction of new plants.

In Vaucluse there is a statue of Jean Althen, who introduced into southern France the dye-plant, madder. At Kamo-Mura in the province of Wakayama, Japan, a shrine was erected to Taji Mamori, who in 61 A.D. was sent on a quest by the Emperor Suinin. After ten years' absence, he returned bearing "eight leafy branches and eight leafless branches of citrus fruits" to lay at the feet of his sovereign. When he found that his emperor was no longer living, in his grief he took his own life. On an old tree in Honolulu, is a tablet commemorating the Catholic fathers who introduced the "Algaroba" (*Prosopis dulcis*) into Hawaii, a tree which has been of great value as a stock food. At Riverside, California, a bronze plaque was created in honor of Mrs. L. C. Tibbetts, who nursed and protected the first navel orange tree from which were grafted the millions of Washington navel orange trees. From this tribute was omitted—by whose oversight I have never discovered—the name of William Saunders, who was responsible for the tree's introduction from Brazil.

Possibly there are other such memorials scattered over the world, of which I have never heard. I sincerely hope that there are. Anyway, this bas-relief at Thebes made a lasting impression, and later I chose it as the design for one side of the Meyer medal to be given "for distinguished service in the field of plant introduction." The noted medallist Theodore Spicer-Simson executed this medal in honor of our explorer Frank N. Meyer, who lost his life on the Yangtze River in China in 1918.

Sir William Wilcox, the most noted irrigation engineer of his time, had told me much about the new dam at Assuan. Three hundred thousand acres of land, formerly watered by the old Egyptian system of great basins, were to be transformed by thousands of canals into land watered by modern methods of irrigation. Sir William felt that a great mistake had been made in putting the dam so far down the river.

At that time the Assuan dam was, I believe, the largest dam in the world. There was much opposition to its being built, because it would submerge one of the most beautiful monuments in Egypt, the temple of Philæ. However, the welfare of thousands must overbalance the sentimental yearnings of the few for ancient monuments.

I was disappointed in the Fayum as an oasis. In common with most Americans, I had believed that from the center of an oasis you saw dunes of white sand undulating away on all sides of you. The Fayum was a large territory and did not give the impression of an oasis at all, but resembled other parts of Egypt under irrigation. I had a letter to an Englishman connected with the Egyptian Markets Co., Ltd., who patiently answered my questions about the region and invited me to dinner at his house. The weather was very warm but I knew that, in accordance with British custom, I must wear my tuxedo. When we went in to dinner, the dining room was as hot as Tophet and had every window and door closed. The perspiration ran down my face in streams, and at last I asked if something could not be opened.

"I'm sorry," my host replied, "we would be glad to open everything if it were not for the insects. They're quite terrible here at this season of the year."

Fly-screens were practically unheard of in Egypt, although they had been used in America for many years. This was hard to understand for there are few places in the world where the flies are more annoying or have dirtier feet than in Egypt.

While I was inspecting the market, a grinning Arab held out a bag of dates. Upon tasting them, I was struck by the unusual flavor; also the mealy flesh, a greenish gold in color, was most attractive. I bought the bag, twenty kilos in weight, and sent it to Washington. As I expected, the dates created a sensation among my colleagues.

The Arab said the dates came from "Wahi," but the word merely signifies an oasis; so I set out to track them down, and discovered that they were grown at the oasis of Siwa. The dates were so exceptional that we were determined to secure suckers of them, and this introduction developed into a long story. Several men were sent on desert journeys but failed to land living suckers for us, and other exasperating misadventures occurred.

After numerous failures to secure the Wahi, a Syrian named Salem appeared in Washington as a graduate from the Agricultural College of Missouri, and presented a letter of introduction from a professor of that institution. The Assistant Secretary wished me to appoint Salem as an agricultural explorer and suggested giving him the job of securing the elusive suckers of the Wahi date. I had met a good many Syrians in Egypt whose sharp practices had given them unenviable reputations, and I was prejudiced against them. Salem claimed that he had lived in Egypt and pretended to know a great deal about Egyptian agriculture.

"How do they pollinate the date palms to make them bear?" I asked him.

His answer proved that he knew nothing about the most common practice of Egyptian date culture. I therefore refused to appoint him. But he was a persistent, plausible talker, and while I was away on a business trip he got the appointment under pressure from Champ Clark, who was then the Speaker of the House. When I returned to Washington, Salem was being equipped as an agricultural explorer to be sent out with full authority of the American Government. He was to visit the oasis of Siwa and procure the Wahi date, and also to send back any other valuable plants which he might find in Egypt.

I was disgusted, and predicted that a man who knew so little about agriculture would make a dismal failure. This he did, most decidedly. His first shipment of "seeds" arrived in the form of a single cantaloupe which had rotted en route and had become a disgusting mass in the bottom of a large bandbox. It was with unholy delight that I put this box on my superior's desk and called his attention to the result of the first appointment made by our Office under political pressure.

Land of the Pharaohs

Salem was not an explorer for long, and all that we have to show for his efforts are a lot of photographs of himself mounted on a camel or wearing a fez and standing beside the sheik of Siwa.

Even after he was dismissed, we had not seen the last of Salem. He arrived in Washington prepared to argue his case for a reappointment, representing to his political friends from Missouri that he had been badly treated. He so infuriated us that Dorsett went to see Champ Clark, recounted Salem's activities and announced that under no circumstances would we have the man on our rolls.

All of this brought us no nearer to obtaining the Wahi date, and it was not until my old teacher, Silas C. Mason, was sent to Egypt to make a detailed study of date culture, that the riddle was solved. With his great patience and flair for detail, Mason classified the date varieties of Egypt, and, in so doing, discovered that one of the date suckers which the wealthy but unscrupulous Greek, Zervudachi, had sent us (and which had since fruited) was the same variety as the bag of dates I had bought in the Fayum. The correct name of this date was the Saidy, and it was one of the finest of all the dates of Egypt, and particularly well suited for cultivation in southern California. Mason secured several thousand shoots and they are now growing at Mecca, California, and Tempe, Arizona.

As I write these lines, a box of beautiful California dates has been put on my desk, and their exquisite flavor and cleanliness would be recompense enough, were any needed, for all the effort we expended in introducing and establishing the best varieties of dates in this country.

Among the interesting Europeans living in Egypt was a Mr. Bayerlé, who had an estate near Cairo where he was using modern farm machinery. The machine with which he was mowing a field of berseem was the only mowing machine in Egypt at the time. The contrast between this and the painfully slow, sickle methods of the fellaheen was appalling. Bayerlé was actually making hay with modern implements on his fields of Alexandrian clover, which he declared to be one of the best crops in the world for irrigated lands.

On his farm, a group of strange, white towers rose fifty feet or so from the level plain. They were cylindrical, tapering to their summits, and dotted with little windows.

"Those are my pigeon towers," he said proudly.

The World Was My Garden

Water bottles of native pottery had been laid in cement in a spiral from the ground to the summit. The mouths of the bottles faced inwards and opened on an air-shaft into which we stepped through a narrow door at the base. It seemed an ideal dovecote, for it enabled the squabs to be removed easily and insured the collection of a large quantity of manure at the base of the shaft. Thousands of pigeons were coming and going, and the nests were filled with little ones. I asked Mr. Bayerlé whether he had to feed the pigeons. He replied that he did not; the birds ranged over the whole country and picked up their food wherever they could.

This incident did not make much impression on me at the time, but later, when visiting the farm of one of my friends in France, I noticed an enormous old dovecote and was told that one of the sources of irritation leading to the French Revolution was the dovecotes of the nobility. The pigeons had fed upon the grain fields of the peasants, but the peasants were never recompensed for the losses caused by the ravages of the birds.

The progress of civilization seems to involve both the creation and elimination of abuses. These dovecotes were an example of the parasitic habits practised by one class upon another.

Egypt is so eloquent of the past, and its agricultural practices are so ancient, that one could spend years in the study of its civilization. One of the greatest of Egyptologists, Professor Archibald Sayce, a dear old Englishman, was a delightful talker and altogether a fascinating person, but had become so absorbed in olden times that he seemed to have dried up like the mummies with which he spent so much of his time. Watching him, I began to fear that, if I lingered there, I would lose my interest in the present and feed only upon stories of the past.

My job kept me too occupied to have time for the museums until just before leaving when I paid my respects to the royal mummies of the Pharaohs, Rameses I and Seti II.

In the Bulak Museum stands the amazing statue of the Sheikh al-balad, carved from a trunk of *Ficus Sycamorus,* called the "Tree of Life" by the ancient Egyptians. This statue is, I believe, recognized as one of the oldest wood carvings in the world. So cleverly is it executed that it is impossible to conceive of it as other than a portrait statue of this sheik who lived five or six thousand years ago.

In one of the smaller cases in the Museum there was a collection of seeds made by two noted German botanists, Professor P. Ascherson and George

A. Schweinfurth. (Schweinfurth was also the first explorer to give the world an accurate description of the pigmies of Central Africa.) These two German professors had made a careful study of all the different seeds which had been discovered in the mummy-cases of the Egyptians. They had taken great pains to eliminate from their collection any seeds which had been surreptitiously slipped into the cases by the tricky Arabs working in the archæological excavations of Egypt. Why the myth of fertile mummy wheat persists, I do not know. Both Ascherson and Schweinfurth dispelled it completely by their investigations fifty years ago. No one who knows seeds would believe that the aged and discolored barleys and wheats in that case could have retained any ability to sprout and grow. Professor Beal of Michigan proved that wheat kernels stored for twenty-five years lose their power of germination.

A full description here of what fruits, vegetables, and cereals composed the menu of the ancient Egyptians would take me too far afield and lead into an almost endless historical discussion. Franz Woenig of Leipzig wrote a volume on the subject in 1886. Barley and later wheat, the horse bean (*Vicia Faba*), the lentil, and the pea, played a major rôle among the grains, as did the onion among the vegetables. It is said that the onion formed one of the principal foods of the hundreds of thousands of workmen who built the great Pyramids. Today the menu of the lower class Egyptians is probably nearly similar. Wheat, beans, lupins, barley, lentils, sesame, and rice, the chick-pea or garbanzo, together with such vegetables as onions, okra, cucumbers, squash, cabbages, beets, and lettuce, about make up the list. Of fruits, the date holds first place. Figs, oranges, lemons, grapes, pomegranates, muskmelons, and of course the olive, comprise the list, although quite recently the mango has been added.

CHAPTER XV

Malta, Tunis, Algiers, and Spain

ORD CAME FROM Washington that Carl Scofield was coming to Algeria as an agricultural explorer in search of leguminous plants, and it was suggested that I join him there in June. Later, upon leaving Algeria, I was to visit the date gardens in southeastern Spain, the only place on the continent of Europe where dates are grown.

I found that I could stop off at Malta and Tunis en route, and engaged passage on a boat sailing the twelfth of May. Just before I left Cairo, there was a "simoon," one of those disagreeable dust storms which make Egypt a place to be avoided in late spring. Similar storms on our great plains do not have the oppressive character of the Egyptian variety. The fine dust sifts in everywhere, and the high temperature, 110° or more in the shade, makes the interior of a house a most uncomfortable place, preferable only to being out in the storm.

Due to the cholera in Egypt, many European ports maintained a quarantine against travellers from there during the summer months. Consequently, on disembarking at Malta, I was told that I must spend seven days at Valletta in a miserable quarantine station instead of enjoying myself studying the agricultural conditions of the island. I meekly submitted to the regulations but asked that the American Consul be informed of the arrival of an official from the United States Department of Agriculture. Mr. Grote was an active and enterprising person, and came down to see me at once.

"It is true that you must go through quarantine," he said, "but it will be easy enough to disinfect you. I'll arrange it immediately."

I had visions of having all my clothes ruined by steam or sulphur, and was much relieved when Mr. Grote reappeared accompanied by an official carrying a tin basin and some fluid in which I was asked to wash my hands. This perfunctory ceremony seemed to satisfy the official, and I soon found myself in a comfortable hotel largely inhabited by officers from the English war vessels in the harbor.

In Egypt, I had received a cablegram from Professor Buffum, of the Colorado Agricultural College, asking me to procure a number of Zarrabi goats from the Khedive's flock. Before the goats could be shipped, I received another cable cancelling the order, as his Board of Regents refused to authorize the necessary expenditure. This episode had made me wonder whether milch goats might be of value to the farmers of our western States, and I watched with interest the goatherds on the streets of Malta driving their flocks along and selling cups of milk direct from the enormous udders of the goats. Even milch sheep appeared in the flocks, and there was a peculiar but palatable cheese made exclusively from sheep's milk. However, already there were rumors that the Malta fever, a serious disease in Malta which often attacked the British sailors, might be caused by drinking goats' milk. Later, investigations by British pathologists disclosed the existence of a bacillus in the milk of the goats and sheep, and proved it to be the cause of Malta fever. This fever became recognized as closely related to undulent fever.

In contrast to the level character of Egyptian agriculture, the rock terraces of Malta seemed particularly picturesque. These terraces, where most of the agricultural plants of the island are grown, are said to have been built by those ancient sea people, the Phœnicians. It makes one's back ache to contemplate the labor which must have been necessary to create them.

One of the crops was sulla, and bundles nearly six feet long lay stacked around the peasants' stone cottages, ready to feed the stock. The roots of this leguminous plant (*Hedysarum coronarium*) were covered with nodules. As it has been cultivated on these limestone terraces for centuries, it must have materially added to their fertility. The peasants recognized that there were short-season and long-season varieties. The seeds were in pods which had to be fermented for fifteen days if the crop were sown in the autumn, or dipped for five minutes in boiling water. A four-year rotation of crops was in practice on these terraces: the first year, potatoes; the second, cabbages or cauliflowers; the third, sulla, followed by wheat.

The World Was My Garden

Some of the Malta terraces were green with fields of cumin (*Cuminum odorum*), the seeds of which are used both for flavoring pickles and in the preparation of curry powder. Unlike the sulla, cumin has a deleterious effect upon the soil and makes it difficult to grow other crops afterward; just why, I have never understood.

The oranges of Malta have a good reputation in the Mediterranean, and I secured cuttings of an Oval Seedless Orange growing in the Governor's garden. However, like most "seedless" varieties, it is not always without seeds. I also secured cuttings of both the Blood Orange and the Lumi-lareng—an orange which is sweet when green.

The Japanese loquats were ripe, and I spent an afternoon watching the owner of a loquat terrace and his family pack these beautiful fruits in baskets made of native cane and cover each basket with the soft green leaves. It was disappointing to find that the trees were all seedlings, no two producing fruits alike, and that the seeds were rather large. For sweetness, however, they excelled almost any loquats I ever tasted.

As Scofield was waiting for me in Algiers, I could not linger to study the curious customs of the people, and cast but a passing glance at the long black headdresses of the women. They might be described as a combination of the black "manta" so characteristic of the Peruvian women, and the sunbonnet worn on American farms. Why this enormous sunbonnet should be black in a climate where the sun is so hot, I cannot imagine. However, this head covering successfully conceals feminine faces from curious bystanders and is even uglier than the cloth masks worn by the women of Muscat in the Persian Gulf region.

I arrived in Tunis on the twenty-fourth of May and had to shift my gears again from English to French. But I always enjoy stumbling along in a foreign language, and much prefer making my own way to depending on an interpreter who seldom understands what you say to him and probably understands the man you are interviewing no better. In fact, an interpreter always seems to come between you and the man with whom you wish to establish friendly relations.

The vast olive orchards of Tunis represent almost a desert culture. They are on such dry land that in order to establish them a special technique was required, the basis of which was the planting of large cuttings, weighing many pounds. These North African methods were so interesting to the pioneer olive growers of California that, when T. H. Kearney was in Tunis studying the date palms, he devoted several months to a thorough investigation of the olives as well, and introduced the best olive, the Baroumi, into California where it proved most successful.

The most amazing sights in Tunis were the enormous Jewesses wandering through the streets draped in yards of brilliant silks. I simply could not turn my eyes away from them and frequently turned my kodak toward them too, although they did not like it.

Many of these women weighed over 300 pounds. Their appearance was made even more extraordinary by their tall, pointed hats of medieval aspect from which floated long, white veils. It seemed incredible that so many women could be so fat. My interest increased when I learned that they were deliberately fattened on fenugreek (*Trigonella Fœnum-Græcum*), a leguminous plant closely resembling the Egyptian berseem. It is fenugreek which imparts the peculiar odor to the conditioning powders used by stockmen when fattening their animals for the livestock shows. Its culture in Asia dates back to prehistoric times and its qualities and odor have been known in Persia, Mesopotamia, and Kashmir throughout the ages. I understand that fenugreek is now grown in California as a green manure; its roots are abundantly provided with nitrogen-gathering tubercles.

Fenugreek seeds, on which the Tunis Jewesses were fattened for the marriage market, contain an alkaloid called choline (*Trimethyloxydahyl-ammonium hydroxide*). According to my informants, it was the custom for a suitor to present a large silver or gold anklet to the parents of the maiden of his choice. If the parents favored his suit, the token was placed about their daughter's ankle. She was allowed no exercise and was fed a mixture of fenugreek seeds, milk, and honey. When her ankle filled the anklet to the satisfaction of the suitor, the marriage took place.

On my return to Washington I showed my photographs of these enormous Jewesses to a young man by the name of Gilbert Grosvenor whom I met in the Cosmos Club. He was building up a little magazine as the organ of the National Geographic Society, and his flair for the spectacular induced him to publish these pictures. They not only brought me a certain amount of publicity but were links in a chain of events which later developed into our becoming not only friends but brothers-in-law as well.

The World Was My Garden

At the hotel in Tunis I met Count Grenfeld and the noted African explorer and botanist, Professor George Augustus Schweinfurth. I had read Schweinfurth's book, *Heart of Africa,* two volumes describing three years of travel and adventure in the unexplored regions of Central Africa, and admired him greatly. I have a cherished photograph of this distinguished man with his plant press under his arm.

The erudition of a real scholar has always fascinated me, and the range of Professor Schweinfurth's knowledge, like that of his friend, Professor P. Ascherson, was amazing. How can a human mind retain the images of over 40,000 species of plants with such clearness that, when dried fragments of any of them are laid before him, he will, without a moment's hesitation, give the scientific name, the name of the botanist who first described them, and often the date when they were named? After an unforgettable day botanizing in the desert with Professor Schweinfurth, I spent a happy evening at the hotel studying his watercolor sketches.

From Tunis I hurried on to join Scofield in Algiers, where he was making a collection of legumes and their living root tubercles.

It had only recently been discovered that leguminous plants have nodules on their roots caused by bacteria which accumulate nitrogen from the air, and that this nitrogen is washed out into the soil and enriches it. A method was being developed by George T. Moore, who later became director of the Shaw Botanic Garden in St. Louis, to grow in culture media the particular species of bacterium which produced the nodules on a particular species of legume. Thus, the seeds of that legume could be treated with the culture of its own specific bacterium before being planted. This at least was the original theory, but it was later found that all the species of clover, for example, would form nodules if inoculated with any species of clover-bacteria.

Scofield had been sent to study under a fascinating French botanist, Doctor L. Trabut. As a young man, Trabut had been consumed by an ambition to publish a description of the flora of North Africa. He therefore elected to perform his military service in Algiers. The region in which he was experimenting with wild species was perhaps more closely similar in climate to that of California than almost any other part of the globe, and his long experience and great enthusiasm were wonderfully stimulating and helpful to those of us who visited him in his enthralling

196

Sulla growing on a terrace in the Island of Malta, where it yields enormous crops on the thin limestone soil. In California, where Sulla might grow, Alfalfa takes its place as it requires less attention, being a perennial.

The enormous udders of the milch goats of Malta give the animals a grotesque appearance.

The Almond kernels we serve on our tables bear small resemblance to the fuzzy fruit from which they come. In Spain, green almonds are a delicacy.

The fat Jewesses of Tunis were creations of an unreasoning marriage custom.

196B

An incredible amount of hand labor is still used by the peasants of Spain to free the wheat of its chaff.

All day long, a man and his wife winnowed the wheat by hand after it had been threshed out by horses on the primitive threshing-floor.

The Jordan Almond trees grew on the dry slopes of the Sierras of southeastern Spain. Men could be seen beating the trees with poles to knock down the nuts.

Malta, Tunis, Algiers, and Spain

Jardin d'Essai at Rouiba, where he had a remarkable collection of leguminous plants from all over North Africa.

Besides the legumes, Algeria had varieties of durum wheats which had not as yet been tried in California and, in the mountains of Kabylia, the natives were known to cultivate excellent figs, including the Capri figs. Experiments were being made in California in an effort to establish the Smyrna fig there, and the various sorts of Capri fig were wanted, as their inedible fruits harbor from one season to the next the minute wasps necessary for fertilization of the fruit.

Scofield and I ate in a café where brilliantly clad sheiks dropped in to sip coffee and gossip with their French friends. We were in a constant state of annoyance, as neither of us had any experience in the art of dealing with French firms, and found them exasperating. We were trying to buy large quantities of the various grains and seemed unable to complete the transaction. Probably we should have invited the Frenchmen to the café.

We finally managed a visit to Biskra and a glimpse of the great date plantations of that oasis, which are the most picturesque and spectacular in the world. Neither of us will forget the suffocating heat of Biskra in June. However, it gave us an idea of the temperature required by the date palm to mature its fruit, a temperature much hotter than is suspected by those who see the palm growing peacefully (and unfruitfully) on the Riviera or in temperate parts of California.

Scofield's meager little authorization was running low, and he had to return to Washington while I stayed on and arranged the details of our shipments. With the aid of the American Consul, I managed to conduct the business in such a way that the Comptroller of the Currency passed my accounts and did not impoverish me by disallowances on technical grounds. By this time, I had learned how to travel for the government without too much personal financial loss!

The Californians had been growing almonds for a good many years, but the New York candy-makers insisted that California-grown almonds became too hard when coated with sugar and candied. They claimed that the public did not like the California nuts, and they continued to send their buyers to Spain to purchase an almond known as the Jordan,

197

a long narrow nut with a kernel of delicate texture and distinctive almond flavor. The Department of Agriculture therefore suggested that I go to Spain, investigate the Spanish almonds, and secure budwood of the best varieties, including the Jordan.

From Algiers I proceeded to Alicante, a town on the southeastern coast of Spain, surrounded by almond orchards, their trees covered with fuzzy fruits already cracking open to show the attractive brown almonds inside. I set up my camera, and began familiarizing myself with those varieties which seemed noticeably distinct and interesting. There is something peculiarly absorbing in the comparison of varieties of horticultural plants. The beauty of form, color, and texture of the creations of nature have an appeal different from that of even the most perfectly constructed articles made by man.

The almond orchards were large and well kept, and some of the trees were very old. The industrious growers were stirring the dry soil about the trees with light wooden plows, primitive of course, but the tiny plowshares had a certain style to them.

Most of the growers alternated their almonds with groves of grafted algaroba trees (the carob of the Italians), some of them two hundred years old. Here I saw a horticultural practice comparable to that of the ancient Egyptians in cross-pollinating the female date palms with flowers from the male tree. Like the date palm, the algaroba is normally diœcious, there being separate male and female trees. In order to insure the pollination of the female trees, the orchardists around Alicante trained up a water sprout from the base of each female tree and grafted or budded into it a scion of a male tree. This sprout grew vigorously up through the branches of the female tree and, when the two sexes flowered, the fertilization of the female flowers was assured. Female trees with trunks three feet through had small male stems not more than five inches in diameter rising from the same roots.

I have often thought that this practice might be extended to other species. According to the Spanish algaroba growers, it has distinct advantages over the ordinary methods employed of grafting a branch with the variety used as a pollinator. This method allows the female tree to retain its shape and makes it easier to control the growth of the male stem. A bisexual sort which I sent to Trabut proved a great success.

The simplicity of these Spanish growers was quite delightful. I have

Malta, Tunis, Algiers, and Spain

always liked horticulturists, people who make their living from orchards and gardens, whose hands are familiar with the feel of bark, whose eyes are trained to distinguish the different varieties, who have a form memory. Their brains are not forever dealing with vague abstractions; they are satisfied with the romance which the seasons bring to them, and have the patience and fortitude to gamble their lives and fortunes in an industry which requires infinite patience, which raises hopes each spring and too often dashes them to pieces in the fall.

They are always conscious of sun and wind and rain; must always be alert lest they lose the chance of plowing at the right moment, pruning at the right time, circumventing the attacks of insects and fungus diseases by quick decision and prompt action. They are manufacturers of a high order, whose business requires not only intelligence of a practical character, but necessitates an instinct for industry which is different from that required by the city dweller always within sight of other people and the sound of their voices. The successful horticulturist spends much time alone among his trees, away from the constant chatter of human beings.

It was hot midsummer and the roads were deep with dust. The dry river-bed had become the marketplace of the town. What could be more picturesque than the brilliant strings of onions and red sweet peppers decorating the market-stalls of Spanish countries in late summer?

We had not yet begun to appreciate the pimiento; it was practically unknown on our markets. Familiar as I was with the sharp, hot flavor of the small red peppers, I was surprised to find fields of a dark red variety the size and shape of small apples. They were being dried to make the paprika with which distinctive Spanish dishes are flavored. A paprika grower took me to a storehouse in which he had a ton or more of the ground peppers. To show me that the powder was good to eat, he put a handful of it in his mouth and I did the same. It was amazingly mild.

Intrigued by the Spanish paprika, I introduced the varieties of the pepper from which it is made, and have always regretted that Americans do not realize how much they miss by not taking advantage of this delicious condiment in the way that the Spaniards do. The recent discovery that it contains ascorbic acid has interested me greatly.

By this time I had collected cuttings of the Alicante varieties of almonds as well as the best algaroba, the "Vera," which has pods so full of sugar that drops of syrup run out when they are opened. The Vera is considered too valuable for horse fodder but is eaten by the people as a delicacy. I also shipped home samples of the best durum wheats from which are made the interesting hard breads of southeastern Spain, and included a sample of the bread for my colleagues to compare with the rather tasteless American bread prepared from excessively bleached flour.

The home of the Jordan almond was in the mountains back of Malaga, and I travelled there by way of the historic town of Granada. Here I paused for a glimpse of the Alhambra and the garden of the Generalife. Although I did not have the courage to admit it, I was disappointed in the Alhambra and thought its decorations too ornate. Years later when I saw the simpler, purer architecture of the Moors in Morocco, I felt that I had been somewhat justified in my disappointment. But my attention was soon distracted from the architecture by the view from the mirador of the Generalife. When I looked down on the Alhambra, there below me stood two giant American trees, a California Sequoia and a *Magnolia grandiflora.* It was exciting to find that the most distinctive trees of this historic garden were both plant introductions from my own country.

Such quantities of Jordan almonds were exported to America that I felt sure the exporters would not be anxious to assist my quest for cuttings. So I called upon Consul Ridgeley and told him of my mission. He was a delightful person and immediately suggested that he go with me to the mountains back of Malaga where he knew the Alcalde of Casara Bonela. We rode on muleback into the hills, and I saw the mountain vegetation dried up as it is in summer in this part of Spain.

As we crossed one of the dry arroyos, the mule driver's face lighted up and he told us proudly that the spring before he had caught dozens of song-birds there by means of birdlime spread on sticks and placed near the pools along the stream-bed.

Theodore Roosevelt and Gifford Pinchot had not yet held their conference on conservation, but pioneers of wild-life preservation, such as C. Hart Merriam and T. S. Palmer, had made me conscious of the destruction of wild life in my own country. The absence of song-birds in Spain was so noticeable that I have never forgotten the mule driver's story of gathering in the little creatures with birdlime.

The Alcalde and his family were delightful people, but I learned little

Malta, Tunis, Algiers, and Spain

about almonds from them. After this failure, Mr. Ridgeley discovered that some of the best Jordan almonds were grown near the little village of Almogia, and I hired a driver who bore an unpleasant resemblance to my idea of a brigand. Although I did not like his looks, he was pleasant enough and during the drive up the mountainside regaled me with talk of the bull-ring. The journey was longer than I had anticipated and we did not reach the tiny village until late in the afternoon. However, there was still enough daylight to collect samples of the soil and to realize that the Jordan almond was grown in a light, sandy clay which it might be difficult to duplicate in California. The Spanish soils had been cultivated for centuries, in decided contrast to the virgin soil of our Pacific coast.

Men and women were beating the nuts from the trees and then spreading them on the floors of their primitive houses. Others were removing the kernels to send to Malaga as shelled almonds. In the plaza children were playing and, as they played, even the tiniest were braiding thin strips of the Mediterranean palm (*Chamærops humilis*) from which their parents made sombreros for the cattle drivers of the Argentine.

The people were miserably poor and I found their food coarse beyond description when I tried to secure something to eat. By this time, the sun was setting and there was no place whatever to stay overnight.

My driver and I turned our faces once more toward Malaga, with our cart packed with cuttings and baskets of nuts. As we approached the narrowest part of the defile where the cliffs shut out even the faint light of the rising moon, stories of the bandits in that region began to surge through my memory. At that moment, my driver reached behind him, pulled out a gun and flourished it in my face saying something in Spanish which I could not understand. It was a perfect spot for a hold-up, and I felt a nervous chill run down my spine. I never carried firearms, but flashed my heavy, nickel-plated tripod in the moonlight, hoping that he would think me armed also. Nothing further happened and before we reached town I decided that, far from his having any idea of robbing me, he had meant to reassure me and tell me that he would protect me if necessary.

At last I had bud sticks of the Jordan almond. But I knew that they were short-lived and, although I packed them carefully in waxed paper and put them in long tin mailing cases, I was afraid to risk sending all of them to Washington. So I shipped half of them to Paris to the firm of Vilmorin and Cie, with a request that they bud seedlings from them and ship the budded plants to America.

I was rather proud of this introduction, although secretly hoping that it would not end the demand for Spanish almonds and destroy the business of the poor growers of the mountains. I wrote a bulletin about the almonds and kept in close touch with the experiments made with them in California.

When the scions I had sent finally fruited in California, the trees proved to be early flowering, blooming in January and February. Consequently, they were frequently injured by late spring frosts and also, for some reason, the quality of the nuts did not equal those grown in their native village of Almogia. So I understand that the candy manufacturers still prefer the Spanish-grown Jordan almonds, and the peasants still have their means of livelihood.

This experience added to my conviction that the quality of an individual fruit, nut, or vegetable is the product not only of the hereditary set-up of that variety of plant, but also of its reaction to that combination of soil and climate which we call environment. As no two spots on the earth's surface have identical soil and climatic conditions, the hope of absolutely duplicating a fruit grown in one locality by growing the same variety in another is a futile one.

CHAPTER XVI

England, America, and West to the Orient Once More

EVERY GOVERNMENT authorization for travel has a time limit and that of mine had almost run out. As there were commissions to be carried out in Paris and London, I left Malaga early in August 1901, with my baggage crammed with almond, fig, and carob cuttings.

In Paris I inquired about the buds of the Jordan almonds which I had sent ahead, and was glad to find that they were all growing. It was during this visit that I met Philippe de Vilmorin, his charming family and his uncle, Maurice L. de Vilmorin. I have already written an account of their outstanding contributions to horticulture in my book *Exploring for Plants*. However, I should like again to acknowledge the debt of gratitude which we owe to the unfailing courtesies extended by the firm of Vilmorin et Cie to the Office of Plant Introduction for over a third of a century.

From Paris I crossed to England to investigate the hops grown in Kent, for they were considered almost equal in quality to the famous Bohemian and Bavarian hops. At the same time I hoped to learn something about the cobnut or filbert industry which also centered in Kent.

The best Kentish strain of hops was originated by a man with a name worthy of Dickens's most brilliant inspiration. Apparently, once upon a time, Philder Fuggle of Fowl Hill had planted a few seeds from his own hops, and the strain now ranked as the best obtainable. Although an excellent variety, Philder Fuggle's Golding Hops were not quite as refined as the Saaz hop which I had collected in Bavaria. They were not

203

seedless, but, for that matter, the value of seedlessness was questioned by the Kent hop growers with whom I talked.

From Fuggle's hop I turned to the problem of the cobnut, and was much disappointed to find that the expert I hoped to see had just left town. This was my only chance to visit Kent before sailing for New York, and I was much upset as I climbed into the train for the return journey to London.

I am often criticized for my habit of talking too much. This day, as usual, I fell into conversation with the only other occupant of the compartment, even telling him of my disappointment in not meeting the man who knew more about cobnuts than any one else in England. My companion proved to be an extremely intelligent man, and seemed well informed upon the subject of filberts. He gave me a great deal of valuable information, which I still have in my notebook as I scribbled it on the joggling train. When we reached London, I thanked him profusely, and told him that he had turned my defeat into success. He then smilingly admitted that he was in truth the man whom I had journeyed to Maidstone to see! The coincidence was a happy one, as many of those of my life have been.

Cobnuts are sold in England in their green husks. The fact that neither they nor the green almonds common in the Mediterranean region have appeared on the American markets is probably an indication of two things. First, our hot summer climate and the leaf diseases of the cobnut combine to make its cultivation a precarious undertaking outside of the Puget Sound region and certain valleys in the Sierras of California. Second, the popularization of even a new nut, new at least in the way it is served, requires a long time. However, there is still hope for the filbert. At the beginning of this century, English walnuts, unshelled almonds, and Brazil nuts were practically the only nuts served, and they appeared only at the end of the meal in a bowl with nut-crackers. Today we have the salted kernels of the American pecan, pinons from the nut pine of the Southwest, pili nuts from the Philippines, pistachio nuts from the Levant, paradise nuts from South America, the cashew nut from the East Indies, and shelled almonds and walnuts. In fact, so popular has the cashew nut become, and so thoroughly has it been distributed, that the growers of California walnuts are finding it a strong competitor.

England, America, and West to the Orient

I sailed from Southampton on the 24th of August, considerably worried as to what the Secretary of Agriculture would do with me when I returned to Washington, for I knew that Jared Smith was still in charge of the office in my old place. Upon landing, I found a telegram from Mr. Lathrop urging me to come to York Harbor, Maine, to discuss plans for another expedition. Labor Day seemed no time to report to the Secretary anyway, so I accepted Mr. Lathrop's invitation and spent ten days with him in a little cottage beside the sea.

York Harbor was quite popular with literary people at that time. Many interesting writers foregathered at the cottage of Mrs. Thomas Nelson Page, Mr. Lathrop's sister, and William Dean Howells had a cottage near by. I renewed my acquaintance with the Bryan Lathrops as well as Miss Carrie McCormick, who had apparently forgiven me the episode of Bat Island in Java. Altogether, I found myself among the most charming people imaginable.

In our first conversation after I reached York Harbor, Mr. Lathrop suggested a trip to China, Japan, and India on the same basis as our previous expedition. Of course I accepted his proposal without a moment's hesitation. From the little I had seen of that part of the world, I realized that from it would come many of the most interesting and valuable plants capable of cultivation in America.

Toward the end of September I went to Washington and had no difficulty in making arrangements for the expedition. The Secretary of Agriculture was enthusiastic about it since it involved no material expense on the part of the government and relieved him of the difficulty of finding a place for me.

We were to sail from San Francisco early in November, and I was soon on my way West, as I wanted to see both the dates and the Egyptian cotton which were being tested in the Southwest.

Plant introduction has two distinct phases. First, the securing of plant material in foreign countries and landing it alive in America; and second, the dissemination and establishment of the plants in the fields, gardens, dooryards, and parks of this country. The first phase is a comparatively simple one, but the discovery of where and how to grow these plants, and how to utilize them, requires years of patient investigation. A pinch of seed may come half around the world for a cost of only five cents. But growing the seed will probably require a flat in a hot-house, followed in sequence by a bench of two-inch pots, a greenhouse of six-inch potted plants, half an acre of rich soil in a nursery, an orchard, and finally an advertising campaign and a selling organization. Most of my activities in the field of plant introduction had been concerned with the easier phase of the two.

There was not time to visit the various State experiment stations, so I made my first stop at Tucson to meet Professor Toumey and see with my own eyes a date palm growing in America. There it was on the college campus, healthy, happy and in fruit! The professor had written to Washington about it and had encouraged us tremendously in our belief that date growing was a real possibility in Arizona. He was a true investigator of plants, and had assembled from the canyons near by a collection of wild species of plants which he thought indicated climatic conditions suitable for the cultivation of the date palm.

In 1901, his date palm was only five years old and yet bore 125 pounds of fresh fruit. We gazed enthusiastically at this tree, never dreaming that a quarter of a century must pass, millions of dollars be spent, and the lives of a dozen horticultural investigators be engaged on the project before date culture was finally an established fact in the Southwest. More actual scientific knowledge with regard to the date palm has been accumulated in California and Arizona than was to be found in all of the Old World.

At Tempe, not far from Tucson, a date garden had been established, in cooperation with the Arizona State Experiment Station. It was considered the most suitable location in the Southwest, and the date suckers sent in by Swingle from the Sahara and my shipments from Egypt had been planted there. R. A. Forbes was in charge and he took me to visit it. Of course the plants were all small, and I remember measuring with my umbrella to gauge their height. The dates had made a remarkable growth, although, according to Forbes, there were indications that the soil conditions were not entirely satisfactory. Later a better location for date growing was found near the Salton Sea.

The next time I visited the Tempe date garden, the palms were bearing. Great bunches of golden-yellow fruit hung thirty feet above my head, and special scaffoldings had been built preparatory to the harvest. By then the whole region was settled with planters.

From Tempe I went to Phœnix to see some date palms of an early introduction from Egypt made by H. E. Van Dieman. Professor A. J. Mc-Clatchie, a botanist of unusual ability, was in charge of them, but his chief interest lay in the Eucalyptus tree, which had already begun to dot the California landscapes. Now I suppose the younger generation believe that they are indigenous, instead of an introduction from Australia.

I next visited the Chandler estate near Mesa, and talked to Doctor Chandler at his house in the plantation. He was still growing cotton from the seed which Mr. Lathrop and I had sent from Egypt, and was convinced that there was a possibility of growing Egyptian cotton as a commercial irrigated crop. He had kept careful records, and told me that cotton seed planted on April 10 came up eight days later, bloomed on July 11, and had set a good many bolls before August 15.

I left him, quite hopeful that something might come out of the introduction. As in the case of the date palms, I had no real conception of the enormous amount of careful pioneer work and plant-breeding investigations which would be necessary before an Egyptian cotton industry was established around Chandler. Nor could I foresee how much the working out of this problem would absorb the lives and interest of my friends—Kearney, Cook, Scofield and others—leading them into the field of genetics, a new branch of science.

Next on my itinerary was Yuma, Arizona, where we contemplated an experimental planting of bamboos. I arrived in the early morning, and found a young fellow who volunteered to take me out to the State Penitentiary where I was to see Herbert Brown, the superintendent. On the way, he enlivened the drive with ghastly accounts of the murders, robberies, and dissipation of the little frontier town, a place largely composed of shacks, saloons and gambling houses.

Brown proved to be a typical, intelligent, broad-shouldered Westerner. He appeared glad to see me, and immediately took me down to the river to show me his garden among the willows on the cracked mud of the flood plain. I was struck by the similarity of the silt there to that of the irrigated basin of Egypt after the Nile subsided each year. But the human element was in notable contrast with that of Egypt. Everything here was in such a crude, pioneer state that I doubted if conditions would be favorable for establishing bamboos, although it had been suggested they might prevent the caving in of the river banks upon which the penitentiary was located.

However, I mapped out an experiment with berseem, horse beans, lupins, garbanzos, durra, and Egyptian cotton, and went back to the penitentiary with Brown to lunch.

We were barely seated when the door of the dining room flew open and a tall, lank cowboy burst into the room. He had a rifle in his hand and the conventional six-shooter on his hip. He was obviously excited.

"That rattlesnake of a Hart has broken jail," he shouted. "He's taken the Mexican with him, stolen a locomotive out of the roundhouse, grabbed all the guns he could get his hands on, and is off down the line. You can see him out of the window."

He waved his gun at an engine and tender tearing across the prairie at full speed.

"The down train is due here at four o'clock, too," he continued. "I've got another engine out, loaded the horses on a freight-car and got a posse together, but we ain't got no guns. Hart took 'em all with him. I came up to see if you'd lend me some of yours."

His posse and rig were waiting outside and I went with Brown to the storeroom and watched him hand out the guns. After they left, Brown explained that Hart was the worst desperado Arizona had seen in years. He had killed half a dozen men, one of them a deputy sheriff, and was such a desperate character that Brown did not want him in the penitentiary. As the trial for his last murder had not taken place, Brown had not been obliged to take him and had insisted that Hart be kept in the town jail which was one of those lockups from which such "killers" as Billy the Kid always seemed able to escape.

"I think they'll get Hart," he said, "but he will make it hot for them. He's a crack shot and desperate, but he wears an Oregon boot."

As I had never heard of an "Oregon boot," Brown explained that it was a heavy metal affair locked to the ankle, which made it impossible for a criminal to travel fast on foot, and also made him easy to track.

As we walked around the walls of the prison and talked to the guards sitting with Winchesters on their knees, we could see the posse, engine, tender and cattle car setting out in pursuit of Hart and the Mexican, who had nearly half an hour's start.

That evening in the hotel bar, the men were discussing the affair and I asked one of them what had happened.

"Well," he said, "it was my engine Hart ran off with. I had had my lunch and was taking a nap on the platform when I heard the clank of Hart's Oregon boot. I woke up and there he was standing over me. 'If you move I'll blow your goddam brains out,' he said. 'Keep quiet!'

"I kept quiet all right, but the fireman got wind something was up and came around the corner of the roundhouse in time to see two fellows making off with the engine. He fired at 'em, and I guess he winged the Mexican, for he dropped."

At the telegraph office, I asked the operator if he had any later news.

"Yep," he said, "they got him and I guess they're bringin' him in."

I had to leave on the night train and did not hear the end of the story until twenty-five years later when, with Mrs. Fairchild and my daughter Barbara, I stopped in Yuma on a motor trip. As soon as I was out of the car, I accosted the first man I saw in the street, telling him that I had been in Yuma in 1901 and asking if he had ever heard of Hart. He pointed across the street to a soda-water fountain which had once been a bar.

"The fellow over there knows all about it. He was here at the time, and up on the wall he's got photographs of the whole darn thing."

Barbara had acted as though she suspected me of inventing the story, so I took her with me. There on the wall were photographs showing the summary execution of Hart and the Mexican, and their burial where the posse shot them. The next time I felt like reminiscing, Barbara greeted my tale less skeptically.

Los Angeles, although a busy place, gave little indication of becoming the great metropolis it is today. Moving pictures had just been invented, and Hollywood was but a bucolic community, hardly even suburban. However, even in 1901 Los Angeles was full of land development companies. Their publicity agents were soon at my heels, begging for favorable comparisons between the fertility of the Nile Valley and the fertility of the land which composed their particular holdings, notwithstanding the fact that much of their land was white with alkali.

In their minds the soil of the Nile Valley was synonymous with an inexhaustible fertility in which the Nile played no part. They had no conception of the effect of the annual floods, and continued to ply me with innumerable questions about irrigation, and crops in Egypt, seeming particularly interested in the berseem, or Egyptian clover. They were an eager, restless lot of pioneers with a flair for publicity and, although they quoted me frequently, there was usually a twist to their stories, particularly when they had to do with the fertility of the Egyptian soil. I do not think that they ever quite believed what I said; certainly they never quoted me when I told them that the fertility was renewed annually and was largely dependent upon the silt brought down by the waters of the Nile; and that the Egyptian soil itself was not to be compared with the deep, fertile soils of the Mississippi Valley fields where wheat had been growing with exceptional yield for twenty-five years without a particle of fertilizer or manure.

On my way up to San Francisco, I stopped at Fresno to see George C. Roeding, who had a large orchard of Smyrna figs. For many years the trees had refused to bear. Suddenly, they began to produce good crops of fruit, due to the introduction of the blastophaga, a tiny wasp. The flowers of the fig are inside the fruit, with but a tiny aperture to the surface. This insect, and only this insect, enters this aperture of the edible figs, thus carrying the pollen from one flower to another; without its services the Smyrna fig refuses to fruit. The romance of the introduction of this wasp, and its establishment in California, is one of the most entertaining stories in the annals of horticulture.

I found Mr. Roeding in his packing house, surrounded by trays of what he had decided to call the Calimyrna figs. Now that his trees were bearing good crops, he had encountered another difficulty when he began to dry the fruit. The figs had thicker, tougher skins than when grown in their native homes. Just as California Jordan almonds were not identical with those grown in Spain, the fig trees did not produce fruit identical with those grown in Smyrna. Although excellent when eaten fresh, and even good when cured, the figs have never, I believe, competed in the Eastern states with the imported Smyrna figs.

Mr. Lathrop was waiting for me at the Palace Hotel in San Francisco, and announced that we were sailing for Hongkong in about a week. He seemed to be looking forward to our trip as much as I was, and had decided that we should spend some time collecting the bamboo varieties which play such an important rôle in Japanese civilization.

In anticipation of the shipments of living plants which I would be sending from the Orient, I spent considerable time making arrangements that Mr. Cooper, the United States Despatch Agent in San Francisco, should be appointed an agent of the Department of Agriculture as well as of the

State Department. California had enacted a State law requiring an examination of all nursery plants coming into the country; so I hunted up Mr. Craw, who inspected all such shipments, and discussed the matter with him in order to insure admission of our plants into the country. I believe that San Francisco was the first United States port to have a plant inspector—a trained entomologist—and the paraphernalia for plant quarantine and inspection service.

My years of travel with Mr. Lathrop had taught me a little about the art of interviewing, and also the folly of poring over notebooks, oblivious of the fact that perhaps the man sitting beside me could give information infinitely more helpful than all the books in the library. Apparently there are three ways of acquiring information: First, from one's own observations; second, from books; and third, from other people. There are good and poor observers. Possibly heredity may partially determine one's powers, but training and self-discipline have much to do with it. Acquiring knowledge from books is an art which definitely has to be learned. Merely rummaging around in a library accomplishes little. Certainly, making the acquaintance of strangers and securing information from them is as much an art as acquiring information from books, and deserves as much study. One must be able to judge character and know what testimony to credit.

Mr. Lathrop had been a successful interviewer during his newspaper days in California, and I used to admire the technique with which he secured interesting information from almost everybody he met. He was as much interested in the steward who waited on him as he was in the captain of the ship, and could start a discussion at the table which would soon involve all the intelligent people sitting around him. On the other hand, I, for my part, would clumsily begin my interview clutching a pencil and notebook.

"That's no way to go about it, Fairchild," he would say. "You cannot sidle up to a perfect stranger, pull a notebook on him and get information that's worth a damn. You scare him. He doesn't know who you are, and he'll string you. Draw him into conversation and find out if he knows any people or places which you know. Talk about your acquaintances, if you have any, in his part of the world. Don't boast, but give him an inkling of who and what you are. Tell him an amusing anecdote; break down his reserve that way. Watch your chance to do him a favor. It takes time to win the confidence of any one worth-while. Also, don't waste time with a man who has no information to give you. Don't let him interview you, or you'll find that he has obtained a lot of information from you and you have learned nothing from him."

I know my friends today will not believe me, but it is true that when I began travelling I found breaking the ice a major operation and desperately hard to do.

Equally difficult for me was knowing how to end a conversation. I once called upon one of Mr. Lathrop's friends in Washington, a delightful woman, and because I had not learned how to leave gracefully I stayed until nearly midnight. When I returned to my room I wrote her an apology for my stupidity. The trick of saying something amusing and leaving in the middle of a laugh is a valuable social asset. In those days I had no "terminal facilities."

I began to practice Mr. Lathrop's theories as soon as we sailed, and found that there were people on board who had lived in Japan and China and knew something about the agriculture there. A missionary's charming wife told me about the Yang mei and the Dragon's Eye, two Chinese fruits now grown in Florida. There was also a kind Japanese who taught me to say "Nan desca?"—"What is it?"—and other indispensable key phrases in Japanese.

Secretary Wilson had sent Jared Smith to Hawaii as the director of the new Experiment Station there. On our day in port, I looked over the site with him, and was much disappointed in the land that had been selected. But even more disappointing were Smith's instructions from the Secretary. He had been told to confine his experiments to such economic plants as might be of interest to the middle-class people of Hawaii; he was not to study plant introductions even though they might later contribute to the prosperity of the islands. Corn and tobacco were on the program, but the mangosteen, lichi, mango and the like were not.

After leaving Honolulu, Mr. Lathrop and I discussed our plans, and decided to push on to India, postponing our visit to China and Japan until we were on our way home.

We ran into Tokyo Bay before a hurricane and we were very thankful when we dropped anchor in sight of Fujiyama with its lovely crown of autumn snow. It was December, and the people were wrapped in their quilted kimonos. No snow had as yet fallen in the lowlands, and the weather was such as one finds in the Carolinas at Christmas. Contrary to the general belief, Japanese winters are not extremely cold; there are no blizzards nor sub-zero temperatures, but much snow, for Japan has a moist climate.

We had only a day in Yokohama, and I utilized my time in making the acquaintance of Uhei Suzuki, founder of the Yokohama Nursery Company, and his son, H. Suzuki. They had offices in New York and London, and were doing an enormous business in lily bulbs, *Lilium auratum, Lilium speciosum,* and *Lilium longiflorum.*

The low, unpainted wooden buildings of the nursery were of charming proportions and pleasing color. The packing sheds presented a beautiful and animated scene, peopled by a hundred or more women dressed in bright blue kimonos with figured blue-and-white handkerchiefs about their heads and white socks and wooden sandals on their feet. Fern balls were then a fad in America; half a million were shipped from Japan to New York and London annually, and these blue-clad women were busy making them. The men were dressed in rather elaborate blue jackets. On the back of each was an enormous white circle in which was stencilled the company's name in Japanese characters.

Bamboo was in use everywhere. The plants in the nursery were tied to bamboo stakes; the man at the wooden pump was drinking from a hand-made bamboo dipper; rows of pleasing bamboo pots contained the seedling plants; the fences were strips of bamboo woven into charming patterns; even the shades which covered the cold-frames were attractive bundles of bamboo twigs.

On the long pine tables were neatly arranged creations of Japanese horicultural art—dwarfed, potted trees, larger trees in figured blue and white porcelain pots (some of them centuries old), and tiny maples in small pots of green porcelain no larger than a teacup. Flat porcelain dishes contained groups representing little garden scenes, or miniature clumps of bamboo which had been growing in the same pots for twenty years or more.

I had a busy time with my camera, photographing right and left. Whenever I looked into the finder, the image made a charming picture. It seemed incredible that a nursery could be so completely picturesque.

I arranged with the Suzukis for a collection of different varieties of bamboo to be assembled for my inspection on our return to Japan in the spring. Mr. Suzuki also suggested that I secure a collection of Japanese flowering cherry trees, and showed me innumerable varieties in their autumn foliage. In fact, that brief day convinced me that I was going to enjoy Japan.

The rickshaw coolie who had drawn me out to the nursery for seventy sen (about thirty-five cents) was asleep in his rickshaw when I finished my visit. As he trotted me down the hillside towards the Bund, I could not resist a little shop where lanterns were being made. As I went into the shop, my coolie put down the shafts of the rickshaw and crossed to a stand which had plants for sale in blue and green pots. When I reappeared, he was gazing with delight at a single little star flower, blooming in a shallow porcelain tray. Here was my "horse," my beast of burden, enjoying a little potted plant like a true connoisseur. I bought it for him as a fitting end to a delightful day.

We passed through the Inland Sea and through the Strait of Shimonoseki, arriving at the mouth of the Yangtse River on December 11. This river and its tributaries promised an endless variety of cultivated plants worthy of introduction into America. Mr. Lathrop and I planned to go up the Yangtse the following summer, and I gazed in fascinated anticipation at its unfriendly yellow waters. In the spring, an epidemic of cholera effectively changed our plans, and I never travelled up the muddy Yangtse but have only received my vivid impressions of it from other men. E. H. Wilson, who became known as "Chinese Wilson" to all horticulturists, was at the time collecting at Ichang for Kew Gardens; Augustine Henry was in the Chinese Custom Service in Szechwan; Frank N. Meyer, at that time assistant to Hugo De Vries of Amsterdam, later spent nine years in China as agricultural explorer of our office, and was drowned in the Yangtse's turbid waters in 1918. From these and others I have learned that my instinct about the river was correct. The list of valuable plant introductions from this region is a long one.

A war junk flying the mandarin flag was anchored at Wusung; coolie women were tilling the fields; in the streets were quaint shops selling the long, celery-shaped cabbages which we later popularized in America as Chinese cabbage; river boats were coming and going up the great muddy river; everywhere one felt the impression of crowded, busy human life.

England, America, and West to the Orient

We continued on the steamer to Hongkong, and I was glad to see the busy little island again with its artificial forest on "the peak"; the street of flowers leading up to it; the shops with dried pigs' heads and dried ducks flattened out like boards and hung up for sale; and the hundreds of junks and sampans bobbing around in the harbor.

Mr. Lathrop spent ten days in Hongkong, but most of the time I was up the river in Canton. The weather was chilly and disagreeable and the sky overcast, but Canton had lost none of its fascination. It was delightful to be there again after eighteen months in the arid regions of North Africa and Spain.

Our Consul selected for me as guide and interpreter Chee Leong, a timid little man afraid of the water. His face grew pale as we embarked in one of the thousand sampans crowded together in the river. Chee Leong had never been on a boat before in his life, in fact I doubt whether he had ever before been that near the water. I was told that even the babies born on the sampans generally grow up and spend their lives on these tiny craft without learning to swim. A block of wood fastened to their backs supports them if they fall overboard, so that they are easily picked up. Later, they learn to hang on to the boat.

The first photograph which I took in Canton was of a particularly picturesque junk coming down the river in the early morning light. Unlike most of the other junks, this one did not seem to be painted, but was coated with an oil or varnish.

After several futile inquiries about this varnish, I went to see Doctor J. M. Swan, a most intelligent and capable surgeon in charge of the Cook-Tau Hospital in Canton. Doctor Swan was much admired by the Chinese and was greatly interested in all phases of Chinese life. He knew the oil which was used to paint the junks, and gave me as much information as he could. Since then of course it has become a common paint and varnish ingredient.

The preparation is called tung oil or China wood oil and is derived from the fruit of the tung tree (*Aleurites Fordii*). The oil, when spread and exposed to the air, forms a thick crust which shrivels somewhat as it hardens. The Chinese use it in the place of what we call "Japan drier." Curiously enough, the oil dries in moist air better than it does in dry air; when it sets, it is waterproof and extremely hard. The Chinese also make a cement from the tung oil. This they use in place of putty and it has the advantage of becoming much harder than any putty. It is always mixed at night to be used fresh.

Doctor Swan showed me the roof of a porch made of shells held in place by this cement and I became so interested in it that I went to Yuen Ching's shop to see it being made from ground-up shell-lime and tung oil. The mixing began at two o'clock in the morning in order to have the cement ready for sale when the shop opened for the day.

I did not see any tung-oil trees, for they grew around Hankow and farther up the river, and I had no knowledge of a report, lying dormant in the State Department, from Consul General L. S. Wilcox of Hankow on the possibilities of tung oil. Had I known of it, I would have made a definite effort to secure seeds. It was not until 1904, when I was in Washington, that we received from Mr. Wilcox about a thousand seeds of this remarkable tree. With this shipment, we started development of tung-oil culture in the United States, a culture which is developing a new industry in the South, in which Florida has taken the lead.

The tung tree belongs to the family of Euphorbiaceæ, as do also the Brazilian rubber tree and the castor-bean plant. It is a rapid-growing, deciduous tree which reaches a bearing age in four years, but never grows larger than an apple tree. The trees are reported hardy to cold (even zero) but sudden drops of temperature damage them. The fruits are borne both singly and in clusters, and are about the size of large plums. Inside their hard, shell-like husks they contain from three to seven seeds, each larger than a castor-oil bean but of the same shape. The shells of these seeds are so hard that one needs a hammer in order to ·crack them open. It is the meat inside these shells which contains the valuable oil.

In spring, the trees are covered with charming pink and white flowers almost as large as those of the cotton plant. I was enchanted when I first saw a tung-oil tree in bloom, but little dreamed that one day orchards in flower would occasion a festival in Florida!

In 1906 and 1907, we distributed about 800 trees to experimenters in the South Atlantic, Gulf, and Pacific Coast States. The cost of the seeds was only $30, but by 1908–09 I estimated that we would have to expend $400 of our slender appropriation for further propagation and the expense of sending some one to inspect the plantings and determine which region seemed best suited to their extensive cultivation.

The silt deposited by the Colorado River, like that deposited by the Nile, dries in the hot sun and cracks to a depth of a foot or more.

216A

Instead of the homely overalls of our country, workers in Mr. Suzuki's nursery wore decorative garments with the firm's name in a circle on the back.

A cluster of Candle Nuts. (The Candle Nut is a near relative of the Tung Oil tree.)

The making of a beautiful dwarf pine is a work of art, requiring great skill in grafting and budding, and an intimate knowledge of grace of form.

One of the first Tung Oil plantations in Florida was planted by B. F. Williamson, near Gainesville.

Chinese Yams carried to market along a narrow path through a peach orchard near Canton.

A field of Water Chestnuts near Canton.

South China hogs have been developed with a maximum of belly, as that part is considered most valuable by the Chinese.

England, America, and West to the Orient

Northern Florida eventually became the choice for serious experimentation. At the present time there are over four thousand acres near Gainesville planted with Chinese tung-oil trees. The Paint and Varnish Manufacturers of America have a plantation there, and I understand that the oil extracted from Florida-grown seeds has proven superior to the Chinese and contains 49 per cent of the oil. But more of this later.

We had one flurry of alarm when some reports came in of illness caused by eating the fruits. (Why any one ate them was not explained.) However, research has proved that, although violently purgative, they are not poisonous. Better yet, cattle seem to show no inclination to feed upon them.

The tung tree has some important relatives. The Mu-Yu (*Aleurites montana*) from South China, correctly classified by E. H. Wilson; the Balukanág (*A. trisperma*) of the Philippines, both of which produce oils allied to the tung oil and are better adapted to cultivation in tropical regions than the tung tree; and the Candle Nut (*A. moluccana*), which grows and fruits abundantly in South Florida and is common in Hawaii, but the oil of which is more like linseed oil and lacks the oleostearic acid which makes tung oil so valuable. Its kernels are good to eat when roasted, but have a purgative affect on many people. Natives of the Moluccas string them on palm spines and use them for candles.

Chee Leong and I next visited a peach orchard on Oxhouse Hill, just outside the great north gate. Near-by were guava and mulberry orchards and patches of Chi Koo, one of the Chinese aroids.

Chee Leong told me that women believed in the potency of the Chi Koo to help them bear male children. Girl babies were not desired, and were often neglected and allowed to die. He chattered away, telling me of the family life and customs of South China. At that time, it was still considered necessary for millions of little girls to endure the torture of having their feet bound. They had no chance of securing a husband unless their feet would fit in shoes four or five inches long. That any convention could so crucify human beings and cripple them for life was something beyond my comprehension. It is hard to realize that man is not essentially a reasoning creature, but an emotional one who perpetually attempts to explain his emotional acts by means of his reason.

This barbaric custom was in striking contrast to the highly developed character of the agriculture which I saw about me. The peach orchard was filled with named varieties, and the names were known and used by the farmers. The peach trees appeared to be vigorous and productive, and were said to represent the result of centuries of selection. Chee Leong particularly praised the Lan-fan, or "soft boiled rice" variety, which he said was a sour, small peach. Doctor Swan had told me that the Chinese often ate their peaches green, a month before they ripened, so I had no way of knowing whether the Lan-fan would be enjoyed by my own countrymen. However, I made a collection of the budwood of these peaches and sent it to Washington for the use of plant breeders.

The San José scale had already appeared in the orchards of eastern Maryland, imperilling the delicious late Crawford variety of Persian peaches which was the principal crop grown there. Chinese types, and hybrids between the Chinese and the Persian, although inferior in quality, had begun to replace the Persian types because the hybrids proved more resistant to diseases and insect pests.

As we walked along the narrow pathways, coolies carrying Chinese yams and taros in baskets slung on long shoulder poles passed us and continued their dogtrot towards the market-place. Beside the path on either side were large shallow ponds filled with pond lilies, grown for their curious rootstocks which are used as a vegetable. There were ponds of Ong Choi, a species of aquatic convolvulus, and growing with them were water chestnuts.

The water chestnut is a rush, like a bulrush. The "chestnut" is the small tuber about two inches in diameter, resembling a montbretia bulb. The interior of the little bulb is brilliant white and is sliced and used as one of the ingredients in chow mein and chop suey. The flavor is crisp and sweet. Quite a large quantity are canned and shipped to this country annually. In China, where there is practically no milk to be had, Europeans complain that the precious milk which they do manage to obtain has often been adulterated by the white liquid from pressed water chestnuts.

Thousands of little gardens crowded each other and covered every available inch of space. Land was so valuable that it must be kept busy producing vegetables every month in the year. Many factors had to be taken into consideration in regulating this agriculture. Perhaps one of the most important is connected with the Chinese fondness for lotus roots, water chestnuts, and Chinese cabbage. Professor King arrived at an interesting conclusion, which he presents in his *Farmers of Forty Centuries*. He says:

Important advantages were gained by the Chinese in the adoption of the succulent forms of vegetables as human food. For at this stage of maturity, the succulent forms have a higher digestibility, thus permitting the elimination of animal food from the diet. Their nitrogen content is relatively higher and this in a measure compensates for loss of meat. By devoting the soil to growing vegetation which man can directly digest, the Chinese save sixty pounds per hundred of absolute waste by the animal, returning their own wastes to the field for the maintenance of fertility. In using these immature forms of vegetation so largely as food, they are able to produce immense quantities which would otherwise be impossible, for the Chinese vegetables are grown in a shorter time, permitting the same soil to produce more crops.*

Thus, through the centuries, these market gardens have become what they are today. Insects and fungus pests are controlled through the continual surveillance of the fields by the sharp-eyed farmers who pick off each insect by hand.

These truck gardens of a city of 2,000,000 people did not contain a single vegetable with which we are familiar in America. The people, apparently well nourished, live on an entirely different diet from that which we consider necessary for health and happiness. Even their grains differ from ours. South of the Yangtse, the people consider their daily dish of rice more essential to their well-being than an American does his loaf of bread. In southern China, Indian corn is practically unknown, wheat a luxury, and, except on rare occasions, no one has either a beef-steak or a mutton-chop. Northern China is not a rice-producing country and there they have some corn and wheat. Throughout China they have pigs, of course, chickens and ducks; but even the pigs appear strange, because they have been bred to be sway-backed and have enormous bellies. This, I was told, was to increase the amount of bacon. Bacon is important as it is an ingredient of the chow mein, the most popular dish of the Chinese menu.

Butter and cheese were unknown outside the foreign settlements, and the only milk sold, except to foreigners, was human milk. There were no beef cattle, and few work animals, buffaloes only, for human labor was cheaper. The sheep were prized for their wool, and seldom killed for meat.

*Franklin H. King: *Farmers of Forty Centuries;* Harcourt, Brace and Company, New York. By permission of the publishers.

The World Was My Garden

Professor King has demonstrated that, according to recent Rothamsted experiments, whereas "only four pounds out of each hundred of the dry substances eaten by cattle are transformed into human food, and five pounds of the dry substances eaten by sheep, eleven pounds in each hundred are transformed into human food by swine." In view of these figures, which have only recently been established as scientific facts, it is significant that the Chinese long ago discarded cattle as meat producers, used sheep more for their pelts and wool than for food, while retaining the swine as the one animal used in the rôle of middleman transforming coarse substances into human food.

Those days in Canton were annoyingly short. My notebook is filled with the names of economic plants such as Chinese persimmons, ginger, olives, dye plants, taros and bamboos, and the descriptions of agricultural practices which I hoped to study later but unfortunately was prevented from doing, as I never returned. However, my contacts with Canton did not end with my visit. I became interested in the activities of the Canton Christian College (now Lingnam University) and have corresponded with Professor Weidman Groff and Professor McClure ever since. This has resulted in an exchange of many agricultural plants of South China with the cultivated plants grown in America. Through all these years Doctor Groff has been unceasing in his interest and efforts as the Chinese agent of this plant exchange. I have a letter from him on my desk today written as he sails from San Francisco accompanied by a shipment of plants which he is taking to China to be planted in the experimental gardens of the college. In time, these plants will find their way into the agriculture of that densely populated region of southern China. Professor McClure has made several collecting expeditions for us and has become an authority on Oriental bamboos.

We left Hongkong on Christmas Day for Colombo and, without stopping at Singapore or Penang, arrived in Ceylon on January 4, 1902. On the boat there was an attractive officer of the 27th Baluchistan Light Infantry, Lieutenant W. H. Maxwell, returning with his polo ponies from Peking where he had been aide-de-camp to the general in command of the British troops during the Boxer uprising. He was on his way to Karachi, and invited me to stay with him there and go with him up to Quetta on the frontier of Afghanistan. He had once been posted at Quetta

Mt. Lavinia, Ceylon, was a lovely beach at the beginning of the Century.

220A

The Cabbage Palm avenue, Peradeniya Gardens, Ceylon.

220B

Lieutenant W. L. Maxwell of the 27th Baluchistan Light Infantry, who sent the Quetta Nectarine (*right*) from Afghanistan to California, where it has become a commercial success.

The Bennett Alphonse Mango tree at Coconut Grove, Florida, grown by Mrs. Haden from the original potted tree which Douglas Bennett sent from Bombay.

In Poona, Marshall Woodrow's native gardeners were propagating from his best mangos by inarching seedlings in earthenware pots to the branches of the trees. To do this, a platform had been built on which the pots were placed.

A carabao or water buffalo bull in Professor Duthie's dairy at Poona.

and told me of many interesting fruits grown there, including a delicious nectarine, or smooth-skinned peach.

After our arrival in Colombo we took our customary drive out to Mount Lavinia to enjoy the loveliness of that immense coconut grove which sweeps down to the beach. As the setting sun shimmered on the myriad palm fronds, the fishermen were beaching their picturesque catamarans.

Surely no tree exceeds the coconut palm in grace of form, and no place is more romantic than a coconut grove when the breezes rustle through the long, green plumes and the breakers roll up the white beach. I would have thought it a dream too perfect for realization had any one told me that one day I would have a home in Florida where the soft rustle of my own coconut palms would lull me to sleep at night.

British tea drinkers had been taught that Ceylon tea was the best tea in the world, and tea-growing in Ceylon was a very profitable business in 1902. Machinery had been introduced, and the method of tea manufacture had already been immensely improved over the old-fashioned days when the wilted leaves were rolled by hand.

Americans are more coffee than tea drinkers, but I knew that propaganda might do much to establish the tea habit in the States. Doctor Charles U. Shepard, a chemist, had begun to plant tea in Summerville, South Carolina, and was trying to train colored children to pick the leaves. As tea growing is not necessarily a tropical industry, I was interested to discover how far mechanical devices had done away with hand labor, and wondered if a picking machine might not some day be invented.

Although Doctor Shepard did produce an excellent tea in South Carolina, and a beginning was made in the evolution of a mechanical picker, his tea had a different flavor from those of Assam, Ceylon, or the green teas of Japan and China. We imported for him the best plants obtainable from the River of the Nine Windings in eastern China, and the Department tried in a feeble way to help him market his product, but the dear old Doctor never had sufficient support to stage an adequate advertising program which might have convinced the American people of the excellent qualities of Carolina tea. I liked Doctor Shepard's tea myself and it had the same amount of alkaloid as the oriental varieties. But the Doctor's funds gave out, his health failed, and the South Carolina tea experiment became classed with the American silk production experiments which were finally abandoned after twenty years or more of struggle.

Milk was difficult to procure in Ceylon, and Mr. Drieberg, director of the Agricultural School, was operating an experimental dairy to see what could be done about it. The Jersey cattle sent out from England languished in the tropical climate and his Jersey bull was a sad-looking sight. However, a Brahmin bull of the Sind breed, with an enormous hump on its back, was in perfect condition.

"If you are going to India," Mr. Drieberg said, "you'll find dairy strains of these Brahmin cattle."

As we sat one evening at the Galle Face Hotel, Mr. Lathrop announced that he felt I had seen enough of Ceylon, this being our fourth visit there. He was perplexed about our next move, for he did not feel up to facing the discomfort and bad food of the hotels of continental India.

"I will go with you as far as Bombay," he said. "From there you can take your choice of spending three months in India or of going up the Persian Gulf to Bagdad. In the meantime I will return to Singapore and cross to Medan on the east coast of Sumatra where I can get the tobacco seed which the Department wants so badly. From Sumatra I will go on to Japan and you can meet me there the last of April. What do you think of the idea?"

The choice between India and Persia was a difficult one, for I longed to visit both. However, I did not know of a single soul who had been to Bagdad, and the date plantations of Mesopotamia and the Persian Gulf region had never been explored. Furthermore, I had the mistaken idea that some day I would have another chance to go through India, while this might be my only opportunity to see Bagdad. So I decided in favor of the Persian Gulf and we left Ceylon the next day on a P. & O. steamer for Bombay.

When we sat down to tiffin in what was considered the best hotel in Bombay, I realized why Mr. Lathrop had mentioned the vile food of India. An English officer was seated beside me and I asked him if this really were the best hotel.

"Well," he said, "I have just come here from the other hotel because my room overlooked the kitchen and I saw the 'boy' preparing my toast for breakfast. He had a tin of butter before him and was spreading it on my toast with his dirty hand. That may be the way my 'boy' does it here, but at least I can't see him do it. In most respects I imagine the hotels are about.alike, both equally bad."

These "boys" did not live in the hotel, but scattered at nightfall to their homes in the crowded city. As there were forty deaths a day from cholera, I felt that one's chance of acquiring that or some other loathsome disease was extremely good. It was rather a blow, as this was my first glimpse of India and until then I had gained romantic and colorful impressions of it from such writers as Kipling and Tagore.

Mr. Lathrop found that I could sail the next week on a British India boat bound for Karachi, a large seaport on the Arabian Gulf near the border of Baluchistan. I needed mailing tins and cotton sacks in which to collect seeds and had all too little time for sight-seeing.

I did go to see a Scotchman, Douglas Bennett, who had organized the market system of Bombay many years ago and had been in charge of the municipal markets of that vast, crowded city ever since. The story of his life was one of self-sacrifice, not uncommon among British administrators in India struggling as best they can with the incredibly unsanitary conditions. Also, the oppressive, hot, almost insufferable climate of Bombay adds greatly to the trials of the men working to give the people a fair deal in the administration of their municipal affairs.

It was winter, and the market contained few tropical fruits: apples from Kashmir, oranges, bananas, and a few jujubes. Mr. Bennett told me that in summer the stalls would be loaded down with mangos.

"The best Indian mangos grow on this side of the peninsula," he said. "I consider the Alphonse, the beautiful Bombay mango, the best of all. If you want plants, I'll give you some cuttings of the finest strain, on one condition. You must name it after me."

I had been disappointed before by such eulogies about plants, and proposed that he show me the tree, pledging myself to name the variety for him if he would let me get some scions myself. With his assistant, I went to an estate near Goregon station. Mr. Cooper, the owner, assured me that few Europeans had ever tasted the fruit of this special variety which had a flesh without stringiness, a small stone, and the best flavor in the world. I secured scions and sent them home. Fearing that they might not reach Washington alive, I accepted Mr. Bennett's offer to send grafted plants which he did later. As a result I have growing by my door in Coconut

The World Was My Garden

Grove a tree of this delicious Douglas Bennett Alphonse mango. Less than half a mile away, in my friend Mrs. Haden's orchard, stands the original mango into which buds were put from Mr. Bennett's little potted trees in June, 1902. Many of my countrymen owe gratitude to Douglas Bennett for this superb variety of a delicious fruit.

Although I would only have two days there, I determined to see Poona, for many years the home of G. Marshall Woodrow, professor of Botany in the College of Science. I had long been the proud possessor of a copy of *Gardening in India,* the book he wrote for use of the British. Unfortunately, he had recently left India, but the native gardeners whom he had trained were still working among his collection of mangos.

Poona has an altitude of 1900 feet and is cool at night (about 56°) and hot in the middle of the day—possibly 90°. The rainfall is about forty inches. The weather was delightful and reminded me of Florida. In fact, Poona, like southern Florida, is visited by frosts from time to time in winter.

The mango is pre-eminently a fruit of British India, and I was at last in one of the oldest centers of mango culture. Woodrow's collection was a selection of those which he described as fiberless and which could be eaten with as much ease as a cantaloupe. Slender, bare-legged gardeners in convention_l turbans were busy watering hundreds of pots arranged on bamboo platforms around a variety called the Pakria. Every branch of the tree had a seedling in a pot attached to it, an "inarch," as such a graft is called.

Woodrow's favorites were the Pirie, the Borsha, the Pakria, and the Alphonse. It has been over thirty years since I sent these four varieties to Florida, and I have found them to be among the best of the eighty-odd mango varieties which the Office of Plant Introduction has brought in and fruited in southern Florida. Of course these represent but a small fraction of the 500 described mango varieties which have been catalogued in India. Our selections were made for us by trained British horticulturists, many of them graduates of Kew Gardens. I had hoped to return to India but the opportunity never came, and, as far as I am aware, no American horticulturist specializing in mangos has ever visited India during the summer fruiting season.

At the Agricultural School in Poona, the wheat and barley fields were in bloom. Professor Gammie was breeding varieties from the great wheat centers of India and presented me with a collection of the Indian wheats for trial in America.

I asked Professor Gammie about the milch cattle and the milch buffaloes of which I had heard while in Ceylon. He showed me through his barns and gave me an opportunity to photograph his prize animals. I soon realized that there were not only beef breeds, but various milch breeds of the sacred cattle of India. These so-called Brahmin cattle had great humps on their backs and were magnificent-looking creatures.

I had made the acquaintance of the buffaloes on the rice fields of Java when I had been chased by the ugly brutes, but I had never heard of their being milked, nor do I think that they are in Java. However, here in the college stables were types from Gujarat which gave twenty pounds of milk a day, milk so rich in cream that it yielded one and a half pounds of butter, whereas the milk of the ordinary cattle in India yields only one pound of butter to twenty pounds of milk. Delhi buffalo had given over thirty pounds of milk per day.

Knowing our Secretary of Agriculture's interest in cattle, I sent my photographs and a write-up back to Washington. They were later published in a little bulletin but, as far as I know, only one cattleman paid any attention to this publication. He was Mr. H. P. Borden of Pierce, Texas, a very resourceful ranchman, who had a Brahmin bull which he had bred with the native stock. He claimed that the Brahmin cattle are not only better range cattle, being greater walkers, but are far less subject to ticks than are our native cattle. Mr. Borden came to Washington and I pleaded his case with Doctor Melvin, chief of the Bureau of Animal Industry. As a result, Mr. Borden was given permission to bring in a shipload of Brahmin cattle. Travellers in Texas today will see Brahmin cattle which date their origin from Mr. Borden's courageous experiment, but, so far as I know, no milch strains have as yet been brought in.

After this interesting interlude in Poona, I returned to Bombay with my mango cuttings, bags of seeds, and photographs, and completed my arrangements to sail.

CHAPTER XVII

The Persian Gulf and Bagdad

O N THE NINETEENTH of January, 1902, Mr. Lathrop saw me off
on a P. & O. steamer on the first leg of my trip to Bagdad. He
sent his best regards to Lieutenant Maxwell, our friend of the
boat from Hongkong, in whose cantonment I expected to stay
in Karachi, and promised to meet me in Japan in April.

Maxwell took me in, gave me an army cot, and introduced me to his
friends. I had never stayed in a cantonment before, but had been with
Englishmen enough to listen appreciatively to army talk. Maxwell was
disappointed when he found that I was seriously in quest of dates and was
headed up the Persian Gulf instead of going with him to Quetta.

"If you show me how to pack the bally things," he said, "I will ship you
some nectarine seeds and cuttings when I get back to Quetta."

He was as good as his word, and on March 24, 1906, there arrived from
Quetta seeds which he wrote were taken from the best nectarine trees in
that frontier town on the border of one of the least explored and most inaccessible countries in Asia, the country of the Afghans. From one of these
seeds originated a splendid variety of nectarine which proved to have such
excellent shipping qualities that large orchards of it have been planted in
California. It is shipped in carload lots across the continent, and I found
it in the markets of New York last year.

After the Quetta became a great success, I wrote the British War Office
trying to find out what had become of Lieutenant Maxwell. By then the
Great War had come and gone, and I was unable to trace him; so he
never knew how great a gift he had presented to the American people.

Karachi fascinated me with its camels and palms, its fakirs and six-foot,
bearded and turbaned soldiers from Baluchistan. Maxwell told me that I
would have to have a "boy," and selected one named Djilsha. We made

up a list of camp gear, cooking utensils and provisions. One of Maxwell's friends remarked, "You'll get sick of nearly everything you've got there except the onions. You will never tire of onions." So I stocked up with twenty pounds, and was grateful for all of them.

In books about India, I had wondered why the writers talked so much about their servants. I began to understand, however, when I looked at six-foot Djilsha and realized that he was to take me to Bagdad, be with me day and night, and constitute my means of communication with the date growers on the Tigris. He was truly amazing; with his enormous white turban he looked nearly seven feet tall. His immense beard was multi-colored, for it had not been dyed for some time and, although the lower half was bright red, the upper half was black, its natural shade.

Wherever the British settle, they seem to create a public garden of some kind, and Karachi was no exception. The superintendent, Mr. Lester, was growing an indigenous variety of grape called the Goolabie, which he considered superior in flavor to the Black Hamburg. It grew remarkably well in a soil which was too alkaline for ordinary European grapes. As there were only seven inches of rainfall at Karachi annually, and the temperature reached 110° in the shade, I was delighted to secure this grape, and the Goolabie made quite a name for itself among the grape growers of our Southwest.

The officers in the cantonment spoke most enthusiastically about the dates of Baluchistan and the cooked dates or karak pokhta which are sold in large quantities in India and form an indispensable part of every marriage feast. Djilsha posted off to the native market and brought back an earthen jar full of the most famous date of Baluchistan, the Mozati. It was a delicious variety and I was determined to get suckers of it. In those days no European dared to enter Baluchistan without an armed guard. Maxwell, therefore, suggested that I send an expedition of natives. While I was hunting date palms on the Shatt-el-Arab, they would make a collection in the Panjgur date region.

I had personal proof that Karachi was indeed on the outskirts of civilization. The doctor of the cantonment warned me that I would have to be vaccinated before I would be allowed on board the ship for my trip up the Persian Gulf. He did not do any vaccinating himself, but said a man named Sunt Singh was superintendent of vaccination in the town. I could not get in touch with Sunt Singh until the day before we were to sail. Finally I hired a rickety old trap and drove to what was called the "plague quarter" of Karachi. The early morning sun illumined the desert through a haze of fine dust, and the camels under the date palms stood out in golden silhouette.

The vaccination station resembled an untidy back yard, but nobody was there. Every house was closed. Not a native was in sight. I drove on for some distance until we met a man who said that every one had fled because of plague, and the entire population was living in a dry river bed a number of miles from town. There I found Sunt Singh, a typical Baluch with bristling, scarlet beard and massive turban. He could not vaccinate me, he said, until the following day. He did not have his tools.

When I arrived the following morning, a calf was standing in the yard. Part of Singh's equipment was a hinged-top table to which he strapped the calf securely and tipped it on its side. Then I realized that he intended to vaccinate me directly from one of the scabs on the belly of the calf, and I began to feel very unhappy. In those days the cause of plague was still unknown, although its contagious character was well understood.

Here I was, baring my arm to the dirty tools of a native doctor. He was going to vaccinate me against smallpox, always epidemic in the Persian Gulf, but, for all I knew, the calf might present me with some other plague. I let the man go ahead but, when I returned to camp, I disinfected the wound. In the light of discoveries of today, this was a stupid thing to do, for I could no more have caught the plague from the doctor's tools, or from the calf, than a person could get malaria by drinking water from a filthy pool, and, in consequence, I might have contracted smallpox, for the vaccination did not "take."

The operation took me so long that I was late for the boat. But the doctor, who was a good sort, held it for me, and, in his anxiety to do me a good turn, he inadvertently let a case of plague slip in among the five hundred native pilgrims joining the S.S. *Pachumba* at Karachi.

I was the only Occidental passenger, but the British captain and officers were companionable fellows, and I was delighted by the prospect of a thirteen-day trip in this little-known part of the world. Our route lay through the Persian Gulf to Basra on the Shatt-el-Arab River, which is formed by the confluence of those two ancient streams, the Tigris and Euphrates.

The Persian Gulf and Bagdad

Our decks were packed with Arabs dressed in embroidered cloaks of camels' hair, Pathans in dazzling white cotton costumes and immense white turbans, and Persians with white felt hats and "puggarees." Among them were three Hindus, one a Brahmin who had tasted neither food nor drink on the voyage, as there was no suitable place to prepare it, no spot where during its preparation the shadow of some infidel might not fall upon the food.

Our five hundred Shiah pilgrims were bound for Kerbela, the burial place of Hosain, son of Mohammed's daughter Fatima. The Shiahs, or Shiites as they are commonly called, are one of the two great divisions of Islam. They really constitute the Persian form of the religion of Mohammed, while the Sunnites are the orthodox Moslems and number 150,-000,000—comprising over half of all Islam. Most Moslems of Persia and India are Shiites, and those of Turkey, Arabia and Africa are Sunnites. The Shiites maintain that Ali, the husband of Fatima, was first successor to Mohammed, but the Sunnites relegate him to fourth place. The Shiites also have many more festivals, a greater cult of saints, and produce most of the dervishes.

As it was stuffy in my cabin, I frequently slept on deck. Every morning I was awakened by one of the pilgrims, a dear old man who came and spread his prayer rug beside my chair and said his prayers as the round disk of the sun illuminated the distant shore of the peninsula of Oman. It seemed to me that he repeated the name "Hosain" at least a hundred times. What effect these prayers to Allah had upon his private life I do not know, but whenever I hear the word "devout," the morning prayers of this old Shiite pilgrim come back to me and I see the gay colors of the costumes, the pink-tinted coast of Baluchistan, and hear again the guttural Arabic, that strange language of the deserts.

It seemed unbelievable that I was sailing toward Bagdad, the city of enchantment. Would I see Ali Baba and the Forty Thieves, the Sultan's eunuchs, and beautiful, veiled ladies? It would be a voyage of adventure, for I was bound for the greatest date gardens in the world. But on this journey I was conscious of more than plants. Among the children of my generation, *The Arabian Nights' Entertainments* held a place second only to *Robinson Crusoe*. The stories were read to us before the open fire on winter evenings, and Bagdad seemed a dream place somewhere in the dim never-never land near Crusoe's tropical island.

The World Was My Garden

Our next port of call was Muscat on the Arabian shore of the Indian Ocean, near the southern end of the Persian Gulf.

About Thanksgiving time, every American grocer used to have solid masses of sticky dates on his counter, which he tore apart with the aid of an ice pick and sold to his customers. These dark-brown dates were mostly "Fards," and thousands of tons of them were shipped each year from the town of Muscat. Fard dates were produced in the various oases of the interior by tribes none too friendly to Europeans. During the season, as many as five boats left Muscat each week, carrying about seven hundred tons of dates apiece, gathered from the regions around Muscat, Basra, and the pearl islands of Bahrein.

The date harvest began in July and August but the shipments to America took place in the autumn. February (when I arrived) was the wrong season to study the dates on the trees, but was the best season in which to get suckers. It seemed possible that dates from here might be even better adapted to the semi-desert conditions of Arizona and California than the varieties which had been sent in from the Sahara and the Nile Valley.

The British Vice-Consul, Mr. McCurdy, considered the Fard the best flavored and best packing date in the region, and offered to secure suckers from the Semail valley in the interior, where half a million date palms were growing. He kept his promise, and I found them awaiting me in Bombay when I returned there from Bagdad.

There were only eight Europeans living in the little town of Muscat and I wondered how they endured the summers. The temperature often rises above 115° in July and the weather is damp and muggy. Captain Martyr told me that when the boat stopped there in summer, he and the officers were in the habit of lowering the gangway into the sea and sitting in the water all night to keep cool.

The Captain took me to call upon the one Englishwoman in the colony, an interesting person with great imagination.

"If I had plenty of money," she said, "I would not leave Muscat. But I would dig a big cave in these burning hillsides, and run a stream of water through it and fill it with ferns until it was the coolest, loveliest spot you ever saw."

In the market place where the great bags of dates are sold, stood the primitive scales on which they are weighed. A haggard, holy marabout was standing by the wall watching me with fanatical eyes gleaming from either side of his big hooked nose, and I managed to include him in one of my photographs.

Our next stop was a desolate place called Jask near the entrance of the Persian Gulf, where the Anglo-Indian telegraph station was located. We had to go ashore in surf chairs borne on the shoulders of half-naked Arab boatmen. Here, in ghastly isolation, lived the telegraph operators, who tried vainly to cheer up their surroundings by little gardens which they watered painstakingly by hand. Jask was bare, arid desert with only a few date palms. When I saw a native boat on the beach, I could not imagine what it had been made from. It proved to be really more raft than boat, and was constructed entirely of the midribs of date fronds; quite as creditable a piece of work, however, as though its makers had been boat builders on the coast of Maine.

Our next port of call was at Bandar Abbas across the Gulf, on the Persian shore. It lay behind a little island where the Queen of Sheba is supposed to have buried her jewels. There was so little vegetation back from the coast that the mangrove bushes, growing in the salt water, were cut and fed to the camels.

We had passengers for Bandar Abbas, and the Governor's boat, a slender, picturesque craft flying the Persian flag, came out to get the mail and take off the Governor's friends. However, before this could be done, an annoying creature came aboard in the person of a half-caste quarantine doctor. He mustered all the pilgrims aft for examination. During the medical inspection they swarmed everywhere, and one especially obnoxious chap appropriated my chair on the bit of roped-off deck where the Captain and I used to sit. When his turn came, the medical officer took an unusually long time to examine him. An altercation then ensued between the Captain and the medical officer, and I began to suspect something was wrong. If the doctor found a case of plague, I felt sure it would be the fellow who had been sitting in my chair. It was.

After considerable discussion, the medico allowed passengers to land, under protest, entering a statement in the log that he had found a suspected case of plague on board. The Captain and I then went ashore to see the place.

Forty miles back of Bandar Abbas, at Minab, there was one of the largest date regions of the Persian Gulf area. Arab sailing craft from all over the Gulf came there to secure dates for the Indian market. The Captain took me to the British agent, hoping that he might get me some palms from Minab. The agent was the only European in the town, and was extremely glad to see us. He complained that the government had suddenly dumped him in this God-forsaken hole with no more ceremony that if he were a bag of dates.

"I've been here several months," he said, "and can't even get a decent flag pole from which to fly the British flag."

Obviously I would get no help from him.

As we were returning to the boat, the Captain said,

"That medical Johnny has messed us up by recording in the log that we have a possible case of plague. The Turkish quarantine fellows in Basra are as likely as not to turn us and our load of pilgrims back and not let us land at all. In that case we'll have to return to Karachi."

As we climbed on board, I realized that Bagdad might again become a mere myth. I might have to spend the next months in India after all.

When we returned to the deck, the Captain and I stood watching the most beautiful mirage I have ever seen. A giant causeway appeared to be stretching for miles along the coast in the brilliant sunset. As we continued our conversation, the second officer came running up excitedly and spoke to the Captain in a low voice.

"The beggar's broken his neck," he said. "I mean the plague suspect. He fell off the bridge. We took him up there to give him a test meal to satisfy the doctor, and he fell off and hit the deck below. He's dead, all right."

The Captain's face was a study.

"That settles it," he said. "It certainly is a good way out of the mess. We can go on up the Gulf now. I'll fix the log all right, and the doctor has no proof that the man had the plague, anyhow."

The burial was arranged immediately. A plank, a heavy lump of coal, and without ceremony the body would be slipped over the side into the Gulf. But something happened. The corpse did a somersault in the air, and an angry cry came from the pilgrims. Their religious feelings were aroused and they were incensed by what they considered disrespect for their dead. The Captain feared a riot on board, and asked me if my man Djilsha could take a hand in quieting the pilgrims. I explained to Djilsha that of course no insult had been intended. The whole thing had been an

At Muscat, on the Arabian coast, passengers are carried ashore on the shoulders of the natives. The heat is terrific, due to the reflection from the bare rocks during the scorching summer days.

232A

At Muscat, a fanatical Fakir leered at us in a sly, unpleasant manner.

Costumes such as these, which reflected the heat, lent a flavor of the desert to the streets of Karachi.

232B

Dhows with lateen sails navigate the Persian Gulf and Indian Ocean, laden with dates for the Indian bazaars.

Some of the Shiah pilgrims lined up to be photographed.

So trustworthy was this old Arab date buyer that the Hills Bros. Company of New York handed him $10,000 in gold each year with which to purchase dates for them.

The residence of Nama, the wealthy date merchant, stood by the river bank, and members of his harem waved covertly from the seraglio as we passed.

232D

unfortunate accident. After an evening of anxiety, we succeeded in quieting the pilgrims.

But we were still on a plague-suspected ship. Even my chair, the only comfortable place I had on board, had held as its last occupant the late Karim Hadji Ramzoon. Although the cause of plague was still unknown, I did know that bacteria could be killed with corrosive sublimate, and I disinfected my chair and washed my hands in sublimate solution for a few days until I forgot about it.

Our next excitement occurred when a tall, aggressive date merchant named Nama quarrelled with a foolish youth from Bagdad and threatened to pitch him over the side. The "Bagdadi" youth came to me in tears begging for protection. He had run away from his wealthy Parsee parents and made his way down to Bombay. Here relatives had found him in a music hall, and put him in charge of our Captain with orders to deliver him back home again. Parsees are followers of Zoroaster, and I had seen them in Bombay walking solemnly along the streets with their characteristic, polished black hats on their heads. This boy had all the most objectionable characteristics of his race, and there were times during the trip up the Persian Gulf when I wished that Nama would make good his threat and throw him overboard. He was a pestiferous, whining youth but, nevertheless, there was something likable about him. When we reached Bagdad, I met his parents, and have a photograph of his sister dressed in a beautiful izar, the heavy silk garment of the Parsee women of Bagdad. I purchased this dress, and years later Mrs. Fairchild wore it at a fancy dress ball where it proved so effective that the *Washington Star* devoted a paragraph to it the next morning.

From Bandar Abbas we steamed west across the Gulf to the pearl islands of Bahrein. In the desert, sixty miles back of Bahrein, lies the remarkable oasis of El-Hofuf. By camel caravan from this oasis comes the superior Khalasa date. The obliging British Vice-Consul secured a number of suckers for me and shipped them to Bombay.

There are probably few oases in Arabia more interesting than El-Hofuf, for it has an ancient culture of its own, with probably the strangest coins in use anywhere in the world. They are about the size and shape of a paper clip and made of a single loop of copper wire pounded flat and turned up at the ends. Called "towela," they are worth a fraction of a cent.

The World Was My Garden

As we approached the mouth of the Shatt-el-Arab River, solid forests of date palms crowded down to the water's edge like coconuts along a tropical shore. The water was fresh here, and there were eight to ten feet of tide in this short river through which the Tigris and Euphrates pour their waters into the Persian Gulf.

Going up the river, we passed between vast plantations of date palms, watered by tidal irrigation. At one point we could see Nama's house, and the women of his harem waved their handkerchiefs to him through the jalousies which hid them from our view.

As we neared Basra, the Captain again said frankly that he did not know whether the Turkish authorities would allow him to land passengers or not. There had been no other case of plague, but the record that the poor fellow had broken his neck was not entirely clear. (I have always doubted that he really fell from the bridge unaided!)

Eventually, after hours of suspense, the Turkish officials compromised by sending us to quarantine for a week. The prospect of detention in a dirty quarantine station was anything but pleasant, but with Djilsha and my equipment I felt that it could be endured. Also, there was an American consular agent in the town and I sent him my card and a letter asking for assistance.

The five hundred pilgrims were landed first, and herded into an enclosure. Next Djilsha and I were sent ashore accompanied by Nama in his goat's-hair abba, and the young "Bagdadi," whose costume by this time had been reduced to a long white nightshirt which badly needed laundering, topped by a conventional straw hat. We were interned in the two-room quarantine house reserved for Europeans. The whining "Bagdadi" was particularly low-spirited. As he jumped out on the muddy beach, his new straw hat blew off and rolled into the water.

"My God, sir, I no like this place," he said in broken English with such emphasis that I laughed until I cried. I did not like it either, but I expected the Consul to help me out. The incarceration was bad enough without the added complication of Nama and the Bagdadi, between whom there was already bad blood.

Before I left the boat, Mr. H. P. Chalk, the agent of the Hills Brothers Company of New York, had come alongside, and the Captain had told him about me. The Hills Brothers had been for years the largest American importers of dates and figs.

234

The Persian Gulf and Bagdad

Mr. Chalk was not allowed to come on board, but he shouted a greeting to me as I stood at the rail.

"I'll send you over some newspapers," he said, "and provisions too, if you need them."

Our consular agent did not pay the slightest attention to my letter and when I met him in the club after my week of quarantine, I refused to shake hands with him. By this time I had learned considerable about him. Incredible though it seems, he was the date buyer for British firms with no contacts in America. Through some stupid injustice of our State Department, he had been made Vice-Consul, and Chalk, who represented the largest date firm in America, had to get his consular invoices from his rival whenever he purchased dates.

After I reported the discourteous treatment accorded me, the State Department rectified the mistake and made Chalk our Vice-Consul. He filled the position honorably for many years, later visiting me in Washington.

The stay in the quarantine station was quite an experience, and, as my friend at Karachi had warned me, I soon became tired of Djilsha's cooking and reduced my menu to boiled onions. Nama and the "Bagdadi" hobnobbed in the other room, apparently forgetting their former animosity. They ate enormous bowls of rice and mutton, shovelling incredible amounts into their mouths with their greasy hands. The nights were noisy with cries of "Allah" and "Hosain" from the enclosure where the Shiite pilgrims beat their chests and yelled with religious fervor until they dropped exhausted. The days, too, were noisy, worse than the nights, for the place swarmed with sparrows. Their chirping drove me nearly mad until I made a slingshot with a forked stick and two elastic bands and amused myself by shooting them. Djilsha cooked some, but balls of india-rubber could not have been more tough.

The days eventually passed, and Chalk came to rescue me. As he took me across the river in his boat, he told me that both he and his wife had just recovered from smallpox and were neither of them feeling very fit. He said that the summer had been extremely hot and moist, 117° on their verandah in the daytime and 100° at night, but the cool weather at the moment was bucking them up. Basra had no hotels or places of any kind in which to stay, and I do not know what I would have done had Chalk not taken me in.

The World Was My Garden

"We always dress for dinner, Mrs. Chalk and I," he remarked as he left me in my room. "People out here get awfully slack if they don't. We find it keeps our spirits up."

They were delightful people and did everything possible to help me get the information that I wanted, as well as suckers of the proper date palms.

Mr. Chalk had a remarkable date buyer, Haji Abdulla Negem, a man of fine personality. He was so reliable that Mr. Chalk would hand him ten thousand dollars in gold coin at the beginning of the buying season with which to purchase dates for shipment to New York. With this man, I visited the plantations and watched the tide-water from the river flow into the network of canals which irrigated the farms. The gates opened only one way, and, after letting the water in, would not let it out again. There was no necessity for pumps or other means of control.

Haji Abdulla knew the varieties of dates as well as an apple grower in western New York knows his apples. He collected early ripening varieties for me, the fruit of which was never shipped to the American market. He also gave me suckers of the standard date, the Halawi, of which there were millions of trees along the river. It is one of the principal light-colored dates imported from the Persian Gulf region to the United States.

The trip to Basra had taken thirteen days, but I still must travel five days by river steamer up the Tigris to reach Bagdad.

A third of a century has passed since 1902, half of my life, and, even with my little red notebooks before me, I find it difficult to give anything but a confused picture of those March days in Mesopotamia. But this is a book of reminiscences, making no pretense to "historical accuracy" (about which I am always skeptical anyway); so I shall cling to my memoranda, and ignore the vast and fascinating literature of this ancient region.

The boat for Bagdad was an old Mississippi steamboat, a decrepit side-wheeler which had been plying up and down the Tigris for years. Nobody seemed to know how such a craft had reached Mesopotamia, but there she was. The boat soon filled with passengers, but again I was the only white passenger beside the Captain, who was a Britisher and a most likable fellow.

Legends hint that the Garden of Eden was located where the Tigris and Euphrates meet. However, had I entertained any hope that the Garden was still in existence, I would have been bitterly disappointed. The junction of the two famous streams was concealed by dusty weeds and a few scrubby trees, and made so little impression that we had almost passed it before the Captain pointed out the famous spot. I fear that attempts to locate the cradle of the human race will be as futile as endeavors to discover the nest of the mythical roc.

The Tigris has been compared with the Nile, and at Basra its broad, muddy waters hinted at its character. But I was unprepared to find it so shallow and so crooked. Nor did I realize that for five days we would see only level, treeless plains, without a stone anywhere large enough to throw at a dog. Somehow, a rockless plain seemed an utter impossibility, and whenever I got off the boat, I hunted for stones, but there were none. The low, perpendicular banks were made up of dark-gray silt, a piece of which seemed to melt in the water like a lump of sugar, turning immediately into thin mud. The waves made by our boat struck the bank, easily undermining it so that great chunks constantly crumbled and fell into the river.

Seen from the deck, the Tigris stretched before us in the curious haze of early morning, twisting like a snake. Native Arab boats were sailing around its many bends, their lateen sails showing above the level plain as though they were a form of covered wagon instead of water craft.

The soil appeared fertile, but the difficulty was to water it. Here and there along the banks, primitive efforts were being made to utilize the water of the river. The crudest contraption was a water-tight basket fastened to the middle of a long piece of rope. The ends of the rope were held by two men. Standing close together, the men would swing the basket down into the river. Then, when the basket was full of water, they would walk away from each other and, as the rope straightened out, up came the basket.

A river trip through a flat country is full of interest. You can study the people without the inconvenience of having to live in their dirty houses and share their food. The region between Basra and Bagdad was not densely populated, but every day hundreds of half-naked, long, slender-legged Arabs ran along the shore, calling for bakshish and holding up their shirts to catch the bits of bread thrown to them.

The days passed without much variety as we paddled noisily up stream. I amused myself by photographing the passengers and villages along the way. One day the falcon-keeper of a local sheik came aboard with his hooded hawks, holding them perched on his gloved hand. I had believed that falcons "went out" with armored knights, jousts and ladyes faire, and was delighted to secure a picture of this falcon-keeper of modern times. Later I used this for a lantern-slide in a lecture before the National Geographic Society. The French Ambassador, M. Jusserand, was present, and wrote me requesting permission to print the picture in a book he was writing on the origin of sports, a copy of which he later sent me.

Within sight of Bagdad we ran aground, and I proposed to the purser that we walk the rest of the way. He carried my tripod, and I my kit, as we trudged along the river bank. Thus I arrived on foot at the gates of Bagdad, instead of on a magic carpet!

Bagdad was a place of narrow, dusty streets between irregular mud walls. Latticed windows of seraglios hid the supposedly fair women of the harem and looked down upon mangy, half-starved dogs snarling on the street corners; swarms of flies; camel caravans crowding pedestrians against the walls; veiled widows in black, passing on donkeyback in groups of three or four; flat-roofed houses on which the family linen hung to dry; date palms raising dusty heads from patios here and there. Everything combined to make the city seem very old and foreign but, to my practical mind, about as unsanitary as any place I had ever been.

Only seventy years before, the Plague had devastated the town, killing forty thousand persons and so terrifying the people that dead women lay in the streets with gold bracelets untouched upon their arms. When I was there, almost every resident had somewhere on his body an immense scar left by the Bagdad boil, a disease carried by flies. There were other human diseases, too—those common to natives of all Arab countries.

Yet, dirt and all, Bagdad was fascinating, and I wandered for hours through the crowded bazaars, studying the Arabs and what they ate and wore. I visited the neglected tomb of Set Zoubeida, wife of Haroun al-Raschid, and was interested to watch the mourners at other tombs. Women in black gowns and masks instinctively turned their backs for fear that I might see their faces. A camel caravan passed—a family moving from one desert place to another, its women hidden from public gaze by an enormous basket-like arrangement covered with gay cloths. I even saw whirling dervishes, whom I had always believed to be inventions of Barnum & Bailey.

Our Consular Agent, Mr. Hurner, proved to be the best-informed resident of Bagdad. He gave me much valuable information and warned me about the mosques which I must not attempt to enter. As a self-respecting American, I resented being looked down upon and considered unclean, but Mr. Hurner said that the prejudices of the people were to be taken seriously and must not be trifled with.

He had once visited Kerbela, the shrine to which the pilgrims on my steamer were going. A sheik there, who was also some sort of attendant in one of the mosques, had entertained him, but as soon as they finished eating, the sheik ordered every dish Hurner had used broken and thrown away. Hurner showed me the cloak he wore crossing the desert from Aleppo in early summer, a two weeks' caravan trip. It was made from the fleece of the merino sheep and was worn with the fleece inside in order to give insulation from the scorching heat of the sun reflected from the desert sands.

Among the many interesting objects which Mr. Hurner had collected, was a jar of barley from the ruins of Nefr, the site of one of the buried towns which have recently been studied by American and German archæ-ologists. The barley was certainly many thousands of years old, and was so charred that I could write my name with it on a piece of paper as though it were charcoal. Every kernel was distinct, and it was Mr. Hurner's belief that it might have been oxidized by age. He gave me some to show my friends who still believed in the myth of mummy wheat. Albert Mann of our Department, who in 1905 began an intensive study of brewing barleys, told me that, in order to keep their grain, the ancient grain dealers sometimes charred them, probably for the purpose of killing the insects. However, it is not clear in my mind how they could ever have eaten this charred barley, and I have wondered if it is possible that they used it as a medicine.

Among the archæological relics of Mr. Hurner, I most coveted his pottery nails, curious spike-like things an inch or more in diameter and six or eight inches long. They had been made of fresh clay thousands of years ago, and had inscribed on them records in cuneiform writing. These pottery nails had been fired and then inserted, driven so to speak, into the mud walls of the houses or tombs. Holding one of these seemed to me almost like shaking hands with the scribe who wrote the message in ancient Assyrian on this bit of soft clay two thousand years ago. Many of these inscriptions have been translated, I understand, but, like tombstones, they give little but names and dates.

The World Was My Garden

At the club in Bagdad, I met a fascinating French engineer. He had been sent out by the Turkish Government to see if it would be possible to reconstruct the Nahrwan Canal, and with it irrigate again some of the vast plain, two hundred and fifty miles long and two hundred miles wide, which stretches along the Tigris—the plain of silt which had worried me because it was without a rock of any kind. This ancient canal had been an immense affair, at places over thirty feet deep and three hundred feet wide, with numerous flood-gates. He believed that it was in operation until about eight hundred years ago when, rather suddenly he thought, it must have been destroyed. He was most interesting when discussing how easily its destruction might account for the decline and fall of the Babylonian civilization. It was easy to postulate that, coincident with an invasion of the Mongols, for example, which disorganized the Government, the Tigris River, forever changing its course, had cut into the canal and during the flood season rushed down it and destroyed it. With the destruction of the Canal, and a disorganized Government, would inevitably come famine, and in its wake deadly diseases such as plague and typhus.

I am afraid that my account of Bagdad may give the impression that I neglected the chief object of my visit. However, the published Inventories of our Office will testify that this is not the case. In fact, I collected every variety of grain I saw: wheat, barley, millet, gram, chick-pea, and even sent in a poor variety of maize to show how inferior it was.

Most of my time was spent securing a collection of the date varieties grown near Bagdad. As I heard that the largest date grower was the Sheik Abdul Kader Kaderry, I called on him with the dragoman of the Vice-Consulate and asked for his advice and assistance. A Turkish soldier had shot him through the arm the day before, but he received me cordially, promised to assist me in every way, and offered me anything I wanted from his date plantations.

Mr. Raphael Casparken, a friend of Mr. Hurner, assisted me materially, even presenting the Department with twenty-four palms of the Zehedi variety. supposed to be the most drought-resistant variety around Bagdad —a so-called "dry" date, the individual fruits of which do not stick together. I became very fond of these Arabian "dry" dates which I could carry loose in my pocket. This type has now become established in

240

The slender "Bellums," poled with consummate skill, formed the chief means of communication along the canals of Basra.

The whining "Bagdadi" dressed up to have his picture taken, clutching a volume of French verse ostentatiously in his hand.

When I photographed Nama, the date merchant, Djilsha held an Indian print as background to indicate his wealth and importance.

Primitive were the methods employed to make the water of the Tigris available for irrigation.

For centuries, this simple tool has sufficed for the culture of millions of date palms grown along the Tigris.

Falconry was still a pastime of the sheiks of the Tigris Valley.

The piles of fresh barley and wheat in the Bagdad grain market were protected by mats from thousands of sparrows and the heavy dews during the dry season.

Like the coracles of old England, the kuffes of Bagdad are baskets, pitched inside and out with bitumen, and are black and smooth as an asphalt pavement. To paddle them requires skill.

Few buildings are more beautiful than the Bagdad mosques covered with mosaics of porcelain.

2401)

Veiled women mourning at a grave in the cemetery near Bagdad.

For generations, Bagdad date growers at dusk have diverted the water from one canal to another.

For five days our decrepit steamer followed the winding Tigris between flat treeless plains.

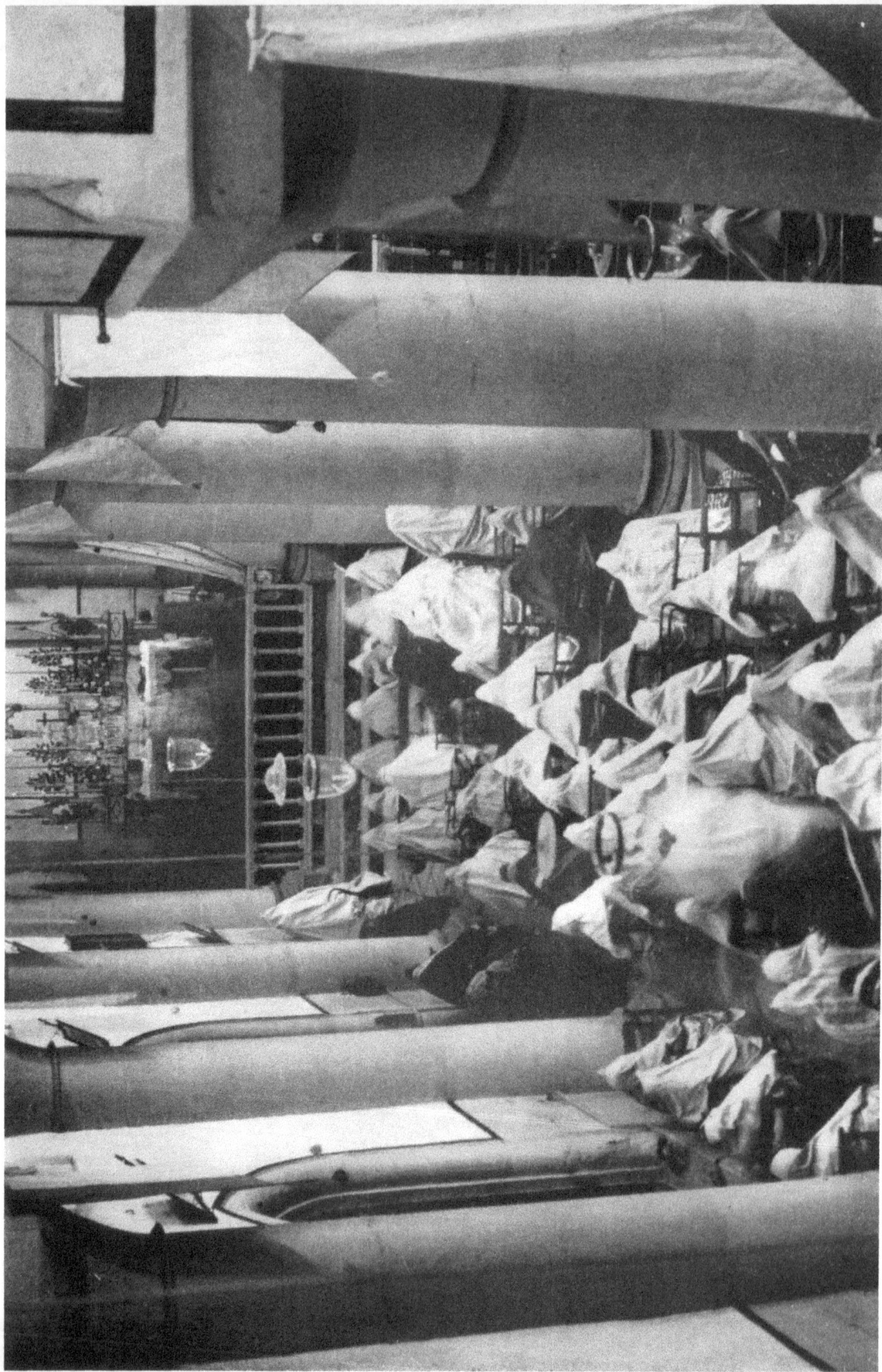

No photograph could suggest the brilliancy of this scene from the choir-loft of a Chaldean church in Bagdad. The izars worn by the women were of heavy, hand-woven silk, each with a pattern in gay colors.

the date gardens of America and I am able to procure them every season. Unfortunately, the dealers at Mecca, California, from whom I buy them, give fancy names to many of their dates, forgetting that the Arabic names would enrich our literature and be more romantic than such supposedly alluring titles as Sunbeam, Sun Loved, Honeydrop, and Bread Dates!

For many years after this, Mr. Casparken posed as an agent of the U. S. Government, and at one time wrote the Department that he had six children, of whom five were adults and one an adulteress!

Tasting the quality of anything as sweet as dates may sound like a pleasant job. However, although eating a dozen dates may be a pleasure, tasting two dozen is less enjoyable, and eating three dozen at once will ruin any stomach as surely as a rough sea voyage.

The date suckers which I collected, each weighing thirty pounds or more, were placed on the aft deck of the river boat, and I had a time of it packing them. To the captain's dismay, I mixed a liquid mud and puddled them. When this dried they were encased in a thick coating of clay. I then sewed each one up in sacking. Sphagnum moss was not obtainable, of course. Upon reaching Bombay, I planned to box all my dates for shipment to New York.

It is difficult in packing a collection of plants to tag them so securely that when they reach their destination the tags will still be attached and legible. Any assistants available are generally so ignorant and unreliable that they are more hindrance than help, and I have always preferred to pack my plants myself.

The Sunday morning before I left Bagdad, I asked Mr. Hurner if there were anything interesting I should see before departing.

"Have you been to a Chaldean church service?" he inquired. "The Patriarch of Mosul is here and you will see some stunning costumes."

I made my way through the narrow, crooked streets to the square where the church stood. It was a morning bright with sunshine, and I found the people gathering for the service. In tropical jungles the unbelievably brilliant Morpho butterflies pass swiftly through the sunlight and disappear in the shadows, and I am sorry for those who have never seen this unforgettable sight. Outside that Chaldean church, hundreds of women dressed in gorgeous izars of heaviest silk, shot with striking patterns of gold, blue, luminous pink, black and green, trooped through the door into the dim church, reminding me vividly of the flights of the butterflies in the jungles of Panama.

With my camera, I climbed into the choir-loft, possessed with the idea of securing a photograph of the colorful scene. The church was filled with women; there were few men. Not a single face was visible. The izar fastens around the waist, the lower half falling to the ground, while the upper part is thrown loosely over the head and held before the face in such a way that the wearer looks out through a narrow slit.

The return trip to Basra was uneventful, and I spent most of my time working on my collections. By the light of half a dozen tallow candles on the aft deck I was able to write up my notes, which were later embodied in a bulletin on the date varieties of the Persian Gulf, a publication of which I have always been proud.

It was the middle of March when I again reached the mouth of the river and found the S.S. *Pachumba* waiting for us. I missed the friendly presence of Captain Martyr, of whom I had become so fond on my outward trip, and learned with deep regret that he had died of fever on the trip back from Basra.

I was worrying a good deal about the expedition to the date region of Baluchistan, which Lieutenants Grant and Maxwell had sent out for me, and which they expected would be completed by the time I returned from Bagdad. Kej lay six days by camel from Gwadar, a little port on the Gulf of Oman, where our steamer was to call, and if the expedition were delayed the date palms would all be lost. When we anchored in the roadstead off Gwadar I was delighted to find a picturesque group of native Baluchistanese in white turbans and loose white costumes awaiting us on the shore. They had succeeded in securing twenty-four varieties including the fine-flavored date which I had tasted in Karachi. Alas, the suckers they brought were discouragingly small. I think that they had selected the smallest they could find in order not to overload their camels. Although I packed them carefully in boxes, few of them survived the nine weeks' voyage to Washington.

When I arrived at Karachi, I found Maxwell had been transferred to Quetta and I never saw him again. His friend, Lieutenant Grant, took me in, and I spent a night in the cantonment.

The Persian Gulf and Bagdad

We dined with Lieutenant Parker of the Twenty-seventh Infantry. They asked about my experiences in Bagdad, where neither of them had been, and I told of a dinner party with an English doctor and his charming wife where an extraordinary hailstorm had interrupted the meal. An orderly entered the room and laid a note beside Lieutenant Parker's plate. He glanced at it and I went on with my story. When we had finished dinner and were having coffee on the verandah, Lieutenant Parker said quietly,

"You were telling us of a dinner party in Bagdad. The doctor was my cousin. That chit which the orderly brought contains the news of his death. I am wondering what is going to become of his wife."

The unemotional, matter-of-fact way in which he announced the death of his cousin accorded with my general feeling that human life hung on a very slender thread in that part of the world. Deaths were frequent, even though the number of European residents was small.

I confided my interest in mangos to an Englishman on the boat, and my desire to get the best varieties from Central India. When we reached Bombay he gave me a suggestion.

"Why don't you go and see Tata?" he said. "He's the wealthiest Parsee in India. He has recently offered a lac of rupees to the Indian Government if they will establish a school of agriculture, but for some reason they have refused it, and he's angry with them. He is really a remarkable man, and can probably help you about the mangos."

It was the end of March. I learned that a boat was leaving Bombay for Colombo in three days and that if I did not take it, I could not make connections and reach Yokohama in time to meet Mr. Lathrop.

Those were three hectic days, which I would not care to repeat. I had to supervise the packing of all my date palms down in the native quarters of Bombay and hunt up a suitable steamer to get them through to Washington. Luck never played a fairer game with me, for I escaped infection, although cholera was rampant in that section of the city, and I had to elbow my way through the crowds and dodge the corpses now and then as they were carried by on the heads of the bearers.

I found a freight steamer which was sailing for Hamburg and arranged with the captain to carry my boxes under shelter on the forward deck and to transfer them at Hamburg for New York. The shipping of living plants is not as simple as the shipping of merchandise. Two careful waterings of the suckers would be necessary during the voyage. Luckily, they were travelling in summer, and would not require protection from the cold.

My last day in Bombay was Sunday, and I determined to take Bottomley's advice and call on Mr. J. N. Tata. I found him with his sons in the drawing room of his marble palace, discussing a trip to America. He was leaving the next day for London and was going on to the United States to inspect some mining machinery in Montana. When I told him that I was looking for mangos, he was much interested.

"I will send you a collection of the best varieties from my mango orchard in Central India," he said.

He then told me of his great interest in agriculture, and I gave him a letter to Secretary Wilson, upon whom he called when in Washington.

Mr. Tata kept his word. On June 7, 1902, there arrived in Washington both my own date shipments and a collection of six mangos which Mr. Tata considered the finest in his orchard. Three of these, the Paheri, the Ameeri, and the Totafari, succeeded and are now grown in the mango orchards of South Florida. They are among the least fibrous and finest-flavored of the mangos there.

CHAPTER XVIII

A Glimpse of Saïgon and a Long Stay in Japan

STILL SOMEWHAT BREATHLESS, I caught my steamer for Colombo, and spent a few days in the Peradeniya Gardens with Doctor Willis, finding him deep in problems concerning rubber culture. I then boarded the steamer *Tonkin* of the Messageries Maritimes for the twenty-eight-day trip to Japan, with a stop-over of three days at Saïgon, the capital of the French colony of Cochin-China. I had never travelled on a French boat in this part of the world, as Mr. Lathrop was prejudiced against them, and I was delighted to have a glimpse of a French colony.

It was a beautiful morning when we arrived in Saïgon, although it promised to develop into a scorching day. I started out early and walked around the town. The cafés on the boulevard were just opening, those informal meeting places which one finds wherever French people live.

The theatre was the most pretentious place of its kind that I had seen in the Orient, completely dominating the shabby streets and native houses of the town. Doubtless it contributed much to the lives of the French officials condemned to live in the continual steam bath of Saïgon.

I presented my official letters and was most courteously received by M. Haffner, the Director of Agriculture, a man who reminded me in some ways of Doctor Trabut of Algiers. He was as anxious to introduce useful plants into the Colonies as I was to secure them for America.

"But my problem is a different one from yours, Monsieur," he said. "The suspicion of these Annamites against the Government is intense. If I propose a better variety of agricultural plant, they immediately suspect that the Government wants to make money out of them by getting them to grow it. However, I have found a way around this difficulty. I introduced a peanut from Java, one with a higher yield which is easier to harvest than the peanut grown here by the natives, but I established Government plantations and put guards over them, instructing these guards under no circumstances to give away any seed. I gave them to understand that the Government is trying to build up a monopoly in this particular peanut. The result has been that hundreds of patches of this peanut have appeared among the native plantations. They have stolen the seeds systematically, as I hoped they would, saying to themselves, 'If the Government thinks these peanuts so valuable that they won't distribute them, they must be worth stealing and growing.'"

Of course my first questions were about mangosteens and mangos. M. Haffner told me of an orchard of mangosteens which had been established a century before by a man who had brought two or three trees from Penang and had planted them at Lai Thiou, where now a plantation of over four hundred trees was growing. "The soil was like that of an old swamp," he said. "Ditches full of water separated the mangosteen trees."

"We have a wonderful mango," he told me, "a sort that is never grafted but reproduces itself from seed. I call it the Cambodiana. You can get it in the market now."

There was no time to be lost. I hurried to the market-place, where I soon satisfied myself as to the quality of this seedling mango. It was delicious, to my way of thinking; an acid type of mango and, while not showy nor as large as the mulgoba, it had a surprisingly thin seed and almost no long fibers at all. I bought every Cambodiana mango in the market that morning, a hundred of them or more, and hurried to the hotel, stopping in a native bazaar to buy six scrubbing brushes, some packing boxes and a lot of charcoal.

When I announced to the head-porter that I wanted half a dozen "boys" who would eat the mangos and scrub the seeds, he had no difficulty in finding volunteers. They certainly made a mess of my room, but they did a good job scrubbing the seeds clean, and I packed them in charcoal and finished just in time to return to the ship. M. Haffner later sent me more seeds, and the fruit became known as the Saïgon Mango.

As I write this account, I see from my window a tree of this mango from which my daughter Barbara gathered delicious fruit last summer and shipped them to our home in northern Nova Scotia.

During the day that the steamer stayed in Hongkong, I secured more acorns of the edible oak (*Quercus cornea,* now *Lithocarpus cornea*) about which I have already written. I also found a beautiful little plum with an aroma which I thought might prove valuable for the experiments which Luther Burbank was then carrying on with Japanese plums.

At Shanghai and Nagasaki we had only a few hours.

The food was bad on the boat and everybody was very tired of it. The evening after we left Shanghai, among the nuts and raisins on the dinner table I noticed some candied fruit which the steward called Chinese dates. The long, pointed seeds and caramel-like texture reminded me of the dried dates which I had eaten in the Persian Gulf region, but there were curious scratches across the fruit which puzzled me because they could not be natural. Also I knew that the fruits certainly were not true dates.

On landing in Japan, I became engrossed with other things, and it was not until five years later that I saw these fruits again and learned their name and nature. One of the shipments which Frank Meyer sent in from the Shantung province of China contained the T'sao or Jujube, and I recognized the "Chinese dates" I had seen on the French boat. I had often tasted Jujube lozenges, but had believed them to be merely medicated cough drops, never investigating their origin or ingredients.

The Jujube (*Zizyphus jujuba*) grew rapidly at Chico, California, and proved splendidly adapted to the climate there. As it flowers later than most fruit trees, danger from late frosts is practically eliminated. It also proved to be a precocious bearer. Trees two years old from the bud frequently bore fruit. Like the grape-vine, the jujube produces flowers on the new wood. The enormous crops never failed to mature, and bent the boughs to the ground with their weight. Ripe jujubes, when eaten raw, are amusing rather than delicious, and have a crisp, sprightly flavor different from other fruits.

Mrs. Beagles, wife of the superintendent of the Chico Garden, began using jujubes both in pickles and put up as a glacé fruit. When Meyer sent in the honey T'saos, they were even more popular. He also sent one of the scoring knives which the Chinese use to slash the tough skin of each jujube before it is stewed in honey or syrup of sugar.

Mrs. Beagles and her friends became so proficient in making good candied jujubes that one autumn Doctor Grosvenor, President of the National Geographic Society, invited us to serve them at the annual banquet of the Society. Swingle helped us write a clever description of the

T'sao, and the fruits proved an unqualified success. Every one ate them and told his friends about them.

The planting of jujube trees has continued, and it has been astonishing to find how wide a range of climate they will stand and how well they endure both cold and drouth. Of course we again ran into the dilemma which faced us with each new introduction—that the Government could not go into the business of advertising or planting orchards for profit. In this case, just as the fruit was becoming established, the War broke out and turned men's thoughts elsewhere.

The jujube has interesting relatives. The common "Lotus" of the Mediterranean region (*Zizyphus lotus*) which is universally used as a spiny brush hedge, is a cousin, and *Zizyphus spina-christi,* which grows in Palestine, is supposed to have been Our Lord's crown of thorns. In the Sudan, a bread made from this *Zizyphus* reminds Europeans of gingerbread.

I had hoped to reach Japan in time for the cherry-blossom season but, when I landed in Kobe on April 26, I realized that spring was over. After the Persian Gulf and Bagdad, it was delightful to return to a country beautiful with green vegetation and flowers.

In Yokohama I found Mr. Lathrop laid up with a cold. When I entered his room, he had a beer bottle in his hand and was pouring seeds of the Sumatra tobacco from it on to newspapers spread on his bed. He had secured the dust-like seeds of the wrapper tobacco from one of the best planters in East Sumatra, and was endeavoring to dry them. In fact, he was so intent upon what he was doing that he scarcely noticed me.

"I had a hell of a time getting this seed, Fairy," he remarked, "and since those fellows in Washington want it for breeding I don't propose to lose it now."

He then told me of his trip from Singapore to Deli on the east coast of Sumatra, the center of the tobacco-growing region. He had happened to arrive there when the Deli tobacco growers were in session, and he had asked permission to address them. This he did, requesting some seed to send to Washington. They flatly refused, on the ground that America was one of the principal markets for the broad, thin leaves of Sumatra tobacco which is used as the wrapper of the best Havana cigars. He told them what he thought of them and left the meeting in high dudgeon.

A Glimpse of Saïgon and a Long Stay in Japan

Later, one of the planters came to him and offered to sell him as much seed as he wanted. Only that morning the seed had arrived in Yokohama.

I mailed Mr. Lathrop's check to the man whose address was on the package. The following March, in Cape Town, the letter and check came back to Mr. Lathrop unopened. On it the postmaster of Deli had stamped a single Dutch word, "Dood." The man had died before the money reached him.

This seed had been especially requested by Doctor A. D. Shamel, who was then breeding special varieties of tobacco for the Connecticut Valley. I believe that its blood, so to speak, today undoubtedly forms a part of the hereditary set-up of the tobacco plants grown there. However, this cannot be proved because, as so often happened, the history of Shamel's work has been lost, and local tradition has it that the Valley tobacco dates back to a package of seed found in some sea captain's attic. I mention this because the same thing has happened with other introductions, and the persons to whom credit is really due are often forgotten.

As soon as Mr. Lathrop could spare a moment from his tobacco seed, he said:

"Well, Fairy, what kept you so long? You missed the Japanese flowering cherry trees. However, I suppose you got a lot of date palms. I would like to have seen Bagdad with you but I knew you would collect more stuff if I were not along. Besides, I imagine the food was pretty bad, wasn't it? Anyway, I've missed you. I've been laid up in the hotel for a couple of weeks now, and if it had not been for my old friends the Eldridges I would have had a tough time of it. The Eldridges were among the first Americans to come to Japan. I want you to meet them."

Mr. Lathrop then continued his dissertation by telling me of the plans he proposed for our immediate future.

"I'm in no shape to go around with you," he said, "you can do better work alone anyhow. We will engage an interpreter, some student from the University. Also, you will have to have special papers permitting you to travel in Japan. You never saw a country so full of horticulture in all your life. You will be worn out trying to see it all.

"One of the main things here," he continued, "is the bamboo. I want to finance a big shipment of the plants to America. We should have them at home. Bamboo is beautiful as well as useful. The Japanese use it for everything. It may take a long time before Americans learn how to use

249

it, but they'll never learn if we do not introduce the plant. Don't spend so much time on other things that you can't study the bamboos.

"There is another thing I want to do," he said, in conclusion. "I'll pay for a special collection of cherry trees, too, if you think they would grow in America."

A youth has little perspective to help him appreciate the importance of an opportunity. As I look over the nine notebooks which I filled with notes, and the photographs which I took during that summer in Japan, I see that I was no exception to the rule. I could not fully grasp the fact that this country I was visiting had one of the most highly developed systems of horticulture in the world, and certainly the most fascinating. Furthermore, I could not of course realize that I had come at the turning of the tide, so to speak.

For generations the feudal lords, or daimyos, of Japan had maintained extensive gardens, and had employed staffs of trained landscape gardeners and horticulturists who lived on these estates and spent their lives in creating an amazing number of new and attractive forms of ornamental plants by breeding, selection, and methods of dwarfing. The living conditions of these gardeners may not have been ideal, but the fact that the daimyos actually spent their time in their gardens and among their plants stimulated the men to do their best and to create new forms; in fact, great rivalry existed among the different noblemen. The spirit of the time resembled that which existed for centuries on the great English estates. Many of our loveliest flowering trees and shrubs and plants trace their origin back to the patient, unnamed gardeners of these country estates both in England and Japan. Unfortunately, the environment of many of the Experiment Stations today does not permit the intimate association between the plant breeder and his plants which existed then. I know that both Van Fleet and Luther Burbank felt that they could never have accomplished what they did had they not been able to live in intimate contact with their plants.

In 1902, the days of the daimyos had passed, but many picturesque gardens were still in existence, containing collections of rare varieties of azaleas, iris, peonies, chrysanthemums, and flowering cherries. Although the age of Japan as an industrial nation was beginning, the awakening of the Japanase to the possibilities of scientific agriculture had also taken place.

A Glimpse of Saïgon and a Long Stay in Japan

Following the lead of America, with the assistance of American experts like Georgesson and Shelton, the Imperial Government of Japan had created a Central Experiment Station with nine branches Beginning in 1890 with a budget of twenty-six thousand yen, by 1902 it was spending almost triple that amount and had a big staff of experts.

There was also a well-equipped Botanical Garden in Tokyo, an Agricultural Institution in Hokkaido, and a Horticultural Institute in Komaba. At all three of these were trained botanists and horticulturists holding degrees from German universities. It was stimulating to watch the rising group of Japanese scientists who were struggling to increase the production of the small areas of land suitable for agriculture. Taken altogether, these areas are only equal to about a third of the State of Illinois.

The day after my arrival in Yokohama, I went out to the nursery to see my friend Suzuki. He had assembled a collection of bamboos and had planted out in nursery rows the most important timber species. Suzuki was, of course, familiar with the varieties in their young stages, but most of them looked exactly alike to me and I realized that it was essential to prevent their becoming mixed. I therefore devised a special lead seal with which I fastened a label onto each plant.

It became immediately evident that the introduction of bamboos into America was a serious undertaking, but I could not foresee that it would lead to the establishment of a bamboo garden with two hundred varieties, and the building of a bamboo museum near Savannah, Georgia, where people would wander in giant bamboo groves as beautiful as those which I now saw about me in Japan.

Before I went to Tokyo for my official papers, Mr. Lathrop took me to dine with Mrs. Eldridge and her two daughters, who had both been born in Japan and spoke Japanese fluently. Doctor Eldridge had been the first American physician in Japan, but had died some years before. The girls called Mr. Lathrop "Uncle," and were ready with suggestions as to what Japanese foods were especially worthy of investigation.

For salad that night, a glass bowl was brought in filled with glistening white shavings resembling celery, though much broader and thicker. I was delighted with its delicate flavor, and Mr. Lathrop, who was fussy about salad and generally insisted on making his own salad dressing, became most enthusiastic, too.

251

"That is something we should have, Fairy," he said. "Let's get some of it and send it home."

"This is a plant which the Japanese call udo," Miss Eldridge informed us. "They grow acres of it in the truck gardens around Yokohama and Tokyo. It is one of the commonest vegetables in Japan."

The brilliant white shavings were quite fiberless and remarkably crisp. They had a slight terebinthine taste, suggestive of the pine. Being fond of tropical fruits, many of which are characterized by this same flavor, I enjoyed the udo immediately. The next day I arranged for a shipment of both seeds and roots to Washington.

Years later Mrs. Fairchild and I grew udo in our garden in the suburbs of Washington. We found that everything depended on having it correctly grown and blanched. If not properly grown and carefully prepared for the table, the flavor of turpentine becomes too strong and is objectionable. Before the roots can be harvested, udo should be grown at least two years in good soil so that it will form a large bush. Once large plants are secured, the trick is to blanch them in the early spring. In the rich, light soil of Japan this is done by hilling the dirt as much as three feet up over the plants. In the stiffer soil of Maryland this was difficult, and we resorted to half barrels, or large drain tiles, which we put over the plants and filled with loose earth in the spring before there was any sign of growth. If the shoots are exposed to the sunlight for even a few days, they turn green and develop the rank flavor which is objectionable.

When udo is to be used in salads, the shoots are peeled, cut into shavings, and soaked in ice-water for an hour or more before serving with French dressing. When they are to be served on toast, like asparagus, the water in which they are cooked should be changed two or three times before the udo is finally put with the sauce. Cream of udo soup is particularly delicious, but the first cooking waters should also be changed.

The plant itself is attractive and useful as a screen in the garden to hide any unsightly corner. The flowers are favorites of the bees and its black berries make an interesting table decoration. Another very great point in its favor is the fact that it is ready for the table extremely early in the spring, at least a week or two before the asparagus.

Mrs. Alexander Graham Bell sent about a hundred plants to Nova Scotia for her garden there. To my great surprise, it has grown there

A Glimpse of Saïgon and a Long Stay in Japan

better than in any of the scores of other localities throughout the United States where it was tested. Not only is it a handsome ornamental, but, best of all, it makes a much-needed early vegetable for this far northern land.

The udo (*Aralia cordata*) has wild relatives which are native in the United States: the wild sarsaparilla, *Aralia nudicaulis;* the spikenard, *A. racemosa,* and the California species, *A. californica.* The spicy roots of *A. nudicaulis* are often used as a substitute for the true sarsaparilla in the making of root-beer—in fact udo roots have also been used quite successfully for this purpose.

Mr. Lathrop and I put up at the Imperial Hotel in Tokyo, presented our gold-sealed credentials at the Embassy, and requested papers from the Japanese Government permitting me to travel. The Minister delayed so long in procuring them that Mr. Lathrop, whose friend, John Hay, was Secretary of State, reported the matter by cable, and we got them next day.

To me, the delay presented a chance to see something of Tokyo and to make the acquaintance of Doctor Matsumura, the accomplished director of the Botanic Garden. I also enjoyed meeting the biologist, Doctor Myoshi, at the University. He had been trained in the methods of microscopical technique in Germany and showed me under his microscope how the fireflies make their light. He told me about the firefly hunters of Uji, who go out at night with nets and catch these pretty, little beetles, popping them into their mouths until they have a mouthful. The fireflies then are blown back into the gauze net and taken home to be put in tiny bamboo cages and sold in the streets all over Japan.

Everywhere in the gardens about Tokyo I saw quantities of the Japanese lawn grass, or Birodoshiba (*Zoysia pungens*), which was used for making beautiful, velvet-like turf among the rocks. I sent a turf of it to Washington; from there it found its way to Florida where I sometimes think it is used too much, as in that sub-tropical climate it has the habit of growing too vigorously.

Doctor Matsumura sent me to a Mr. Takagi, who was a great authority on flowering cherries. This charming little man, an artist to his finger tips, received me most cordially and produced his water-color sketches of the cherry-blossoms. We sat together on the spotless matting in his simple little house, and he explained the drawings to me one by one. I have rarely been so thrilled, for I had had no idea of the wealth of beauty, form, and color of the flowering cherries. In Washington there were a few trees of the weeping, single cherry (*Prunus subhirtella*, variety *pendula*), and a tree or two of what was then called *Prunus Pseudocerasus,* the Japanese flowering cherry. Also, there was a large tree on Fourteenth Street, of what was said to be *Prunus Chinensis,* supposed to have been introduced from China, and in the Arnold Arboretum were some specimens which I think Professor Sargent brought back from Japan but which I had never seen in bloom.

Mr. Takagi's water-color sketches were so beautiful that I spent the morning poring over them. With his help, I selected thirty which he considered the finest and used them as a basis for ordering a shipment. In the corner of each sketch was the name of the variety in Japanese characters. This name was a part of the picture, and I appreciated for the first time how ugly and pedantic the scientific names at the bottom of our prints or wood-cuts seem in comparison with these strokes of the artist's brush.

Japanese cherry trees had been grown to some extent in England, and were advertised by English nursery firms like Veitch, usually with such titles as *Floro-pleno Veitchii,* but our collection was to carry the Japanese names. However, trouble developed in the translation of the Japanese characters into an English equivalent. I did the best that I could, but as the years passed and new books were published, much readjustment of these English and Japanese names has taken place. There may have been other collections of Japanese flowering cherry trees introduced into America, but I believe that this one of Mr. Takagi's and mine, which arrived in Washington in 1903, was the first one in which an attempt was made to feature the original Japanese names.

There is a great deal more to this story of the introduction of the flowering cherry into America, including the history of the now famous trees in Potomac Park. But these accounts must wait until later, for it was a good many years before a single cherry tree was planted around the Tidal Basin in Washington.

Unfortunately, in 1902 the Office of Plant Introduction had no place to plant the cherry trees which I collected except in the recently established garden at Chico, California, and the whole shipment was sent there. The burning summer sun of Chico beat on them unmercifully and, to my bitter disappointment, most of them were eventually lost, but not, however, before a considerable number were propagated and distributed to experimenters in other portions of the country.

There was a gigantic gingko tree standing in the Botanical Gardens at Tokyo. Doctor Matsumura told me about the servant in his laboratory, a man named Hirase, who had learned to use the microscope and came to him one day asking if he could not have a problem—a piece of research to do. The Director pointed to this gingko tree and told Hirase that the microscopic history of its fertilization was as yet unknown. As the tree was close at hand, and fruited every spring, he suggested that Hirase work out the details of its fertilization by means of microtome technique.

The gingko is a tree of ancient lineage, so ancient that the other members of its tribe have disappeared and it alone remains, the sole representative of a large family which in earlier geologic times was abundant throughout the world.

Hirase set to work and, to the amazement of the botanical world, discovered that the gingko's process of fertilization resembles that of the cycads, a class of palm-like plants which was scattered pretty generally over the surface of the globe during the coal age but which today is represented by only a few species.

Hirase's name is now in every text-book on botany. I knew of his discovery, of course, and went to call upon him. He was a teacher in a secondary school at Hikon, and received a salary amounting to six hundred dollars a year. Upon this magnificent sum he depended to bring up a family of six children. We had luncheon together in a tea house and he gave me, as a souvenir of that afternoon, a microscopic slide on which was mounted a section of the gingko flower, one of those which first revealed to him the motile character of the male nucleus, a characteristic of cycads which was also disclosed somewhat earlier by H. J. Webber's work on the Zamia in Florida.

With my papers at last secure, I hired as guide an interesting young student named K. Yendo, who spoke good English, and together, in rickshaw and by train, we travelled through Japan. I look back with peculiar pleasure on those days spent in the small villages of central Japan. I learned to appreciate little gardens, and studied methods of landscape gardening and flower arrangement which have had a profound influence on all the gardening which I have done since.

Yendo and I started out at once to see some of the bamboo groves and to meet the actual growers who were selling the products so immensely important to Eastern people.

The Bamboos compose a natural division of plants classified by botanists as a subfamily of the grasses. They are in fact the most spectacular of all the grasses, ranging in height from a few inches tall to canes a hundred feet high. Bamboo has been used by man since prehistoric times. It has been suggested that tools of bamboos were used by savage races in the Orient during the Stone Age. The hollow stems of bamboo construct the homes of millions of people and furnish the material from which countless handmade utensils are fashioned. Last but not least, bamboos are among the most graceful plants known to man.

Yendo and I first visited a commercial grove near Kyoto, where the Moso bamboo, then called *Phyllostachys mitis,* was grown for its edible shoots, a great delicacy of the Japanese menu. The plantation was on a hillside, and the late afternoon sunlight streamed through the tall, green stems and delicate, leafy canopy as into a vast cathedral. The wind was blowing outside the grove, but a curious stillness prevailed within. Even our footsteps made no sound on the thick matting of brown leaves. If we can have groves as beautiful as this, I thought, it is worth any effort to introduce the bamboo.

The owner of the grove showed us how to dig the giant shoots which were coming up here and there between the tall, green culms. The speed with which the shoots grow in spring is almost unbelievable. Careful measurement has shown that they grow thirty inches in twenty-four hours—an inch and a quarter an hour—in California; in Georgia they average about six inches a day. The smooth, tight sheath which protects the young shoot as it pushes up through the soil is easily peeled off, leaving the tender core which is edible throughout. This is sliced into chips which should be boiled for twenty minutes in salted water and then allowed to simmer in a frying pan. Bamboo shoots are firm in texture and never become soft or mushy. Their taste reminds me of sweet corn with an oriental flavor which I cannot well describe.

This small grove was the owner's only source of income. However, bamboo culture was one of the most profitable plant industries in Japan.

A "back yard" garden in Tokyo; the apparently careless arrangement of rocks and plants is the result of centuries of culture.

The T'sao, or Chinese Jujube, fruiting in America. (Page 247)

Barbour Lathrop and Mrs. Graham Bell in 1922 examining her Udo patch in Nova Scotia. The plants are still vigorous after fifteen years of growth, and each year provide a crop of blanched shoots for the table.

Blanched Udo shoots.

256B

A fresh shoot of edible Bamboo ready for market.

A Bamboo lumber yard in Japan, with thousands of poles in assorted sizes displayed for sale.

The soft sheaths, like corn-husks, which cover bamboo stems, when dried, folded, and sewn together, make the smooth covering for the Japanese sandals.

Mr. Tsuboi, the authority on bamboo, enjoying his dwarf plants in charming porcelain bowls and trays.

256D

Hillsides, from which the Cryptomeria forests have been cleared, are covered with the Mitsumata shrub. From its bark, some of the softest papers of Japan are manufactured.
(Page 253)

At right, one of Frank Meyer's bamboos from Peking proved perfectly hardy in Maryland, and has been grown in gardens as far north as Connecticut.

Boards on which fresh paper pulp had been spread were drying along the main street of this mountain village.

A Soy sauce factory at Ichang. Fermentation vats with their bamboo covers.

A prolific Soy Bean plant ripe for the harvest.

The closely trimmed hedges of the tea bush in Japan made the idea of a mechanical tea-picker seem less impracticable than had the individual bush plantings of Ceylon.

The small paddy fields of Japan were really "pony size."

256G

Days are spent plucking and pruning Japanese trees to maintain the desired form and size.

A sacred Plum tree stands near the graves of the Forty-seven Ronins.

The Graves of the Forty-Seven Ronins, who held honor dearer than life, are shaded by venerable Cryptomerias. They form one of the most sacred shrines of Japan.

Aside from the shoots, the hollow stems are valuable for timber, and are cut when four years old and marketed in Kyoto, where the timber yards keep great stacks standing on end.

Even in our country, bamboo has a thousand uses. It is peculiarly fitted for light-weight ladders; baskets of every description; porch screens or blinds; fishing poles, of which several million are imported into the United States annually; phonograph needles; flower stakes, chair seats, and certain types of wicker furniture. One use, which I had never considered, was called to my attention by the conductor on a local train in Florida. He addressed me as follows: "Say, kin I get some of those canes you fellers are growin' to the 'speriment station? I want 'em fur my chicken yard. The hawks is mighty bad and I see they don't like to git into the thick brush of them canes. The chickens'll love them fur the shade, and I just want to plant the hull dern chicken yard to them." Certainly there could be no more attractive or safe place for chickens in the South, than in a grove of madake bamboo.

I had been fascinated by the black bamboo, which the Japanese call Kuro-Chiku (*Phyllostachys nigra*). Its stems have an ebony cast and grow about twenty feet tall. Yendo and I visited a grove of this handsome species and found the owner of a higher social order than the grower of edible bamboo. This man sold his canes exclusively to the basketware and fancy article market, and as we walked through his grove, he showed me specimens of rare species in which he took particular pride.

The complexities of bamboo culture seemed overwhelming, and I longed to talk to a real expert. Yendo therefore wrote to Mr. Tsuboi of Kusafuka, a recognized authority. He was a member of the samurai class which in feudal times were permitted to wear two swords, and was a charming gentleman of means and culture.

We were invited to spend a week-end at his country place, and wandered through his groves, learning all the details possible regarding the culture of timber bamboo. One evening, with Yendo acting as interpreter, Tsuboi told me that he remembered the days of Admiral Perry's arrival in Japan in 1853. He well recalled the talk in the village when it was rumored that these giants of the western world drank the milk of cows. His father and some of the wise men decided that if these big Americans lived on milk, it must have something to do with their stature and strength. Therefore, they themselves would also drink cow's milk. None of them had ever milked a cow and the only cows in Japan were work cattle with small udders and small teats. However, they managed it somehow and tasted the milk.

"You know," Tsuboi said, lifting a glass of milk to his lips, "I've drunk cows' milk ever since those days. I take a glass every night. I don't like it, but I drink it. I think it must be good for me."

Our visit to Mr. Tsuboi also gave me an opportunity to study the Japanese foods served in a home of wealth. The menu was completely new to me and I had considerable difficulty in digesting the raw fish and the seaweed, but became very fond of the egg dishes, the bean cheese or "tofu," and the universal soy sauce.

At Mr. Tsuboi's table, I made the acquaintance of the giant Japanese radish, which sometimes weighs twenty pounds or more, and is salted and pickled before serving. But the dish which I afterward served in Washington was a delicate variety of the soy bean. The young, flat, rather hairy pods—much smaller than the lima bean—are boiled in salted water and served in the pod in little lacquer bowls. I enjoyed opening the pods and taking out the delicate green beans, which have a pleasant, nutty flavor. Some Americans complain that it takes too long to empty the pods, and of course there is no reply to such an objection.

In Japan, meals are delightfully quiet. There is a general use of wooden chop-sticks and almost noiseless lacquer trays and bowls. Therefore, there is no clatter of dishes, or noise of knives and forks on china plates.

The Westerner feels that the use of chop-sticks requires great skill. It seems inconceivable that any one could pick up food with such slender implements. But, after two or three amusing experiences, I found them simple enough, and my children have all learned to use them. I have been asked many times how the Japanese would eat beefsteak or a mutton chop. The answer is: the meats and other foods are prepared in the kitchen, and cut into pieces of proper size.

The Japanese have their own technique of food preparation. Their cuisine is conspicuous for the absence of gravies, for which they substitute a sauce made from the soy bean by a complicated process of fermentation. Almost every meal includes boiled rice, often served with delicate, soy bean curds; vegetables (generally eaten either raw or pickled);

258

chicken or fish instead of beef, mutton or pork; and, as sweets, wafer cakes and pretty bits of colored soy bean candy.

Japanese women are supposed to know how to prepare rice perfectly. They cook it in a kettle with a wooden top, but the real secret of their success is the fact that they wash the rice so thoroughly, changing the water five times as they stir the grains about. This grinds the rice until the powdered starch on the kernels has been washed away. Then a scant inch of cold water is added with a pinch of salt, the lid is put on, and the kettle is placed over the fire. When it boils over, the fire is immediately reduced, but the cover is never removed until the rice is to be served, thus conserving the moisture. The Chinese, on the other hand, put their rice into boiling water. When it is cooked, they dip it out in a sort of basket and wash it in cold water, re-heating it before serving.

From Mr. Tsuboi's delightful home, we journeyed to a branch experiment station at Anjo where there were fields of the chief agricultural crops of the country. The soy beans particularly interested me, as in 1898 I had imported both beans and some soil from a Tokyo soy bean field. Waite and I discovered that root nodules could be induced on soy bean plants by growing them in Japanese soil, but, when they are grown on new soil, they will not develop the nodules unless the soil is inoculated. In Virginia, Maine, Michigan, and Kansas, a general interest in the soy bean and its possibilities had developed among the experiment station workers, although it was only used as cattle food.

I hope that a book will be written describing the vast soy bean industry which has developed in America since those early beginnings, and the important part played by the varieties introduced by Frank Meyer and later by W. J. Morse and P. H. Dorsett. Morse has devoted the best years of his life to a careful and penetrating study of the soy bean, and he should receive the full measure of credit for his contribution to an industry now measured by millions of acres under culture in America each year.

One of the most fascinating of the Japanese industries was paper. Every house was protected by treated sheets of soft, white paper made from the bark of the paper mulberry. Paper stretched tightly on sliding screens was often all that kept out the cold winds of winter or the prying eyes of the passers-by. I had thought of Japan at first as a bamboo world,

259

but I soon realized that it was, more strictly speaking, a land of paper. Even the umbrellas were paper. Paper fans and paper lanterns were already familiar in America, but the Japanese had paper clothing, paper handkerchiefs, waterproof paper cloaks, and paper bags tanned with persimmon juice, in which they stored their rice.

Our next objective was the nursery region of Ikeda among the hills back of Kobe. It was the plum season, and I enjoyed collecting the best varieties although really all Japanese plums have an unpleasantly flavored zone of flesh around the seed.

A beautiful fruit called the Nagi (*Myrica rubra*) was also growing at Ikeda. The dark red fruits were covered with tiny protuberances and had a slightly acid flavor and bright crimson juice. The juice makes a brilliant drink and cooked nagi reminds one of stewed mulberries. They belong to the small, "left over" order Myricaceae, a remnant, as it were, from the Tertiary times when its plants were widely scattered over Europe, and even as far as Greenland. In America, the genus is represented by the bayberry (*Myrica carolinensis*) from the wax of which candles are made. I introduced this little fruit and we grew it in California. Later, Frank Meyer found large fruited varieties in China, much superior to the Japanese. These plants have been fruiting for a good many years at Chico, but their slow growth has kept them from becoming popular with the nurserymen, even though nagi can be grafted easily on the southern bayberry. In appearance they resemble the Pittosporums in both leaf and habit.

In a nurseryman's tiny house, Yendo showed me a diploma hanging on the wall; there had been an exhibition of home-made charcoal, and the nurseryman had taken first prize. We were sitting with the men around their picturesque hibachi (a sort of brazier), watching the glowing coals of cherry-wood charcoal as we knocked the ashes from our delicate Japanese pipes by tapping them against the upright piece of bamboo placed for that purpose. The continual filling of the little pipes with tiny pellets of shredded tobacco, and the sharp noise made by the pipe against the bamboo, were an essential part of the lives of the people until the cigarette habit supplanted the hibachi and its interesting paraphernalia. The bamboo which served as an ash tray in a hibachi was grown in the mountains, and had a tough, hard wood which withstood the continual hammering of the pipes.

A Glimpse of Saïgon and a Long Stay in Japan

The raw fish, which I had such difficulty learning to appreciate, was nearly always accompanied by "wasabi," a grated, green horse-radish which I really enjoyed. Probably my previous interest in horse-radish made me particularly notice this vegetable, and before returning to Tokyo, Yendo and I visited a wasabi grower in a mountain valley back of Shedzuoka. Both the wasabi (*Eutrema wasabi*) and our own horse-radish are members of the same family, the Cruciferae. The large, heart-shaped leaves, rising on slender stalks, covered the sides of the narrow valley wherever the ground was saturated with water from the many springs which trickled down the slopes. The short rootstocks are produced in a single season, and from them is made the pungent condiment. I tried to grow this in America, but found it difficult to discover the right conditions for its culture.

After our return to Tokyo, Doctor Baelz, a noted Austrian physician, invited me to his home which was in an old daimyo's garden. As I stood on the porch, I watched four men perched in a beautiful pine tree (*Pinus densiflora*), the "character tree" of the charming landscape. When the Doctor appeared, I asked him what the men were doing.

"They are pulling out the needles and pinching back the small buds," he said. "We do that every year. That is the way we keep the picturesque shape of the tree."

Doctor Baelz had saved the life of the young prince and, as a token of appreciation, the ladies of the court presented him with an elaborate, miniature landscape. In it were minute pines and maples, a tiny azalea garden, and a little rivulet with sandy bed. If any one would understand the Japanese, he should study their fondness for little gardens, not only the miniature gardens of dwarf plants, but the back-yard gardens, where old men and women of the better classes sit in contemplation after they have stepped aside to give their sons a chance to carry on.

One of the favorite fruits of the Japanese is the loquat (*Eriobotrya japonica*) which belongs to a small genus with only a few species. It is one of the few early spring fruits of the temperate zone, and is now grown all over the world wherever the climate is suitable. However, for some reason, very little has been done toward its amelioration. As far as I know, there have been no attempts to hybridize it with other species, and there are only a few recognized varieties.

The loquat has decorative, attractive foliage, and the delicious fruits are a rich, golden yellow. In northern Florida the trees have proved somewhat susceptible to the pear blight bacillus, but in South Florida the fruits set well except in particularly warm winters. Bordeaux mixture on the flower clusters has proved conducive to a better setting of the fruit, and arsenical sprays are also helpful in preventing insect attacks on the fleshy flowers.

During my stay in Tokyo I called upon a horticulturist named Tanaka, who had originated an especially fine variety of loquat. He proudly showed me some of the largest fruits I ever saw, which were also much rounder than the ordinary fruit. He gave me scions from this tree, and some years later I had the pleasure of eating Tanaka loquats from grafted trees in California.

Before sailing for home, I went with Suzuki to visit one of the most historic spots in Japan, the burial place of the forty-seven ronins. I had read the amazing story of these men and, as I stood beside the forty-seven simple gravestones, I felt that this spot epitomizes the true spirit of old Japan. Forty-six of the forty-seven were samurai who had deliberately committed hari-kari after avenging an insult shown to the daimyo who was their lord. The forty-seventh grave was that of an official who felt himself disgraced because he had misunderstood the actions of one of the forty-six, and had spat upon him in disgust.

In *Tales of Old Japan,* Mitford gives a picture of the civilization of old Japan which is important and absorbing to any one desiring to understand the Japanese character. Among my souvenirs was a hari-kari knife with short, thick, polished blade, and sheath of unpolished gray wood. It lay for years on my table reminding me of the men with a love of honor greater even than their love of life.

I was sorry, very sorry to leave Japan, and I have often wanted to go back. But I hear that the rickshaws are gone, that motor roads have been built everywhere, and the picturesque costumes of the men have disappeared. I could not bear to see the workmen wearing ugly American straw hats. I would rather keep my memories of old Japan as it was at the beginning of the century.

CHAPTER XIX

I Visit Luther Burbank and
Circumnavigate Africa

I EXPECTED TO BID Mr. Lathrop good-bye in San Francisco and to return to Washington where the "Section" of Plant Introduction had now become a larger unit with the title of "Office." But Mr. Lathrop began talking of another expedition.

"You've seen Egypt and Algeria, but how about South Africa? You don't know a blessed thing about the 'veldt,' or Natal, or the unusual vegetation of the Cape. You really should see Africa before you settle down in Washington. I'll look up the steamers and if you can get away we'll leave some time in October." Again I accepted.

I spent ten days in San Francisco, talking bamboos, and making arrangements with people who had horticultural training and facilities for the testing out of foreign plants. It was vitally important to find the proper persons to grow the plants we introduced.

Before going East, I made a trip to Santa Rosa to see Luther Burbank, and spent the night in his modest little home in front of which stood his famous hybrid walnut, a cross between the black walnut (*Juglans nigra*) and the California walnut (*Juglans Hindsii*). I have a photograph of him standing beside this "Royal Black Walnut" (as he called it) which was only twelve years old at the time and already forty feet high.

The magnitude of his breeding operations was bewildering, and I was fascinated by his quiet talk about them. There was the old apple tree behind the house on which he said that a thousand sorts of seedling apples had been budded or grafted. There were rows of Japanese plums, 40,000 in all, different seedlings which he had grown; a bed of native flowering bulbs, 200,000 of them; his new crimson, winter rhubarb from Australia;

his hybrid Japanese-Spanish chestnut. He was already besieged by visitors who wanted to see him, and who took up much of his time. He told me that he was weary of trying to produce new varieties of plants to sell in order to support his extensive operations.

As we drove in his little buggy from Sebastopol, where his nurseries were located, to Santa Rosa where he lived, we talked over possible plans for a foundation which would insure him financial support and would relieve him of the necessity of the nursery end of his business. Eventually plans did mature, in which I had a hand, which for a while gave him $10,000 a year. But this support from the Carnegie Fund was later withdrawn, and the dream of a scientific institute of plant-breeding at Santa Rosa never materialized.

At times, I have felt inclined to criticise Burbank for this collapse of our plans, and for allowing those around him grossly to exaggerate the claims of his new creations which frequently proved failures when grown away from the climatic conditions in which they were produced. However, it should not be lost sight of that Burbank grew up as a commercial nurseryman and spent his life in an atmosphere where plants meant money, and where exaggerated claims such as those in many nursery catalogues were the rule, not the exception. This early training unfitted him in many ways to handle scientifically the large amount of plant material which he grew and imported. Consequently, I am inclined to believe that he had little control over the circumstances resulting in the exploitation which brought him a certain degree of disrepute in scientific circles.

On the other hand, had Burbank's large-scale methods of plant selection been emulated more widely, I believe that the orchards of this country would have many more fine varieties in them than they have today.

As we passed a little school-yard, bare of any sign of tree or shrub, he checked his horse long enough for me to take a photograph while we discussed how strange it was that the majority of our schoolhouses surrounded the children with an environment containing nothing beautiful.

I was surprised and nonplussed to find that Burbank believed in clairvoyance. He told me of a coincidence which happened at the time of his father's death, when his mother had a vision of the accident that had caused it.

When we reached the house where his secretary, who afterwards be-

Luther Burbank and his hybrid walnut tree in front of his house in Santa Rosa, California. According to Burbank, this hybrid between the California wild species (*Juglans Hindsii*) and the Black walnut (*Juglans nigra*) grew twice as fast as the combined growth of both parents. *Right:* Burbank and a fair neighbor.

From the terrace of the old monastery near Taormina, Mr. Lathrop gazed at the snow-covered peak of Mt. Aetna.

Mr. Lathrop admiring a gnarled Olive tree near Syracuse, which was believed to be 1500 years old. (Page 271)

The medieval town of Sfax in southern Tunis was designed to withstand attacks by the desert tribesmen.

Accounting stick of the Sicilian peasants.

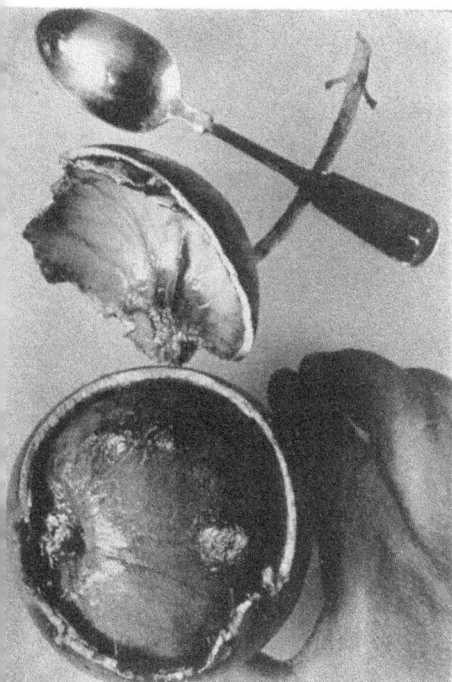

The amazingly round fruit of the Kaffir Orange contains glistening pulp within a hard shell.
(Page 273)

Harbors are scarce on the east coast of Africa, and passengers are lowered by twos and threes in a hamper to the deck of the waiting tender. (Page 273)

Professor Macowan introduced me to the
Cape Town Botanic Garden. (Page 276)

The Kaffir coolies at Durban deck their heads
with polished horns embellished with gay
feathers. (Page 275)

The first Carissa hedge in America; in the Plant Introduction Garden in Miami.

came Mrs. Burbank, was busy writing up his notes, two pretty girls came to call, and we had a jolly time, sampling some of his creations. He was at heart an altogether lovable person, whose intuitive sense with regard to plants was most extraordinary. The wave of interest which spread through the civilized world as a result of his activities at Santa Rosa had a real and lasting influence on people who had never before thought of plants as capable of change and improvement.

One might describe Burbank as like Tolstoi, in that, when one was with him, one felt the strange force of his simplicity and his profound confidence in his own abilities. But, on leaving him, the impression faded, and one began to wonder wherein lay his power, for his results did not quite seem to justify his claims.

A letter which I wrote to Cook after this visit seems worth quoting from:

There are those who say he (Burbank) is not scientific. It is true only in the sense that he has tried to do so much, and has been so fascinated by the desire to create that he has not always noted and labeled the footsteps which he has taken. I mean that he cannot always tell you the parentage of some of his creations; neither is he always quite sure that his pollinations were the only possible ones— bees might have followed and vitiated them.

. . . .

He is interested in so many things, I could write all night and not give a correct idea of what he is doing. He sees the particular characteristics which are desirable and talks of adding them to his hybrids as one might talk of adding ingredients to a pie.

Although not at all apropos, I cannot resist quoting the last sentence of this letter written in 1902:

By the way, Burbank has a graphophone, and so has President Jordan of Leland Stanford University and Professor Hilgard.

On the first expedition with Mr. Lathrop, I had sent four forms of spineless cactus (Opuntia) to Burbank. One had come from the Gran Chaco of the Argentine, another from the calcareous hillsides of Sicily, the third from Ceylon, and the fourth from the plantations of spineless cactus which had been in operation for many years on the coast of Tunis. In Tunis, I had spent some time investigating the use of the cactus by a dairy firm there. Some years later, on my return to America from one of my trips, I heard of the "creation" of a spineless cactus by Burbank, and soon convinced myself that he was announcing that he had made the *only* spineless form in the world. Naturally, the esteem in which I held him suffered a shock from which it never recovered.

One reason that I felt so keenly Burbank's abandonment of scientific horticulture, and his tie-up with journalists and exploiters, was the fact that, at a critical moment in the negotiations, I had appeared before the Directors of the Carnegie Fund urging them to support him. I had even suggested that one of my most intimate scientific friends should be sent out to study his results from the genetic standpoint. Needless to say, my disillusionment was a severe blow. I never saw Burbank again and, now that he is dead, and so many years have passed, I prefer to think only of his lovable nature and those charming days with him in his little house in Santa Rosa, as the sun shone upon his garden friends.

To summarize my impressions of the whole Burbankian episode in American horticulture, I should say that he was the pioneer in large-scale methods of selecting seedlings from beds sown with millions of cross-bred seeds, and that his work marked a distinct advance over the old method of chance discovery of "sports" in hedgerows and dooryards, and thus he blazed the way to much larger operations than had previously been thought necessary for the discovery of new varieties of fruit trees.

I left San Francisco on September 12 and went straight through to Washington, where I spent ten hectic days attempting to see my friends and make arrangements to leave in October for South Africa. I had much to tell about my trip to Bagdad and the shipments of plants which were already coming in. I went up to New York to spend a few days with my mother and sisters in Brooklyn, and attended one of the first meetings of the American Breeders' Association. Among those present was the great English genetist, William Bateson. The science of genetics was emerging at this time, and Hugo De Vries of Amsterdam had startled the botanical world by his theory of mutations. O. F. Cook read a paper at the meeting on what he called "The Network of Descent" and its effect on the persistence of those groups of interbreeding individuals called species. Our dreams with regard to what man could do to change and create forms of plants may have been a bit fantastic, but I am amazed when I study the vast accomplishments which have come in the field of

I Circumnavigate Africa

agriculture since then through man's understanding of the laws of heredity. Hybrids have been produced in great numbers, but the power to predict the results of hybridization has not come to us in quite the way we thought that it would thirty years ago.

On October 19, I awaited Mr. Lathrop on the deck of a North German Lloyd steamer bound for Naples. Nine years had passed since I met him on the *Fulda* of this same line off the Azores. By this time my relation to him had ceased to be that of a mere investment. We were friends.

The captain proved to be a rather stupid fellow and Mr. Lathrop's attempts at conversation inspired nothing interesting in reply.

"This is an eleven-day trip," he said to me after a day or two, "and I don't propose to be bored all the way across just because I am supposed to sit at the captain's table. There are some vacant places at that table over there," and he indicated one at which sat two attractive ladies. "I am going over there and you're going with me. I don't care what the captain thinks."

Mr. Lathrop arranged the transfer with a minimum of friction and we appeared next morning at the other table. In the meantime, he had learned that the ladies were Miss Lucille Hopkins, daughter of the president of the Diamond Match Company, travelling with her chaperon, Mrs. Collins. Both proved to be charming and intelligent.

It was the fashion at the time for women to wear rats, artificial somethings over which they combed their hair in front, pompadours I think they called them. Miss Hopkins had an enormous one which threw her pretty face completely out of scale. Mr. Lathrop, whose philosophy of life included frankness at all times, blurted out,

"I don't see why the girls of today ruin their looks by such ugly things as these pompadours. I think they're hideous."

The young lady's face turned scarlet, and I felt thoroughly ashamed of Mr. Lathrop, and exceedingly embarrassed as well. But I did not know as much about women as I thought I did, for long before we reached Naples both ladies were captivated by Mr. Lathrop's wit and humor and even enjoyed his utter and devastating frankness.

We went to the Hotel Royal in Naples, and Mr. Lathrop invited our table companions to dinner. The orchestra played those lovely Italian airs which bring memories of sunny hillsides, pine trees, blue waters and gay, happy people. Mr. Lathrop seldom showed emotion, but at the first bars of "La Paloma" his eyes filled with tears.

"This is my favorite of them all," he said. "The Emperor Maximilian loved it too. His last request was that it be played for him before they shot him like a criminal at Querétaro."

His sentimentality was over in a moment. He cleared his throat and made plans for the next day.

My first visit, naturally enough, was to the Stazione Zoologica in the Villa Nazionale, where I found Paul Mayer still in the same little laboratory surrounded by vegetable dyes and piles of literature, everything in perfect order. He took me to the library and ran his hand affectionately over nine large volumes of the *Zoölogical Year Book*, each containing hundreds of closely printed pages. I do not think that he quite approved of all my travelling, although he did not say so, but he was glad to know that I was not married. He thought there was still hope of my settling down and doing some scientific work again.

The Director, Anton Dohrn, invited me to dinner and showed me his collection of Phyllocactus, a genus of plants of which he was particularly fond. At his suggestion I called on Carl Sprenger, a brilliant botanist who had established a nursery, and gathered together an interesting collection of plants from many parts of the world. Sprenger was very deaf and difficult to talk to, but he was one of those real plantsmen who both know the names of plants and how to grow them. At that time he had taken a particular fancy to the Yuccas and was busily hybridizing them. His name is known to American gardeners through his introduction of *Asparagus Sprengeri*. Possibly a million people have grown pots of it on their window-sills, but it is doubtful if a dozen or more have ever wondered about the man who introduced it into cultivation and after whom it was named.

Many letters passed between us in later years, some after he became superintendent of the Kaiser's Corfu garden. His most dramatic epistle described the appearance of his Naples garden after an eruption of Vesuvius had buried his plants under volcanic ash, destroying hundreds of his best specimens. I lost track of him completely during the war, and it was not until 1924 that I heard from no less a person than the Kaiser himself that Sprenger had been interned by the Italians and had died in prison.

I Circumnavigate Africa

Since my former stay in Naples, the American macaroni manufacturers, although improving their product by the use of Durum wheats, had not yet succeeded in producing a first-class product. I had heard that in order to taste the finest macaroni in the world I must visit Gragnano and sample the "pasta" fresh from the mill.

Gragnano is a little town on the Bay of Naples, back of Castellammare. My route took me through Pompeii and Herculaneum, and I strolled for awhile in those cities of the dead to see if by chance they might give me some perspective on the cultivated plants which had grown around the Bay of Naples almost two thousand years ago. Perchance some had been preserved in the ashes and mud hurled from the crater of Vesuvius. Mills with which wheat had been ground were in the museums, but I could find no evidence that the art of macaroni-making had been known. In a glass case in Pompeii, near those tragic forms of prostrate men and women, were the remains of a few figs. I had bought some Neapolitan figs that morning and had them in my pocket. I have always liked the way they are prepared in pairs with a walnut between. Amazingly enough, the figs in Pompeii had been prepared in this same way, nearly two thousand years ago.

The owner of the mill at Gragnano gave me all the information I desired and, at the little trattoria across the street, I enjoyed the most delicious plate of macaroni I have ever tasted; as it came so fresh from the mill, it had a delightful flavor of its own. The manufacturer told me that in making this particular type, he used Saragolla wheat from the province of Apulia on the Adriatic coast, where only eighteen inches of rain fall during the year. While it had not the strength of the Russian hard wheats and did not keep as well, it had a better flavor and produced a macaroni which sold at a premium. I imported a quantity of this wheat, hoping that somewhere in America it could grow under conditions similar to those of eastern Italy. But I was disappointed, and people who want Gragnano macaroni must still go to Italy.

After three short weeks in Naples, we pushed on to Reggio. I had heard that the best English briar pipes were made, not from the sweet briar or eglantine rose, but from the roots of an Ericaceous plant (*Erica arborea*) which grew wild in Calabria near Reggio, and I made a trip into the mountains to investigate this industry.

The World Was My Garden

Great gnarled roots of this Erica, dug from swamps back of the little town, were piled in great heaps. They were kept alive for months in moist sheds, and sawed into pieces when the wood was still green. These pieces were then boiled eight to twelve hours to prevent their cracking open. Five hundred factory hands were employed, with thousands of peasants who dug up the briars in the swamps. The industry amounted to over two million lire a year, a substantial sum in those days. The men who worked the dangerous saws received five to six lire (one dollar to one dollar twenty cents) a day, high wages as compared with one lira fifty (thirty cents) a day paid to common labor on the farms.

Inside the little factory, men sat before small buzz-saws and "blocked out," as they called it, these great pieces of root into smaller pieces ready to be shaped later in England into conventional briar pipes. I noticed that most of the men had mutilated hands and I asked the proprietor about them. He explained that the Government had put a price on each finger, and that it was not uncommon for a worker who had run into debt to saw off one of his fingers and get the bounty.

I was not disappointed in Taormina. Who could be? Its beauty cannot be described in words, and the view from the old Greek theatre lives in my memory as vividly as the first day I went there. It seems amazing that people spend thousands of dollars on trivial pleasures and yet pass their lives in commonplace surroundings, content never to see Mt. Ætna from the Greek theatre in Taormina.

The last time that I was there, I was pushing off from the beach in a rowboat after giving the fishermen a liberal tip. One of them pretended that he was dissatisfied.

"We are poor," he said. "Noi siamo poveri."

"Yes," I said, "I know. But I have come three thousand miles to spend a day in this spot where you live, because it is one of the loveliest spots in the world. And you are privileged to live here year after year. You have something which costs you nothing, which it costs me hundreds of dollars to see."

He had not thought that I could speak Italian, and was rather taken aback, but he smiled and said, "Ah, yes, that's true." And he waved his hand to me as I pulled away from the shore.

The poverty of those days was appalling. Twenty thousand able-bodied workmen were leaving Italy each month and emigrating to the United States. Once there, accustomed as they were to the frugal fare of Italy, they could live so cheaply that they were sending back more than half their wages to their families. In 1902, the principal bank in Naples distributed over a million lire a month, sent from America.

It was somewhat chilly in Taormina, and we moved on down to Syracuse, where we spent a week while I filled my note-book with facts and figures which I intended to use later in a comparison between the living conditions in Sicily and the United States.

The illiteracy was amazing; only three men among the servants in our hotel knew how to read or write. Less than 20 per cent of the population could understand the printed word. Catania, a city of over one hundred thousand inhabitants, had no daily paper.

In the market-place I watched people keeping their accounts by means of notches cut in sticks. I secured one of these curious devices. The marketman cut notches with his knife to indicate how many dozen eggs, chickens, or broccoli had been sold to him by a peasant. Then he split the stick in two, gave one-half to the peasant and kept the other himself as his record of the transaction. When the peasant returned with more eggs or vegetables, he presented his half of the stick, the two were fitted together, and more notches made.

The olive trees about Syracuse were the largest I had ever seen, some of them eighteen feet in diameter at the roots, and so old that no one living there had any idea how ancient they were. A French expert, M. Hugon, whom I met in Tunis, believed them to be at least fifteen hundred years old. Who can say?

From Sicily we sailed for North Africa and spent a day at Sfax, a fascinating old town whose crenelated walls seen from across the desert reminded me of the woodcuts in Abbott's Histories, where crowds of warriors were shooting from the battlements with bows and arrows at other warriors below who had only shields for protection. In Sfax I again saw enormously fat Jewesses wearing peaked caps and white silk robes.

After a few days in Tunis, we sailed on a stormy Christmas Eve in what Mr. Lathrop called a "futty" little French boat. It was one of those boats that would "roll your inwards out," as he used to express it, and everybody, even the captain, was sick.

The World Was My Garden

We reached Port Said on January 1, 1903, and made a hurried trip to Cairo to lay in an extra stock of linen for the forty-nine day trip down the east coast of Africa. In those days no laundry was done on passenger steamers.

The trip from Port Said to Aden, eight days through the Red Sea, had begun to be an old story to both of us. The rocky, desert shore is about as interesting to a botantist as a tropical jungle to a geologist. The boat was comfortable and there were pleasant people on board, including some officials bound for the German East African colonies.

As we steamed along the desolate coast of Somaliland, touching at Jibuti, and rounding Cape Guardafui, I began to realize the harborless character of the east coast of Africa. Five days it took us to steam the thousand miles from Aden to Tanga, a tiresome trip had it not been for the company on board.

I remember nothing of the day at Tanga, a hurried call of only a few hours. But, as we neared the island of Zanzibar, the air was filled with fragrance, for the clove industry, transplanted from the Spice Islands of the East Indies, had already begun to be profitable there. Zanzibar has always had the reputation of being hot, and so we found it. There was not time for much, but I did see a sight which amazed me—a date palm in full fruit. That a palm adapted to withstand the bone-dry desert air, where not over 2 or 3 per cent of moisture occurs, should be living and fruiting in Zanzibar in an almost saturated atmosphere is certainly an indication of the toughness of its constitution.

From now on the trip seemed annoyingly hurried, a day on land and then a day or two at sea. At Mozambique, Portuguese East Africa, I went ashore late in the afternoon. It was a miserable place where evidently the white man found it difficult to live, and every one seemed shaking with malaria. I wandered about the town until dark, and on the edge of the village came across a group of black men dancing in the moonlight on the dusty roadway. They were abandoning themselves to a kind of instinctive rhythm as they clapped their hands and stamped their big bare feet on the hard ground as naturally as the monkeys in the forest swing by their tails from the forest trees. I, with generations of Calvinism and Quaker creed behind me, stood fascinated, understanding for the first time that perhaps the instinct to dance is so deeply rooted in the make-up of the human animal that the puritanical idea of eliminating it is more unnatural than the dancing.

Until this trip, I had never comprehended what a handicap the lack of harbors can be to a country. Open roadsteads are most unsatisfactory substitutes. When we reached Chinde, at the mouth of the Zambezi River, the crew set out a big wicker basket on deck, resembling a giant clothes hanger with a door at the side. One of the ship's derricks was unlimbered while I stood wondering what they were doing. At first I thought that they were joking when they explained that passengers must land in these cages. The weather was frightfully rough and the tug coming out from the shore was bobbing up and down in an alarming fashion. It came as near as it could, and the passengers who were landing were put into the basket three at a time and hoisted over the side onto the deck of the lighter. The weather was so rough that our steamer only stayed off Chinde for an hour or two, so that all I saw of the Zambesi was a stretch of greenish gray water where it emptied into the sea. Later, at Durban, where we did land, we had the experience of scrambling for places in the basket ourselves.

At Delagoa Bay, our boat pulled into a respectable harbor, and we spent a day at Lourenço Marques. I made a short trip out through the parched, brown country-side, and found the vegetation entirely new to me. Aside from a few grasses and wild species of garcinia, there was little of promise to collect. I was much disappointed, for I had hoped to find something worth-while, and felt sure that there would be many things if we had time enough to get into the interior.

As we were leaving the residence of Consul Hollis, I noticed a low, rather scraggly tree, with grayish-green foliage and rough bark, on which was hanging dark-green fruit about four inches in diameter, each one round as a ball, in fact the roundest fruits I have ever seen. Hollis explained that they were Kaffir oranges and said that the Kaffirs ate them, but that he did not care for them at all. I took one on board with me, and forgot it for several days until I became conscious of a spicy perfume in the cabin, an odor of cloves, and traced it to the Kaffir orange which had in the meantime turned light yellow. I was delighted by the fragrance and immediately tried to open the orange, but the rind was hard as a shell and I finally had to get the carpenter's hatchet. Inside, it was filled with a brown, glistening pulp of the consistency of a very ripe banana, and a pleasing aromatic taste.

My reference books stated that the Kaffir orange is *Strychnos spinosa* and, aware that the drug strychnine is produced by a species of Strychnos, I had some misgivings about eating the fruit. However, the consul had said that the Kaffirs ate it, and I took the risk. Later I learned that the seeds alone contain strychnine, and an analysis showed that there was only a trace in them.

I sent the seeds to Florida where they did well and fruited heavily. When Mrs. Fairchild and I bought our place on the shore of Biscayne Bay, the Kaffir orange was the first tree we planted. The late George P. Brett, President of The Macmillan Company, who owned a place near mine, used to serve me what he called his "strychnine" cocktail, flavored with the Kaffir orange. Last Christmas, our table decorations were gilded Kaffir oranges arranged with branches of the Caribbean pine.

At Durban, the capital of Natal, we spent a delightful week. After a month at sea, life on shore appealed strongly to a landlubber like myself. I found it a great relief to look at a horizon not forever shifting its levels, and to sleep at night without being continually tossed from side to side in a narrow bunk.

The jinrickishaw had been introduced into Natal, but the Kaffir coolies had more imagination than the Chinese or Japanese, for they decked their heads with polished cattle horns embellished with feathers, and, although they did not trot through the streets with the speed of the Chinese in Singapore, the bobbing of their horns was most amusing.

Durban possessed a good Botanical Garden and I hired a rickishaw to take me there. At the entrance, the coolie lowered the shafts beside a dense hedge of glossy, dark-green leaves. Scattered here and there were white, star-like flowers and brilliant scarlet fruit the size of bantam eggs. To my delight, the flowers proved to be jasmine-scented, and the coolie made signs indicating that the bright red fruits were good to eat. When I picked some, the man ate one to prove that it was all right, and I followed suit, finding that the fruit had a delicate, refreshing flavor. The plant had long, sharp spines which forked at the end, making a Y. Here was an ideal hedge plant, impenetrable to cattle, and having beautiful foliage, white perfumed flowers and an excellent fruit. In all the category of hedge plants I doubt if another such combination of desirable characteristics has been brought together. The director informed me that this *Carissa grandiflora* was a food plant of considerable importance in Natal, and that fruits of it appeared in the markets and were enjoyed by Europeans. I imported a large quantity of seeds which were grown at the Plant Introduction Garden in Miami and widely distributed in southern Florida. I believed for years that mine was the first introduction of the Carissa, but T. L. Meade, the noted horticulturist of Oviedo, Florida, appears to have introduced it in 1886 by correspondence from somewhere in Australia, and I am delighted to yield to him the honor of introducing this successful and popular plant. The State of Florida owes a very real debt of gratitude to Mr. Meade for his many introductions, which have been distinctly to the advantage of the State.

There are about twenty species of the Carissa, distributed from West Africa to Australia, and the remarkable characteristics of the genus would seem to entitle it to more study than has been given it. I hope the time may come when many hybrids between the species will be created, and the genus will be even better known. It not only is a splendid, subtropical hedge plant but is sufficiently salt-resistant to grow near the ocean.

I had read about the black Wattle of Australia (*Acacia mollissima*) and of its cultivation on the plains of Natal. I therefore made a trip to the Hilton Wattle estate and found thousands of acres of this lovely acacia growing in the center of a treeless prairie. I was reminded of the shelter belt plantings of cottonwood and box elder on the plains of Kansas in the early days, with the difference that these Wattle forests are unusually profitable because of their excellent tan-bark. Although I was feeling "pretty rocky" at the time, coming down with a touch of fever, I secured some good photographs showing the methods of tan-bark preparation, and wrote a full report which I sent to Washington. It was published as a little pamphlet, but I was unprepared for the repercussion it caused in Hawaii, where, in order to test the bark, Jared Smith violated the sensibilities of the old residents by cutting down a small forest of this species which had been planted years before on the Punch Bowl. My pamphlet certainly got him into trouble!

We went to Johannesburg by rail through Pietermaritzburg and Lady-smith, but I recall little of botanical interest on this journey to the greatest gold-mining town in the world. I always have had a dislike for mining towns, the great dumps they produce are so hideous. The gigantic size of those of Johannesburg put it in a different class from other mining towns, but even this could not make the place beautiful in my eyes. Of course the eucalyptus had been introduced and was everywhere, as was also the Monterey pine, *Pinus insignis*. Both were growing splendidly on that high, dry plateau.

On the journey down to Cape Town the waiter in the dining-car served each of us an individual pineapple prepared by simply cutting off the "eyes" and rough outer covering. The fruits were about the size of an ordinary drinking glass, and were served with the "topknot" of leaves still attached. I had never seen the fruit served except in conventional slices with a hole in the center. In this case, we tore the pinapples to pieces with our forks, and, to our astonishment, found scarcely any core. Also they were far sweeter than any we had ever tasted, and had practically no fiber. I saved the tops and sent them in, and went to a good deal of trouble to secure more suckers from Natal and from Trapps Valley in the Cape Colony.

Webber and Swingle were then deep in the problem of breeding pineapples, and, upon my return to America, I took up the cudgels with Webber for the "individual pine." But he drowned me with arguments against it, claiming that the market wanted the largest ones it could get. He enthused about his delicious hybrid pineapples until I did not care whether they weighed ten pounds or more.

However, I still believe in the appeal of the "individual" size, and maintain that I like a pineapple of my own, just as I enjoy my own orange or apple. At "The Kampong," I grow a small plantation of Natal pines which gives us every year some really sweet pineapples in strong contrast to the acid, commercially grown "pines" which usually come on to the market.

Cape Town was a city of about 50,000 inhabitants and, like most pioneer towns, it had been built in a helter-skelter manner. The houses were of indiscriminate architecture but many had charming grounds. The first morning I gravitated naturally to the Municipal Gardens, where I found Professor Macowan, a charming old gentleman of great dignity and keen sense of humor. He had been largely responsible for the garden and shared my enthusiasm in regard to plants. At the time, he was collecting species which might be useful in holding the shifting sands of the coast.

Almost the first plant which caught my eye was a curious native shrub or small tree which I had also noticed in Natal. It was *Portulacaria afra,* a succulent, known as Spek-Boom by the Afrikanders, and has but a single related species in Africa. Professor Macowan said that it is a favorite food plant of the elephants, and is enjoyed by other animals as well. Its branches and stems have the puffy appearance of the arms of a doll, and you could easily bend a tree several feet high until its tip touched the ground. It was altogether a weird thing, reminding one of a giant purslane. In my imagination I could see this succulent growing as a forage plant in our Southwest, and I wrote a glowing account of it in which, perhaps because of its uniqueness, I overemphasized its usefulness to the elephants. My chief, Doctor Galloway, whose sense of humor never failed, was entertained by this feature of my description and remarked, "It's all very well to introduce an elephant fodder tree, but I'd like to know where the elephants are to come from."

The Spek-Boom grew, and there are still plants of it around San Diego, California, where there should be elephants to eat it but are not. I had overlooked, however, a most important difference between the climate of the Cape of Good Hope and that of our Southwest. The rainfall around Cape Town is only about twenty-one inches, but it occurs in the warm season, whereas in general the rainfall of California occurs in the cool season. This difference may not seem great, but constitutes a material variance, and, as a result, certain plants from the Cape seem unable to accommodate themselves satisfactorily over here.

In the garden at Cape Town there were some beautiful eucalypts with gorgeous red and yellow flowers. Professor Macowan helped me to collect a quantity of seed, some of which was sent to the Golden Gate Park in San Francisco, where another fine Scotchman, John MacLaren, had already reclaimed the sand dunes and had begun to create a magnificent park. Years later Mr. MacLaren took Mr. Lathrop and me to see these splendid eucalypts in bloom.

I was on the lookout for any native forage crops, and pricked up my ears when Professor Macowan spoke of a native bush called the "Karoobosch"

(*Pentzia virgata*). He said that it furnished excellent sheep pasturage and that its slender branches took root, bound the sand, and prevented gullying and also withstood drought unusually well. I sent seed home and the last time I saw this plant was near Superior, Arizona, where Doctor F. J. Crider of the Boyce Thompson Southwestern Arboretum was experimenting with it in an effort to check erosion. The plant is a composite with small heads of yellowish flowers, a rather nondescript-looking little thing, but its habit of striking roots wherever the tips of its branches touch the ground is extremely valuable.

Professor Macowan was also proud of a row of shade trees which he had planted near the Parliament building in Cape Town in an exposed situation, where they were whipped by the continuous winds blowing down from Table Mountain. The tree is known as the Kaffir Plum (*Harpephyllum caffrum*) and is distantly related to the pistache and mango. The seeds which I sent in grew well in southern Florida and became attractive trees which resisted the hurricanes well, but the plums were disappointing. I have one on my own place, and when I put my hand on its strange, rough bark, a picture comes back to me of the days I spent with the dear old Scotchman in his gray top-hat wandering through the Cape Town garden.

Cecil Rhodes had died before we visited South Africa, but his will provided that his residence, "Groote Schuur," be preserved as a residence for the Premier. It was a fine old Dutch country house with well-kept grounds. Behind the house was an amphitheatre of considerable size, entirely filled with the most gorgeous blue hydrangeas I have ever seen.

I wandered around the place until I found the gardener who told me of Mr. Rhodes' interest in pasture grasses. Several years before his death, he had made plantings of *Chloris gayana,* from which he had distributed quantities of seed throughout the colony. The gardener believed that the grass was unusually drought-resistant, although I did not consider that the conditions under which it was growing really indicated this. However, I was grateful for the bag of seeds which he gave me and sent them in to Washington for trial. C. V. Piper interested himself in the distribution of this grass, and it became well established both in Florida and in California, and showed possibilities as a hay crop.

Twenty years later I was invited to make a speech at a little town in the Imperial Valley of California, before a group of settlers, men, women and babies. The speaker who preceded me gave a long and most enthusiastic account of "Rhodes" grass for meadows on irrigated lands. He did not know where it came from, and was much surprised when I prefaced my own speech by describing Cecil Rhodes' country house and the planting of Rhodes grass from which had come the first seeds twenty years before.

I spent an afternoon on Table Mountain among those unusual plants, the Proteas. The most familiar one is the Silver Tree (*Leucadendron argenteum*), called by the Dutch "witteboom." Both upper and lower surfaces of the leaves are densely covered with white hairs, giving the tree a very striking appearance. This species has been a favorite in our greenhouses but, like others of the *Proteaceae,* requires special treatment for its cultivation. The family is mostly confined to the southern hemisphere, to those land masses nearest the Antarctic. The Grevillea and Macadamia have done surprisingly well in Florida and California, but, considering their beauty, the Proteas are sadly lacking in our gardens.

I ended my stay in Cape Town by making an excursion to see the sand-dune vegetation near Sea Point, where the dunes, like shifting snowdrifts, were covering every tree or shrub except those capable of binding the sand and growing up through it. There is a fascination about sand dunes and the types of plants capable of growing on them. Although living within sight of the ocean, these plants must have constitutions similar to the plants of the desert. The root cells must allow the molecules of pure water to enter their protoplasmic substance, but at the same time shut out the sodium chloride or salt which, except in small amounts, is toxic. This is a matter of the molecular permeability of the cell membrane. In my book *Exploring for Plants,* I described extensive attempts to bind the sand dunes near Mogador on the coast of Morocco, and therefore will not repeat the account of this fascinating field of research.

Autumn was approaching and I wanted very much to experience a winter in the southern hemisphere. Also, I was fascinated by this glimpse of a vast country nearly half a million square miles in extent. However, on the seventeenth of March we sailed for Lisbon.

Mr. Lathrop's cousin, Charles Page Bryan, had recently been transferred as United States Minister from Brazil to Portugal. When we entered the harbor of Lisbon, the rain was coming down in torrents, which was most unusual, for Mr. Lathrop had phenomenal luck in such matters and the sun always seemed to shine when we came into port. We were looking over the rail, wondering how we could protect our baggage in the open boat in which we were supposed to land, when we saw a launch approaching bearing a tall man in a black Prince Albert coat and top-hat. It was "Cousin Charlie," the American Minister, who came on board drenched through, but with the same unruffled dignity which he always displayed.

Mr. Bryan took charge of us and installed us at the Hotel Braganza, the most fashionable hotel in Lisbon. While Mr. Lathrop hobnobbed with the dignitaries, I visited the markets and shops to study the food plants of the people.

In a delicatessen store I saw some candied prunes which differed from the ordinary glacé fruits, being drier and less sticky. They were called "Ameixa doces d'Elvas," and I bought some to show to Mr. Lathrop. They appealed to him as unusually delicate confections and he thought that the California prune growers would be interested; in fact, he had a friend in the Santa Clara Valley who would like to have the recipe for their manufacture. So I went to Elvas and spent a day with the prune growers. There I discovered that only two varieties of the plum were grown, the Reine Claude, and the Green Gage. The method of candying was not entirely responsible for the quality of the ameixas, for the skin of the plums was unusually delicate and permitted the infiltration of the syrup in which the fruits were cooked.

There was considerable hard work involved in candying these plums, as each one of the fruits had to be handled separately. Since the women and girls who did the processing in Elvas were paid next to nothing, the ameixas could be sold at a price which was not prohibitive. Mr. Lathrop's friend to whom we sent the recipe had to pay a high price for labor and I do not believe ever attempted to do anything with it. I have seen the ameixas for sale in London, and still think them among the best of sweetmeats. The industry brought in thirty-five thousand dollars a year to the little town of Elvas, and I hope still persists there, for I believe in agricultural monopolies which are built upon the quality of the product produced.

In Madrid we stayed at a hotel in the Puerta del Sol, which seemed a noisy place but was the hotel where all the foreigners stayed. Outside, on the street corner, a nondescript fellow with dark glasses was selling

"Groote Schuur," the home of Cecil Rhodes near Cape Town, South Africa. Behind the house, the hillside was covered with a forage grass which has come to bear his name. His manager gave me a bag of seed from which grew the Rhodes grass now in America.

Garbanzos in bowls of water are in the markets of most Spanish countries.

Seed pods of one of the Medicagos—leguminous cover-crop plants which add to the fertility of the soil on which they grow.

Spanish peasants are still turning over their soil with old-fashioned mattocks.

papers—or rather I should say "the paper," for apparently there was only one published in Madrid at that time, and it was called the *Imparciál*. With the regularity of the fever-bird in India, he yelled at the top of his penetrating voice, "Imparciál de hoi," "Imparciál de hoi," "Imparciál de hoi." When the street noises quieted down in the evening, his cry rang out upon the night air, and its deadly monotony was terrible. Sleep became an impossibility, particularly for us, as our windows were almost over the newsboy's head. Mr. Lathrop remonstrated, but discovered that the hotel management was helpless, as the bespectacled "newsboy" had held his position on this street corner for nearly a quarter of a century.

When I visited the markets I discovered that middle and lower class Spaniards were essentially garbanzo-eaters. Although they consumed quantities of bread made from durum wheats, their principal dish contained this excellent legume, which they much prefer to corn. The markets were filled with the largest and finest garbanzos I had ever seen, and I sent some home, hoping that somewhere in the United States this legume would find a congenial home. It seems to prefer a saline soil and is capable of withstanding long periods of drought. Today the garbanzo is grown in America wherever Mexican or Spanish taste prevails, but it still lags as a major crop owing to the lack of machinery with which its cultivation could be made profitable in a country of expensive labor. In Spain, a hundred thousand acres of garbanzos were planted annually in Castile alone, and large quantities were imported from Mexico.

Through my studies of the garbanzo I learned something of the agricultural conditions of Spain, and of the incredible taxation, amounting to 23 per cent, under which the peasants struggled. The landed gentry owned enormous estates; it was said that the Duke of Medina Celi owned half the province of Cadiz. So obsessed were the peasants with the idea that labor-saving machinery would take the bread out of their mouths, that they smashed improved gang-plows and put stones in the threshing machines. Instead, they prepared their wheat fields by working them over by hand.

I returned to Spain twenty-five years after this first visit and found gangs of peasants still turning over the soil of their fields with mattocks. The adjustments which the introduction of labor-saving machinery would have made possible, and the resulting transfer of labor to the other industries which sorely needed development in Spain, had not yet occurred.

The World Was My Garden

While there certainly are tragedies connected with the introduction of machinery, there are no such persistently drawn-out tragedies as those connected with the refusal to adopt it.

I spent some time in the Botanic Garden of Madrid, where the director had assembled an unusual collection of species of Medicago, a genus of leguminous plants strangely absent from the native flora of the American continent. This genus includes the alfalfa (*Medicago sativa*), which was introduced early into the Western Hemisphere and is easily the most important single fodder crop in the world. He also had a large number of the bur clovers which throughout the Mediterranean region play an important rôle in adding nitrogen to the soil, and furnishing forage for cattle. I was delighted to obtain seeds from this collection to add to the other medicago species of which Roland McKee later made an exhaustive study.

The first question asked of any one who had visited Spain was whether he had seen a bull-fight. Consequently, Mr. Lathrop insisted that my education would not be complete without a visit to one. We joined forces with Mr. and Mrs. James Harlan who were staying in the hotel, and Mr. Harlan and Mr. Lathrop engaged front seats while Mrs. Harlan and I modestly preferred to sit farther back. However, I found it the most exciting spectacle I had ever seen.

The entrance of the bull into the ring gives a somewhat similar thrill to that of the line-up of two contending football teams in our big college stadium, but it is more spectacular and colorful. I cannot agree with those who look upon the bull-fight as a degrading sport, but I must confess that I left the bull ring after seeing horses gored and bulls killed, feeling about as joyous and happy as after a visit to the stock-yards of Chicago or after a tragedy in the theatre.

I prefer the Portuguese bull-fight in which the bull's horns are capped with rubber balls and the horses are ridden by their owners, who are proud of them, and who train them as carefully as polo ponies. This spectacle ends, not in the death of the bull, but by his being overpowered by the toreadors who, after vaulting over him with long bamboo poles, throw him down and sit on him.

CHAPTER XX

I Meet Alexander Graham Bell

AFTER A FEW DAYS in Paris, we crossed to London and I spent a
day with George Harrow, who was in charge of the trial
gardens of Messrs. Veitch and Son, a family firm like that of
the Vilmorins of Paris, and a worthy rival in the number of
generations of horticulturists who have maintained it at a high, scientific
level.

This firm had commissioned E. H. Wilson, then only twenty-four years
old, to go to China and procure seeds of *Davidia involucrata*, a tree
which had been reported by Augustine Henry and others as being one
of the most beautiful trees in the world. When in bloom it is covered by
immense, white leafy bracts as large as one's hand. Wilson had been in
China three years and, in addition to the Davidia, had sent in seeds and
cuttings of twelve hundred different species of Chinese plants. Harrow
showed me the whole collection, which included maples, pears, Vitis and
Prunus. Among them were seedlings of the Davidia, and I naturally
wanted to find out what the chances were of its growing in America.
Mr. Harrow had been much disappointed to find that a temperature of
twenty-three degrees (or as the English express it, "nine degrees of frost")
had killed the tips of the young seedlings even after they had dropped
their leaves. Since temperatures below 23° occur throughout the United
States with the exception of limited sections of California and South
Florida, the chances of the Davidia becoming naturalized in America
seemed rather slim. As the years have gone by it has, I believe, found a
congenial home in a few places in America, and has proved hardier than
these first observations seemed to indicate.

We had ten days in London and I spent three of them securing information regarding the rôle of the horse bean (*Vicia Faba*) in the daily

ration of the horses which drew the hansom cabs and carts of London. It is with a touch of melancholy that I remember this episode which seemed an important investigation at the time. Certainly nobody dreamed that a horseless era was at hand. None of us could believe that the occasional, noisy contraptions which sped through villages at the ungodly speed of twenty miles an hour would multiply and overrun the earth, supplanting the intelligent horses behind whose well-groomed flanks millions of people were riding in style.

Doctor Augustine Henry, the great authority on Chinese plants, had returned to England and was living near Kew Gardens. He was a fascinating person, rather small in stature, with the nervous manner which I have often noticed in scientific explorers who are more accustomed to plants than people. He received me most cordially and helped me to fill one of my notebooks with information, even writing in it many Chinese characters and notes of his own. During the twenty years which he had served in the Imperial Chinese Maritime Customs, he had spent a large part of his salary on his scientific studies of the plants of China. He gave me a pamphlet which he had published and distributed in 1893 called "Notes on the Economic Botany of China." It began with the following significant sentence:

"Missionaries and others living in the interior are often in a position to make inquiries concerning the natural products of China, the results of which would be of great service to science." In the pamphlet he encouraged his readers to send him specimens of dried plants, drugs, woods, dyes, etc., which he would forward to Kew. It was a remarkable little book, giving the scientific names and Chinese characters of a large number of plants, together with discussions regarding their uses and habitats.

I was delighted to possess this article, which was out of print, and later it proved so useful that the Department of Agriculture had it reproduced photographically for me, so that the library could retain the original copy on its shelves.

My main object in calling on Doctor Henry had been to discuss exploration in China. I asked him whether he would consider a proposition to return there. He replied that Professor Sargent had already offered him a thousand pounds a year and his expenses if he would go back and collect for the Arnold Arboretum, but that he had written saying that he was tired of China and did not want to return. He was, however, most enthusiastic about the possibilities of the western provinces, particularly Yunnan and Szechwan. He declared that these provinces "are immense plains irrigated by many rivers; seven crops a year are raised; and no botanist has touched this region." Doctor Henry also told me of wild pear and peach varieties, hardy bamboos, eighty species of the genus rubus to which the blackberry and raspberry belong, and a persimmon in Peking, which the dealers ripened by puncturing the skin with a small stick to let in the air. I was particularly anxious to know if he had seen Tung oil trees growing. He knew it as *Aleurites cordata* and I took down the following note at his dictation:

"The Tung oil tree (*Aleurites cordata*) is cultivated about Ichang as an orchard tree and grows on waste land where nothing else will grow. A strictly mountain plant, stands the snow, but is generally grown in the citrus belt. The rains come in April, May and June. Hardier plants of the species are found high up in the mountains, where it is much colder."

He called my special attention to a tree called *Eucommia ulmoides* which he said grew like a poplar in England. He advocated planting it because of a gum which was contained in the leaves and bark. Later we distributed this tree throughout America, but, although the gum was so abundant that you could pull the leaves apart and the individual pieces would remain attached by delicate strands of latex, the gum has not been found to have commercial value. He disappointed me by saying that the home of the evergreen oak (*Quercus cornea*), which I had seen in Hongkong, was unknown.

The industry of this unusual man was evidenced by a manuscript of his which he showed me—a dictionary containing 20,000 Chinese names of plants, including references from the principal Chinese botanical books. Moreover, he told me that he was working on a dictionary of the Lolo language, the Lolos being a simple, independent people living in the interior of China, about whom little was known.

That afternoon spent with Doctor Henry had far-reaching results. The information which he gave me made a great impression upon me, and subsequently, when I was directing the plans and policy of the Office, it determined me to send explorers into China.

On my way to Washington, I stopped at New Brunswick to see Uncle Byron. When I left home he had been primarily concerned with the diseases of farm and garden crops, but I found that he had transferred his interest to the new subject of genetics. At the moment, he was engaged in breeding varieties of sweet corn and had made crosses between the Black Mexican, the sweetest of all corns, and the Country Gentleman, already a standard variety. As I ate the black and white kernels, I did not suspect that the corn plant would prove to be one of the best subjects with which to study the laws of heredity. Some of the most brilliant minds in America were soon to be engaged in an attempt to solve the problem of the origin of corn, and the laws which govern the inheritance of its characteristics.

It was the end of August, and Washington was boiling hot when I returned there, again an unsalaried employee of the Office of Seed and Plant Introduction. During my absence the organization had come under the direction of A. J. Pieters when Ernst Bessey had resigned to go to the Michigan Agricultural College as Professor of Botany. I became one of the Agricultural Explorers attached to the Office, and settled down in a small room in the ugly red brick building which was torn down as soon as the first wings of the present white marble "Palace of Agriculture" were erected.

The Department of Agriculture had grown enormously since 1889 when I first climbed the stairs to the laboratory under the mansard roof. Instead of a few hundred, there were now over four thousand employees working in the various sub-divisions of the Department scattered all over the city. In fact, this branch of our government had become one of the greatest research centers in the world.

It is an interesting thought that our Department of Agriculture had its origin in the far-sighted wisdom of two of our greatest patriots. George Washington first suggested the organization of a branch of the national government to care for the interests of farmers; and Benjamin Franklin, even when in England as agent of the Commonwealth of Pennsylvania, sent home silkworm eggs and mulberry cuttings in an effort to start a silk industry here. After the Revolution, many representatives of the newly formed United States Government followed Franklin's example by sending home plants until, in 1839, Hon. Henry L. Ellsworth induced Congress to make a small appropriation to attempt the distribution of the seeds and plants thus received, with information concerning them. This experiment was successful, and such activities increased until, in February,

I Meet Alexander Graham Bell

1889, the Department of Agriculture was raised to first rank and accorded a seat in the Cabinet.

In Washington's last message to Congress in December, 1796, he said,

> In proportion as nations advance in population, the cultivation of the soil becomes more and more an object of public patronage. Institutions grow up supported by the public purse. . . . This species of establishment contributes to the increase of improvements . . . by drawing to a common centre the results everywhere of individual skill and observation and spreading them thence over the whole nation.

But to return to 1903!

Even when I had first come to Washington in 1889, the Department had different ramifications and divisions. Later these were subdivided until a maze of organizations resulted. Below the Secretary and his purely administrative force, came the great scientific Bureaus, such as the Bureau of Plant Industry, of Animal Industry, Entomology, Forestry, Weather Bureau, etc. These in turn were composed of "Divisions," "Offices" and "Sections." The competition on the part of heads of the respective units was rather keen at times in their desire to see their unit raised in social standing from a Section to an Office, or from an Office to a Division. Increases in the appropriations for each unit were either asked for by the Secretary, or by some Congressman in a special bill.

When I first joined Doctor Galloway's group, it was too new and small to be called either a Division or an Office; so it had been named the "Section of Vegetable Pathology." Only one group, that of Animal Industry, had been elevated to the status of "Bureau" in 1889; all the other units were still Divisions, Offices, or Sections. One by one these increased in importance and new activities developed side by side with them.

The Weather Bureau was transferred from the War Department to the Department of Agriculture in 1891. Six hundred new stations were added within a short time, and plans were perfected as rapidly as possible for increasing the usefulness of the Bureau by extending the system of frost, flood and storm warnings.

The Bureau of Chemistry grew from a small beginning in the basement to an important organization employing scores of chemists in an effort to solve problems connected with our various plant industries, such as the sugar beet, cane sorghum, cotton, and a dozen others. This Bureau also became involved in the pure food campaign to abolish the use of harmful preservatives and coloring matter in food products. Their investigations culminated in the Pure Food Laws to whose passage Doctor Harvey Wiley's "Poison Squad" contributed largely by bringing the facts before the public.

The laboratories of the Bureau of Animal Industry were evolving new vaccines while the Bureau was also working over the study and elimination of hog cholera, bovine tuberculosis, and Texas cattle fever.

Milton Whitney had built up a Bureau of Soils which aimed at a complete analysis of the soils of the country and the mapping of them so that every farmer would know what type of ground he was farming. Although this elaborate charting proved too general in character to aid individual farmers as much as had been hoped, it served to educate the public mind about the existence of soil types, and was therefore ultimately of real service.

The Bureau of Entomology was assembling the greatest collections in the world of insects injurious to farm crops and to man himself. At the same time, the entomologists were educating the public in the use of insecticides and were studying the fly and mosquito plagues; they also had taken up the question of beneficial parasites which would prey on the injurious pests. The difficulties encountered in attempting to prevent introduced insect pests from sweeping through the country finally crystallized in the quarantine laws and their administration through a Federal Quarantine Board.

In fact, all of the units composing the Department were doing valuable work, and a volume could and should be written describing their activities. It should be interesting for taxpayers to know more about where their money goes!

Of course, to me, the most important of all the units composing the Department was (and is) the Bureau of Plant Industry, which was made up of various divisions, including the Office of Seed and Plant Introduction about which my life and thoughts revolved. Other separate units included in this Bureau dealt with fruits, their growing, shipping and storage; with cereals, forage crops, vegetables, plant breeding; tropical plant investigations, including cottons and Mexican corn; systematic botany and a dozen others.

The old Division of Plant Pathology and Physiology, with which I made my debut, now formed the center of a new Bureau with Doctor

I Meet Alexander Graham Bell

Galloway still as the chief. The scope of activities had increased tremendously since the days when only one bacterial disease of plants was known—pear blight. New diseases caused by bacteria or fungi seemed on every hand and the technique for their detection and diagnosis was being rapidly improved. More efficacious methods for their treatment and prevention were also being devised, although many of the diseases proved baffling. The problem of controlling their spread by regulatory methods had appeared on the horizon, and was forcing the pathologists—unwillingly, it should be stated—to join the entomologists in quarantine procedures which we of the S. P. I. regretted, as we felt they would affect the spirit of the whole organization.

Soon after my return to Washington that fall, I met Gilbert Grosvenor, the young editor of the magazine published by the National Geographic Society. He had heard of my trip up the Persian Gulf and asked me to address the Society on my expedition to Bagdad. I enjoyed his amazement at finding that I did not possess a long beard. For some reason, he had confused me with Professor A. S. Hitchcock, the grass specialist.

I was much flattered by this invitation to lecture, and spent anxious hours arranging my lantern slides and preparing my notes. I do not remember much about the actual speech except my pleasure in hearing that Alexander Graham Bell was present in the audience. Mr. Bell was Ex-President of the Society and his older daughter was married to Gilbert Grosvenor. Although I could not know it, it was to be my good fortune to join this interesting family circle. The chain of events which led from Bagdad to Washington and my meeting with Gilbert Grosvenor has continued throughout the rest of my life.

Soon after the night of my lecture, I received an invitation from Mr. Bell to attend one of his "Wednesday Evenings"—probably the most interesting social events in Washington at the time. There I met not only Mr. and Mrs. Bell and Mrs. Grosvenor but, even more important to me, their younger daughter Marian.

I have always been grateful that I first met Mr. Bell at one of his "Wednesday Evenings." These affairs had an intimate quality, and were more a forum than a mere reception. The handsome library created a suitable background for the distinguished men who assembled there. The walls were lined with books of travel, biography, and general literature, and the woodwork was of carved teak, brought from India. A stunning painting of the Temple of Isis at Philæ was lighted from the frieze by concealed lights—a lighting effect quite new and original then.

Mr. Bell was at his charming best on these occasions, for he enjoyed his guests, drawing them out with courteous and interesting questions. You were conscious of his dominant personality the moment he entered a room; his thick grey hair curled back from a high, sloping forehead, and he had a full beard and extraordinary eyes, large and dark, under heavy eyebrows. Mr. Bell was tall and handsome with an indefinable sense of largeness about him, and he so radiated vigor and kindliness that any pettiness of thought seemed to fade away beneath his keen gaze. He always made you feel that there was so much of interest in the universe, so many fascinating things to observe and to think about, that it was a criminal waste of time to indulge in gossip or trivial discussion.

On Wednesday evenings, when twenty-five or thirty scientific friends had assembled, Mr. Bell, with tender courtesy, would escort his aged father to a comfortable chair in the center of the room. The old gentleman, Mr. Melville Bell, lent an added dignity to the meeting as he sat there smiling benignly.

On my first Wednesday evening, after Gilbert Grosvenor introduced me to Mr. Bell, I settled myself contentedly in a corner. I remember feeling immediately at ease, as one does with any really great and simple character. As the years rolled by, this belief grew to a happy conviction. To this day, the mere thought of Mr. Bell stimulates and enheartens me. It was one of the great joys of my life that I had the privilege of knowing him so intimately.

The next link in the chain of my romance was the result of a streetcar ride to Chevy Chase which I took one Sunday morning in order to enjoy a walk through the woods. On the car, I found myself beside the noted horticulturist, Peter Barr, a burly Scotchman known in gardening circles as the "Daffodil King." Better yet, he possessed the heartiest, most infectious laugh I ever heard.

Peter Barr was on his way to the Hubbard estate to see Peter Bisset, the head gardener there. "He knows more about water gardens than any one in America," Barr said. "You had better come along and meet him."

I had been in Washington so little that I had not heard of Gardiner

I Meet Alexander Graham Bell

Greene Hubbard, whose daughter married Alexander Graham Bell when he was a young inventor struggling to gain recognition for his invention of the telephone. Mr. Hubbard had great faith in Graham Bell, and gave him both financial and moral support, founding the first telephone company in order to bring the invention into practical use. Mr. Hubbard had died several years before.

After a walk through the garden, Peter Barr and I went to the house and were greeted by Mrs. Hubbard, a tall, graceful woman with a lovely, intelligent face. Mrs. Hubbard had heard of the "Daffodil King" and was charmingly cordial to us both. It was a most delightful way to meet one's future grandmother-in-law, and I quite fell in love with her even before my affections became the property of her granddaughter for life!

Altogether, it was a wonderful time for me, that spring of 1904. Washington was full of interesting people and events, and I was fortunate enough to have the privilege of enjoying both.

Theodore Roosevelt was in the White House and his famous Tennis Cabinet was the talk of Washington. Men interested in such fields as mine were delighted to have as Chief Executive a man who thought in biological terms and was familiar with the importance of preserving wild life. We realized that he would take a sympathetic view of our problems, many of which had been brushed aside by previous Presidents who had no conception of biological matters.

Both the young Civil Service Commissioner, Alford Cooley, and Gifford Pinchot were members of the Tennis Cabinet, and through Cooley I came into contact with the President.

Some time before, at Secretary Wilson's request, I had investigated Gifford Pinchot's qualifications for the position of forester, and had reported favorably about him. During my absence from Washington, Pinchot had started his conservation propaganda to save the forests of America, and I was much interested in the quite spectacular success he was having. In fact, I was somewhat overawed by the rapid growth which the Bureau of Forestry had made under Pinchot's direction, and was also much disappointed to find that his interest in the purely scientific phases of forestry did not seem very deep. On the contrary, the men he was gathering around him were not as much interested in the trees which composed a forest, as in the propaganda for their preservation. There was already considerable friction between the Bureau of Forestry and other more scientific Bureaus in the Department of Agriculture. I was particularly distressed to learn that Pinchot had placed the Populist professor of Economics, who had nearly wrecked the Kansas State Agricultural College in my father's day, as editor of a forestry magazine.

The Forestry Service was not the only division of the Department involved in a controversy. As a result of my investigations in Bavaria, quite an argument was going on among the brewers regarding the comparative values of different types of barleys and hops in their effect on the quality of American bottled beers. It therefore became advisable for me to go west to meet representatives of the Schlitz and Pabst brewing companies in Milwaukee, and of the Anheuser-Busch Company in St. Louis, to discuss the Hanna barley from Moravia, the Bavarian hops, and the procedure advisable in order to introduce them into cultivation in this country.

At the time, our brewers were using the six-rowed barleys because the American farmers found them better for the general market. Barley for making beer was merely a by-product with the farmers. It had only recently occurred to any one to study the differences between the types of grain. In fact, I found that, although this idea had occurred to the Department of Agriculture, it was still beyond the understanding of most of the brewers themselves.

When I was in Sweden in 1900, Doctor Hjalmar Nilsson demonstrated to me the great difference which there is botanically between two-rowed and six-rowed barleys, telling me that they were classified as distinct botanical varieties, but that only a study of the kernels would show the difference. I soon understood why the European brewers preferred the two-rowed type. In the brewing process, there was an excess of certain proteid substances in beers made from the six-rowed grain. These substances were difficult to eliminate and caused an objectionable cloudiness in the beer.

When one looks at a two-rowed head of barley (*Hordeum vulgare* var. *disticum*) from the end, one sees but two distinct rows of kernels. With the six-rowed types (*Hordeum vulgare* var. *hexasticum*), there are two large side rows of straight, full kernels and four smaller rows of curved kernels. In a handful of barley the presence of crooked kernels immediately proves the presence of six-rowed barleys. However, I found that up to this time the brewmasters had never looked critically at their grains through a hand lens. They felt that they knew everything there was to know, and could judge the quality of the grain by appearance and feel and by a chemical analysis of the starch and protein content, regardless of the hereditary uniformity of the barley or the proportion that the scutellum bore to the whole grain.

In the making of malt, the barley is sprouted on special floors kept at a certain temperature. As soon as the little rootless and sprouted stems reach about half an inch in length, the barley is "killed" by roasting. After water is added, the malt-sugar is extracted and under treatment becomes the thick, molasses-like substance called "beer wort." On this beer wort, special yeasts feed, producing the alcohol and flavor of the beer. The amount of sugar in the roasted barley kernel is in proportion to the amount of starch in the live barley.

The controversy in the brewing world therefore resolved itself into the question of the relative starch content of the two- and six-rowed varieties of barley. I had discovered that the superiority of the Hanna barley was recognized in Europe, and had sent quantities of it home, and it had been distributed. But in America there was no region specializing in growing brewing barley. Our farmers grew the grain primarily for feed, and continued to prefer the six-rowed types, for they contended that there was more feed value in the high protein barley than in the two-rowed, which had more starch.

Eventually, in 1903, the shipment of Mariut barley, which I had arranged for, arrived from Egypt, where it grew on the saline soils about Lake Mariut (or Maryut) which had been flooded by Napoleon in 1801 during the siege of Alexandria. This barley became a commercial product of our Southwest, where it has held its own for thirty years in competition with all sorts of selected types, and now is grown on 233,000 acres of land, mostly in California.

From 1903 to 1906 I spent much time and thought on the brewing industry. With Doctor Albert Mann and H. V. Harlan, who later became the world authority on barley, I studied such matters as cell structure and cell chemistry in an attempt to understand the function of the scutellum of the barley kernel. The scutellum is the shield-shaped organ which fastens the embryo to the kernel. When the barley begins to sprout, the inner layer of the scutellum pours out diastase, which converts most of the starch of the grain into malt-sugar. This is very important, because, unlike starch, sugar is easily soluble in water, and can be extracted; furthermore, it is fermentable by the yeasts of beer. In other words, while the barley is on the sprouting flour, the diastase manufactured by the scutellum converts the starch into sugar, which can be made into the molasses-like beer wort which is the basis of all beer. Hops are used merely in the rôle of preservative and to add a bitter flavor.

During Prohibition, I imagine few of the younger generation thought much about barley and the part which it has always played in the life of man. I remember, in the ruined palace of Knossos, looking down at the huge earthen jars which had been used by the kings as receptacles in which to store their scorched barley; in the Museum that morning I had seen ancient Cretan jugs ornamented with designs of barley and at my feet were dry hills of barley stubble grown by the Cretans of today. If we could trace their ancestry, barley and beer would probably take us to Neolithic man.

After many exhausting conferences with the brewers, I attended a meeting in St. Louis of the Association for the Advancement of Science, of which I had long been a member and where I met many old friends.

I have never been an organization man, preferring to spend my free time with a few congenial individuals instead of in the continuous hand-shaking necessary when one frequents crowded places. At this St. Louis meeting the atmosphere became hectic from the intense personalities present. Almost against my will I was drawn into the meetings of the new American Breeders Association, and found myself in a kind of whirlwind of optimism which seemed to surround the personality of Professor Willett M. Hays, whose work in the breeding of cereals had already attracted wide attention throughout the United States. He was a large man, and his mind was so full of ideas that they sometimes seemed to tumble over each other when he attempted to express them.

I took but a minor part in the meetings, however, for the plans of the American Breeders Association were on such an extensive scale that I felt some skepticism with regard to their success. To Professor Hays the outlook seemed entirely clear, and I listened until early in the morning to his world-wide schemes "for the improvement of plants and animals by the development of expert methods of breeding."

Wherever these plans dovetailed with those of the Section of the S. P. I. for the introduction of new varieties of cultivated plants, I entered into them most heartily. The securing of plant-breeding material for members of the Association, as for all plant breeders, became an important feature of our plant-introduction program.

William Saunders, head of the Canadian Department of Agriculture, was also in St. Louis and I spent much time with him. He had begun his work on the improvement of the Red Fyfe wheat, then the popular wheat in the great Northwest, and had introduced and bred with it the Ladoga wheat from Russia and the Gehun from 11,000 feet elevation in the Himalayas. He had already produced his "Preston," precursor of the famous "Marquis" wheat which his son Charles later developed in the breeding plots which his father had established in Ottawa.

Professor Saunders was a most congenial personality, and our meeting in St. Louis led to the freest possible exchange of introduced plant species between Canada and the United States.

I had expected to find it difficult to settle down to a desk in Washington, but on the contrary it proved so fascinating that there were not hours enough in the days or nights in which to accomplish all there was to do. Beside each day's mail, and routine, and reports on my travels to be written for future reference, there was the world-wide field of plants still waiting to be introduced.

I had fondly imagined that I would have time to sit down, write bulletins and arrange my photographs. Instead, a stream of visitors began to pour into my office, a stream which increased into a flood during the next twenty years. One of my first visitors was Frank Carpenter, and he was the cause of many who came later. He was beginning his syndicated letters, a new idea in journalism out of which he made a fortune and a reputation. He wrote dozens of articles about new plant introductions and I gave him free access to my collection of photographs. His articles were headed with catchy titles, such as "How the dimple got into the plum," and they soon attracted public attention and brought us bags full of "fan mail."

The Weather Bureau was the product of the genius of Cleveland Abbe, who had the reputation of being able to dictate two letters at the same time to two separate stenographers. He had already gathered together an immense amount of data on the climate of the country. It seemed to me that this information could be charted on maps which would be of great use to us in distributing plants. As it was, the data were so scattered through their files that I had to spend many hours at the Weather Bureau to get information about any particular locality.

Professor Alfred J. Henry, the weather forecaster of the Bureau, listened sympathetically to my complaints and to my proposal that a bulletin be prepared summarizing not only the usual data with regard to temperature, precipitation, mean humidity, sunshine, etc., but also including the dates of the occurrence of extreme temperatures, particularly the extreme low over a long period of years. After much urging, the Chief of the Bureau, Willis L. Moore, consented to the preparation of this publication, and "Bulletin Q" was the result. Contrary to Mr. Moore's prediction, this bulletin was one of the most popular publications issued by the Bureau. It proved of the greatest use to us, and frequently prevented the distribution of plants to regions where the climate would have proved totally unsuited for their culture.

CHAPTER XXI

A Grand Tour of These United States

I N VIEW OF the numerous introductions which were coming in to
be tested in different parts of the country, it seemed advisable for me
to spend the summer in an inspection tour through the West. I
planned to visit the experimenters who were already testing plants,
and also to make arrangements with other experimenters who had the
facilities necessary for such work.

When I left Washington in June, one of my main problems was to find
some one suitable to send as an explorer to China, and my first stop was
at Boston, where I found that Professor Sargent had already engaged the
services of E. H. Wilson to collect trees and shrubs in China for the
Arnold Arboretum. I made tentative arrangements for Wilson to co-
operate with us in the exploration of that vast country. Sargent's interests
were chiefly in the wild plants, whereas we were equally interested in
varieties of the cultivated crops.

On my way west, I spent a memorable day with Mr. Bryan Lathrop,
who was the head of the Chicago Park Board. He had excellent taste and
was convinced that America should develop its own types of landscape
gardening along naturalistic as opposed to formal lines. He introduced me
to Mr. O. C. Symonds, who was in charge of Graceland Cemetery and had
accomplished some of the first successful moving of large trees. It was
his contention that, in landscape gardening, the plants native to a region
should compose the major part of any landscape, and this idea sank deeply
into my mind and affected everything that I have done in the way of
landscape planting.

Mr. Bryan Lathrop showed me a friend's place at Highland Park where
a bit of open prairie had been transformed into a landscaped estate. It
was the first demonstration that I had seen of a small place so cleverly

landscaped that it appeared to be several times larger than it really was. Some exotic plants had been used, but the colors of their foliage were carefully blended into the predominating background of native species.

This small "oasis" at Highland Park created a dissatisfaction in my mind with the formal gardening then practised in America, using unrelated collections of garish shrubs or trees in artificial "beds" and specimen plantings. I had noticed while at the Arnold Arboretum that Professor Sargent disliked variegated plants, and I found Mr. Lathrop shared his abhorrence and would not have one on the place. I was much impressed by his explanation that spots of white in the landscape attracted the eye and were as bad as a stray piece of paper in the middle of the lawn.

After a brief stop to see the wheat-breeding experiments of Professor Willett M. Hays, and make arrangements for testing the varieties of grain then coming in from different parts of the world, I continued west.

One of our correspondents, to whom we had been sending high-altitude grains to test, was a German named Meyers. My train drew into the little town of Montpelier, Idaho, late in the afternoon, and I was met by a stocky man with a bushy beard, who was waiting with a team and wagon. He said it was only a short distance to his place, but he drove on and on for hours, assuring me, each time I asked, that we were approaching our destination. Finally, as the sun was setting over the mountains, we drew up in front of the little log cabin which he and his sons had built. His wife had dinner waiting for us, and she too spoke little English. Meyers proudly displayed the year book of the Department of Agriculture, and a Sears Roebuck catalogue; patting them both affectionately, he called them his bibles.

On the long drive through the mountains I had been depressed by the dilapidated, shanty-like homes and run-down, untidy appearance of the places which we passed, but here was a well-kept homestead with flowers growing at the doorway; a neat vegetable garden; rows of currant and gooseberry bushes; and ornamental shrubs planted here and there. Meyers and his wife were happy and interested in their home and took me out in the dusk to show me their barley, alfalfa, and oats, and a little hop yard which they had set up. They were particularly proud of a home-made fanning mill with which they winnowed the grain after they threshed it with hand flails.

A Grand Tour of These United States

As we sat chatting about the supper table, enjoying some home-made beer, Meyers told me how he had happened to come to this lonesome spot in the mountains of Idaho. He had been a weaver in Dresden, he said, and on two separate occasions some Mormon missionaries had come there and talked to him about this wonderful country, urging him to emigrate there. Their persuasion finally prevailed, and he packed up his belongings, which fortunately included hand tools and seeds from his Dresden garden. When he reached Idaho with his wife and three young children, he was bitterly disappointed to find that his neighbors were stock raisers instead of farmers, and were convinced that crops could not be grown at that altitude. Furthermore, the neighbors were not in the least interested in attempting to grow anything.

The first year after Meyers' arrival, he and his wife spaded up an acre or so and planted it with barley, during which process they had to endure the ridicule of any neighbors who drove by. They harvested the grain with a cradle, such as was then universally used in Germany, and during that first year subsisted on the crudest of barley cakes and the few vegetables they had managed to grow. Soon they acquired an old horse and harness by working for the settlers, and progressed little by little until they now had what was easily the most valuable farm in the community, in fact the only one which could be called a farm. The check which they had received from the Department of Agriculture in payment for the high-altitude barley they had grown and sent to Washington was the only money they had seen for years, and they were overwhelmingly grateful for the assistance it had been.

When I inquired about their sons, I was amazed to hear that one of them was in Florence, Italy. I could not imagine what a son of this poor family was doing in Florence, and it took me some time to discover that he was there (largely at his father's expense) as a missionary sent to convert the Italians to the Mormon faith. Mr. Meyers explained that his son had been chosen for this position by the church. It was a great hardship for Meyers to lose the services of his son on the farm and to have to support him in Italy, but he was obliged to do this, as his neighbors were all Mormons and it was expedient for him to remain in the Mormon church, for the present at least.

I spent that night and the next day with the Meyers family, fascinated by the charming home which they had created with their own hands.

The World Was My Garden

A stream ran through the place, furnishing them with delicious mountain trout, and their pantry was filled with jellies and jams from the
berries which Mrs. Meyers grew in the little garden.

Through my good fortune in being able to speak German, I was able
to become more than a mere passing acquaintance. Meyers drove me back
to the railway station, and when he said good-bye, his voice broke and
tears came into his eyes. I was the only person he had seen since he came
to America who had taken the slightest interest in him or his labor in the
mountains in Idaho.

I should like to erect a monument to those pioneers whose intelligent,
self-directed toil transformed the great western ranges into a land of
farms and gardens. Surely such lives deserve memorials as much as those
of men killed in battle.

Seattle was still in its boom days in 1904, and everybody seemed to be
selling land. The climate of the Puget Sound region greatly resembled
that of Japan, and I got in touch with a Japanese nurseryman hoping that
I could promote the cultivation of the Japanese vegetable udo and in this
way encourage its general cultivation around Seattle. It grew well
enough, but, once again, there was no adequate organization for its popularization.

One of the main objects of my trip was to visit the Chico garden while
it was in its infancy, and to settle the numerous problems connected with
its maintenance and personnel. Dorsett was in charge and Chico soon
became the most important of all the plant introduction gardens.

The land had been donated by the citizens of the little town, and they
were very proud that the rich alluvial land, irrigation facilities, and
climate of that particular section of California had been chosen as a site
for an introduction garden. However, the climate in summer was distinctly hot, 117° in the shade at times, and while this temperature was
suitable in some respects for a place in which to propagate plants, it was
not by any means ideal for the personnel who had to live there.

On August 7, I reached Chico and found Marks, Beagles, and Edward
Goucher, one of the best plant propagators I have ever known, there with
Dorsett. They were all working with feverish activity to put the land in
order, construct the buildings and greenhouses, erect the pumping plant
and equip the garden to accommodate the stream of foreign seeds and

plants which had already begun to pour in on them. These men paid little attention to hours of work. Dorsett paid none. The terrific heat often lasted until late at night but seemed only to stimulate their activities. The week I spent with them was a strenuous, exciting experience. We all believed in the importance of the farm crops, garden plants and avenue trees which were being sent in, and we keenly felt the responsibility of developing this place in which to test and care for them. A volume could be written about Chico full of both plant and human interest.

Mr. Bidwell, one of the early pioneers in California, lived near by and had preserved a large tract of virgin land containing many magnificent trees, including an oak supposed to be the finest in California. It was called the Hooker Oak in honor of the famous botanist, Sir Joseph Hooker, who had once visited Chico and expressed admiration for the tree.

Mr. Bidwell had introduced various interesting plants, among them the cassaba melon. In 1904 this fruit was just beginning to appear in the markets in California, and since then it has become a feature of the autumn and winter markets throughout the country. The success of this melon encouraged us to introduce all the varieties we could find from the Caucasus and the oases of western Asia, where it is native. The winter melons on our markets today represent a product of the breeders' art to which strains of these varieties contributed their part.

In the mountain town of Nevada City, a nurseryman named Felix Gillet was writing articles for *The Pacific Rural Press* on filberts and almonds. Early one summer morning I arrived there, and walked out to Barren Hill Nursery as the world was waking from its slumbers. Gillet proved to be a real connoisseur of plants, an active little Frenchman with piercing black eyes and a perfect command of English. His laboratory was filled with almonds, filberts, walnuts, and chestnuts, and he had an interesting story to tell of the history of each fruit or nut tree growing in the garden. We spent a happy day among his plants, which included numerous hazelnuts and Jordan almonds sent him by the Office of S. P. I. As we sipped our wine in the cool of the evening, he told me his history.

When he first came, he had run the only barber shop in the place, but he had a soul above shampoos and shaves and, as soon as he had saved up enough money, he bought a completely barren piece of land on the outskirts of the town. Then, with the confidence of a man who knows that he can grow plants, he sent an order for $3000 worth of stock to a nurseryman in France. He had no money for an irrigation system and, when there was a spell of dry weather, he nearly killed himself working night and day watering his plants by hand from a well which he had had dug on the place. In the course of a few years his bare, ugly hillside became a paradise of trees and shrubs, and he began distributing rare varieties of nut trees up and down the Pacific Coast.

I became very fond of Felix Gillet and corresponded with him until the time of his death. Since then I have visited his Barren Hill Nursery several times and am glad to say that it continues to be a center of horticultural interest.

I stayed with Mr. Lathrop in San Francisco at the Bohemian Club, although most of my time was spent with the botanists and horticulturists who were beginning to assemble about the University at Berkeley. My old friend Osterhout was there, and Hilgard, the soil chemist, and Wickson, the horticulturist, each contributing his part in the beginnings of what has become a truly gigantic institution of learning.

It was the time of the "Midsummer Jinks" of the Bohemian Club, held annually in their Redwood Grove in the mountains. We sat on logs under the redwoods, below a terrace on the hillside which served as a stage. It was during the August moon, and silver light poured down through the great trees. At midnight, dressed in long priests' robes, and carrying lighted candles, we formed a procession and wound through the forest to another grove, where we set fire to a great funeral pyre symbolizing the "Burial of Care." I shall never forget those hours when Druids and Hamadryads wandered among the redwoods in the unearthly moonlight as soft music echoed among the giant trees.

As I had seen cork oak plantations in Spain, Italy, and Corsica, I visited the Santa Monica forestry station of the State University, but found that the trees were making slow growth compared with those in the parks of Fresno, and Agnew, California.

This interesting tree furnishes raw cork for which a substitute has never been discovered. Although it grows perfectly well in California, I believe that, to this day, no cork oak forest has been planted there. In fact, little seems to be known about the industry, and I was shocked to read some official correspondence which passed over my desk many years after this California visit. A man had written the Department of Agriculture asking whether a plantation of cork oaks would be profitable. In reply he was advised that, "inasmuch as not a great deal was known about the profits of cork-growing," he had better plant a few trees first and find out. If the man knew anything about oaks and their slow growth, this official advice must have revealed to him an abysmal ignorance in the Department with regard to a forestry problem. And how could a man find out by planting a few trees whether a forest would be profitable?

On September 2, I left the train at a God-forsaken little station called Mecca, California, sweltering in the desert valley between the San Bernardino and the San Jacinto mountains. It was blistering hot but nevertheless, due to irrigation, there were settlers living there and experimenting with alfalfa, melons, and barley as well as asparagus, figs, eggplants, and sweet potatoes. The Mecca Land Company was selling land at seventy-five dollars an acre and advertising the profits to be made from early melons. I interviewed all the better-class pioneers in this desert settlement, where later we established a date garden. As the years went by, many kinds and conditions of people have been attracted there by the lure of the date palm, and today superb dates are grown there on a commercial scale for the American market. There is certainly a dramatic contrast between the Mecca of today and the scribbled description in my notebook written over thirty-one years ago.

After a day at Tempe, Arizona, to see how the date plantation was progressing, I pushed on to southern Texas, which was still pretty wild country in 1904. I remember some strange personalities such as Lon Hill, whose revolver had notches cut in its handle to indicate the number of men he had killed; and Mr. R. J. Kleberg, manager of the King Ranch, where I spent a night and filled my pockets with rattlesnake rattles from a bushel bag full of them in his office.

The King Ranch was supposed to contain a million acres. I was told that King and a friend of his named Kennedy had owned a blockade runner during the Civil War, and had smuggled goods into the country through the port of Brownsville. At the close of the war, they had a tidy profit and each man established himself on a big tract of land in that vast unsettled region. Eventually the King Ranch increased to a million acres, and the Kennedy became three-quarters of a million in extent.

Mr. Kennedy, a son of the original owner of the Kennedy ranch, was a man much interested in plants. He had discovered artesian water on his place, and showed me a small fresh-water lake which had been formed by his flowing well and in which fish had begun to appear. He did not know where they could have come from.

I was surprised to find a grapefruit growing near his house, and asked him how it stood the winters, for he had said that the temperature dropped to sixteen degrees Fahrenheit. He explained that, although it was frozen to the ground, it had come right back again. The weather records ran back only about ten years but recorded a low of twelve degrees above zero, even as far south as Brownsville. I therefore felt safe in predicting that South Texas could not become citrus country. However, I had overlooked the fact that, in the rich alluvial soils of southern Texas, citrus trees make a phenomenal growth and, after being killed back by a "norther," grow up again quickly and bear unusually large crops. Twenty-four years later I photographed the same tree beside which we were standing when I made my rash prediction. In the meantime it had been killed back several times, but was again full of fruit. During the intervening years a grapefruit industry has grown up which is so vast that it may be said to be measured in square miles.

Mr. Doherty, superintendent of the new Corpus Christi-Brownsville Railroad, invited me to go down the line as his guest. He was much interested in our plans to introduce foreign plants into the great new region which his railroad was opening up.

There were practically no towns on the coast then except the old settlement of Brownsville. Corpus Christi consisted of a little shack of a hotel, a few dozen houses, and a sandy beach on which grew some scraggly tamarisks.

The line between Corpus Christi and Brownsville ran through an utterly wild landscape of scrubby mesquite and nondescript vegetation. At the end of our trip was Brownsville, which seemed to me about the most miserable shack town I had ever seen. The place was insufferably hot and incredibly dirty, while the hotel was unspeakable and the water not fit to drink. Yet this jumping-off place was filled with men determined to do something with the land and in some way make a living out of it. I again travelled over this same route with Gutzon Borglum, the sculptor, in 1927, when he and the authorities of Corpus Christi and Brownsville had conceived the idea of planting the entire stretch of 150 miles between the two places with palms and ornamental trees in order to make of it "the most spectacular coastal driveway in the world." Moreover, both places ranked as cities by then.

In 1904, I spent only one day at Brownsville, and hoped I would never have to spend another. As things turned out I spent many days there, for we took over the old fort and attempted to make a plant-introduction garden out of the reservation. But that is another story.

When I left Corpus Christi, I went to another small town, a place called Nursery, Texas. I had heard of a genial Dutchman named O. Onderdonk, who had a nursery there. He was delighted to see me, for he seldom met any one interested in plants. He had grown up in New England, but his health failed and in the early eighties he had bought a horse and carriage and driven south until he reached this dreary spot in southern Texas. The land was all owned by the cattle barons, who were greatly annoyed when they saw Onderdonk plant his place with trees and nursery stock.

The struggle between the farmers and cattlemen was as bitter here as it had been in California. The ranchers refused to sell their land to bona fide farmers, and I think Onderdonk voiced the sentiment of most of the new settlers when he said,

"The cattle barons are the curse of the country."

Onderdonk was a true plant enthusiast at heart and had gone into Mexico and brought back sixteen varieties of Mexican peaches, which he thought might prove better adapted to conditions in Texas than the American sorts. In fact, some of them have persisted there. Later, the Office sent him back to Mexico in search of more peaches and other useful fruits.

Houston was suffocatingly hot when I arrived there the 18th of September. It was the type of town which had hotels, but where every one seemed to take his meals in cheap "hash house" restaurants.

At the Texas Agricultural College, David Houston, its president, assembled the members of his staff in order that I might tell them something about cottons and the various other foreign plants in which they were interested. He also entertained me in the College dining-hall and showed me the usual courtesies extended to a visiting scientist. I greatly enjoyed meeting him and was more than usually impressed by his quiet, forceful personality. Houston had been born in North Carolina and was a graduate of Harvard. He seemed to me to be a person having great stability of character, one not easily stampeded by a new idea. Nine years later he wired my friend Waller Page, a Harvard classmate of his, asking Page to engage rooms for him in Washington. Waller stuck the telegram in his pocket and forgot it until he overheard some one remark that a man named David Houston had been chosen by Woodrow Wilson to be Secretary of Agriculture. Waller made this discovery barely in time to meet the new Secretary at the railway station.

For seven years Mr. Houston was Secretary of Agriculture, in my estimation one of the most forceful men who ever occupied that position.

The discovery that there was an almost unlimited supply of water available from artesian wells along the coastal plains of Texas and Louisiana had led to extensive rice plantations using new methods of pump irrigation. In 1898, Doctor Seaman A. Knapp had been sent to Japan to study the rice industry there, and shipped back ten tons of short-kernelled rice, mainly from the island of Kyushu. By 1904, this rice had largely replaced the long-kernel rice we had grown heretofore, and had brought about a great saving to our growers. The long kernel had broken to pieces on the power mills then in use, causing an estimated loss of over two million dollars a year in consequence.

The short-kernel Japanese rice proved more vigorous and better suited to light soils, out-yielded the Honduras rice, made a better straw for hay, and had a thinner husk. As a result, these newly irrigated regions produced so much more rice that, whereas in 1898 we grew only half enough for our own needs, it was not many years until we produced enough for ourselves and were also exporting a considerable amount.

From Houston I went to Crowley, the real rice center of Louisiana. As I was sitting in the hotel lobby the evening of my arrival, I could not help overhearing an animated discussion between some rice buyers seated near by. One of the men represented the National Rice Milling Company of New Orleans. I had never been in an American rice mill, but I had taken it for granted that the milling process would be superior to that in the Orient. I was therefore considerably surprised to gather from the conversation that in the United States not only was the rice "buffed" until the delicate outer skin was removed, but the rice was then rolled in great iron cylinders which coated it with paraffin to give it the polished appearance which the public demanded.

I pricked up my ears, for I had just been in Japan and had heard that, when rice was polished, certain valuable phosphoric compounds were removed. (Vitamins at that time were unknown.) During the Russo-Japanese War, the Japanese troops in Manchuria were fed polished rice and developed beriberi, a disease which they did not have at home when living on *unpolished* rice. Now, in this hotel lobby, I was learning that American methods of milling removed this vitally important part of the rice kernel and left an inferior product. I already knew that American processed rice was inferior in flavor.

The men beside me were complaining that the American public ate too little rice—only seven pounds per capita a year. It seemed small wonder to me, when not only were the most nutritious cells removed, but the grains were rendered even more tasteless by covering them with paraffin. I became quite wrought up about it and, when I returned to Washington, collected all the available facts and had enlarged photographs made showing polished and unpolished grains of rice. Thus armed I made a special trip to Philadelphia to present the matter to the editor of *The Saturday Evening Post.*

The editor was reluctant to publish my story, for he had recently printed an account of a four-headed wheat which the originator claimed would produce four times the yield of ordinary grain. The story proved to be a hoax, and consequently the editor was shy of all agricultural articles. However, he finally relented and published a half column entitled "The Fiction of Polished Rice." Coming as it did in the early days of the pure food campaign, the article stirred up a wide interest and brought me into contact with the pioneer agitators for pure-food legislation.

Later, in 1908, H. W. Fraser and A. T. Stanton, scientists employed by the English government, read a paper at the first meeting of the Far Eastern Association of Tropical Medicine at Manila, announcing the cause of beriberi. It was proved that the outer coating of the rice grain contains an element essential to good health if rice is to be the constant and exclusive diet. In other words, when only rice is eaten, if it has been polished and this element removed, beriberi is the result.

The contrast between the optimism of the rice planters in Texas and the pessimism of the planters of South Carolina was striking. I had not realized that the development of the Texas and Louisiana rice fields was destroying the century-old Carolina rice plantations. By the use of machinery and artesian water on the virgin, alluvial soils of Texas, rice could be grown much more cheaply than by the old-fashioned methods of the Carolinas. Already the historical "rice plantations of the South" were practically abandoned.

These ruined rice planters had appealed to the Department of Agriculture for another crop with which to replace the rice, and A. J. Pieters had begun an investigation of the rushes from which Japanese matting was made, as this formed an important article of export from Japan at the time. General E. P. Alexander came to Washington and asked me to visit his plantation on South Island near Georgetown, South Carolina. He already had several hundred acres of a native species of rush growing wild, and he thought if this were cultivated it might grow tall enough to use for mattings. The old General was a charming Southern gentleman, with courteous manners and a keen knowledge of the agricultural conditions of his State. We discussed the possibilities of rushes as a substitute crop for rice and I arranged to carry on experiments on his plantation.

This proved to be the beginning of a long and discouraging struggle against numerous unforeseen difficulties. At the start, the prospect seemed rosy, for a Mr. R. H. Sawyer of Kennebunkport, Maine, had invented a power loom with which he felt sure that he could manufacture cheaper and better floor matting than that made on the hand looms in Japan. Therefore the possibility of rehabilitating the Carolina plantations and, at the same time, developing a manufacturing business in Maine spurred us on, and we attacked the problem in as systematic a way as possible.

The Japanese rush is closely related to the common American variety except that the American rush is shorter. We hoped that by cultivation we could make our rushes grow to the length required for manufacturing purposes. In the meantime, we sent John Tull to Japan for plants of a strain of *Juncus effusus*. These we planted in the Carolinas, beginning on General Alexander's plantation and increasing the plantings as fast as possible. At the same time, we experimented with some rushes, including the Samar, a giant sedge from the irrigated lands of the Nile Delta.

Our efforts proved futile, for, with all the skill at our command, we could not make any rush grow sufficiently long for matting purposes except the Samar, which had to be split before it could be used. This presented technical difficulties which we could not overcome with the machines available.

Sawyer really had a wonderful machine and was prepared to import the straw from Japan, since it could not be grown here. However, in order to do this, he must have a reasonable tariff and this no amount of lobbying seemed able to obtain. As a last resort, we attempted to help him utilize the common reed of our marshes (*Phragmites communis*) to manufacture coarse roll matting which could be used instead of lath to plaster on. In Sweden there are houses over a hundred years old in which this reed was used in place of wooden lath under the plaster. The idea therefore seemed feasible, but before proceeding further we made a survey of the supposedly "immense areas" of the reed growing in this country. To our surprise we discovered that the supply was smaller than expected and quite inadequate. Of course this was all long before any one had thought of manufacturing metal lath.

I became so fired with enthusiasm that I built a house with my own hands and lathed it, floor, roof and walls, with matting which Sawyer manufactured. I applied cement plaster over it, and felt that the house was a credit to both Sawyer and myself. It stood for ten years, a shining example of the possibilities of rush matting, until, suddenly one summer, the floor gave way and the trunks and other effects of a Dutch diplomat who had rented my house were deposited in the cellar—much to my chagrin and his.

From Carolina I returned to Washington in October, having made almost a complete circuit of the United States. I had met some of the most interesting plant enthusiasts in the country, men with whom I was to be in almost constant correspondence for years, but I had not found any one with the necessary qualifications for a plant explorer to send to China.

The newspaper correspondents in Washington continued to find interesting stories in our work, and while they flattered us with their attention, they also embarrassed us. Every article they published increased the flood of correspondence which was taxing our slender force of stenographers to the utmost. Occasionally this correspondence put us in touch with intelligent experimenters, but most of the letters were from people more curious than interested.

About this time, Walter Hines Page was in Washington looking for articles for *World's Work,* of which he was editor. He spent an afternoon with me, going over the details of plant introduction work in his usual enthusiastic, thorough way. At the close of the interview he made some remarks about the work which I remember as vividly as I remember his optimistic personality.

"What you need in this plant introduction business, Mr. Fairchild, is a conspiracy. People won't eat fruits and vegetables which they have never heard about. The newspapers and magazines will publish articles about these introductions, for they know people like to read about them. They're news. You have the facts and the photographs, and if you let the magazines have them they will give you the free advertising which you must have in order to induce people to buy your new things as they come on the market. You could write these articles yourself providing you can present them to the public from an interesting angle. It isn't what you say, Mr. Fairchild, but the way you present it, which determines whether or not people read what you write."

I saw a good deal of Mr. Page in the years that followed and he invited me to his round table at Garden City, where I served the dasheen (taro) at a luncheon of the editors and assistants of the *Garden Magazine.*

When Wilson was elected, it was rumored that Walter Hines Page might become Secretary of Agriculture. The day after the election, I happened to meet him on the street, and told him that my colleagues and I all hoped that the rumor was true. He shook his head and said, "No, I'm not to be your Secretary."

The next day his appointment as Ambassador to the Court of St. James was announced. I never saw him again.

CHAPTER XXII

Mostly Personal

AFFAIRS IN THE OFFICE were exciting enough, but other things were happening outside which seemed far more important to me. Early in November, I received an invitation from Mrs. Grosvenor, asking me to dinner. It was a small party, and I found myself seated beside Miss Marian Bell, who had recently returned from New York, where she had been working in the studio of Gutzon Borglum. Our conversation was largely on art, about which I knew nothing but could talk a good deal, having travelled with Mr. Lathrop, who was a real connoisseur. It was the first chance I had had to talk to Miss Bell, and I was fascinated by her.

Just before the party broke up, I realized from something which Mrs. Grosvenor said that her sister was not engaged, as I had assumed inevitable. I left the house, my mind in a whirl, a whirl which has really never stopped since. It was the beginning of a part of my life which has been completely different and vastly more beautiful than anything I had dreamed possible.

Concentration proved difficult, for I no longer had only a single master, Plant Introduction. Miss Bell honored me by visiting my office and brought her grandmother, Mrs. Hubbard, to see the first flowering of that gorgeous tropical creeper (*Camoensia maxima*) named after the Portuguese poet Camoens. The holiday season arrived with much gaiety. There was the Christmas tree at Mrs. Grosvenor's, where I ruined a new coat with candle wax; a performance of "Parsifal"; a concert by Paderewski; weekly lectures of the National Geographic Society in Hubbard Hall; my collection of Javanese batiks, which Miss Bell wished to see; and experiments with vegetable dyes.

February opened with an ice-storm, a gorgeous event which transformed Washington into fairyland, and brought out all the old-fashioned

311

sleighs. For a few days the jingle of sleigh bells tinkled throughout the Capital. Just after this Miss Bell told me that she was going away.

Charles Lang Freer had offered his magnificent collection of Oriental and American art to the Smithsonian Institution, asking that a committee be sent to Detroit to examine it. Instead of the artists and art connoisseurs whom Mr. Freer very naturally expected, the Smithsonian decided to send its Executive Committee. Secretary Samuel P. Langley, Mr. Bell, and ex-Senator John B. Henderson were to judge the value of old Chinese pottery and Whistler paintings.

Mr. Bell did not pretend to have a knowledge of such things as Oriental porcelains, and had asked his daughter to accompany him. Marian was keen to go, as she realized that this particular committee would probably fail to appreciate the value of the collection, and might even refuse to accept it. In this event, Mr. Freer would give it to New York. The Smithsonian was strictly a scientific institution, and although the National Museum came under its supervision, the Museum had not yet taken over to any extent the custody of works of art.

The account Marian gave us of the visit drew a picture both humorous and pathetic. Mr. Freer proved to be a courtly, sensitive and gentle man, his life completely centered in his beautiful art treasures, while his three distinguished visitors were entirely out of their element. The scientists sat wearily hour after hour as each picture or objet d'art was brought out and displayed separately, Japanese fashion, for their inspection. The opinions of all three regarding the priceless examples of ancient Chinese art were expressed in a remark of Senator Henderson that "The things were all very well of their kind—but damn the kind!"

Marian proudly treasures a telegram which Mr. Freer sent her, indicating his appreciation of her influence in bringing about the decision of the committee to accept his collection, which is now housed in a beautiful marble building on the Mall.

Marian was returning to Washington on March 1 to be present at the Cabinet dinner given by the Secretary of Agriculture. I met her in New York, and the trip to Washington proved an eventful one indeed.

Of the hectic days between March 1 and our wedding on April 25, I remember little except one lovely evening at Twin Oaks when Mrs. Hubbard, Marian and I saw the moon through the branches of a drooping cherry tree in full bloom, as the Japanese love to do, and a tea party which

Charles Marlatt gave, in Japanese style, for an old double-flowered cherry tree in his yard.

There were few automobiles in Washington at the time, but Mrs. Bell gave an electric to Marian and we made expeditions out to the country at incredible speed, twelve miles an hour, I believe it was.

The wedding was to be an out-of-door affair between the twin oaks at Mrs. Hubbard's place, and since it was the season for the "cherry-blossom trees," as Marian chose to call them, I proposed that the bride's bouquet should be a spray of these blossoms. The spring advanced more quickly than I had anticipated, and I was obliged to have the flowers sent down from Boston on ice, as the few trees in Washington had dropped their flowers before the twenty-fifth. I was much disappointed about this, but it did not matter in the end, for Marian forgot to carry them anyway.

We returned from our wedding trip to live with Mr. and Mrs. Bell, for we were a vague young couple and did not know exactly where we wanted to settle.

In these first weeks of our married life I came to know Marian's mother as I had not been able to before, and many were the long talks we used to have over bowls of blackberries in her sitting room. We had many interests in common, and I think she was the most sympathetic listener I ever knew. Although Mrs. Bell had been totally deaf from an attack of scarlet fever when she was five years old, she was an accomplished lip-reader and had no difficulty in following a conversation with almost any one. She was full of initiative, and as I look back to those years I realize more and more how much Mrs. Bell's enthusiasm encouraged us at each turning point.

Mrs. Bell agreed at once that we should not plan to spend our lives on a city street. We both wanted to be as far out in the country as possible, so we began looking for a place beyond the suburbs of Washington.

Mrs. Hubbard's gardener, Peter Bissett, knew of a tract of land near Kensington, Maryland, through which Rock Creek flowed. It was ten miles from the Capital and a mile beyond the end of the Chevy Chase street-car and the road was terrible. He took us to see it and we fell in love with the place at first sight and bought it that summer.

Just as we began to plan its development, Marian developed appendicitis and had to have an operation, and her grandfather, Mr. Melville Bell, was taken ill and died.

The World Was My Garden

It was the thorough training in the mechanism of speech which Melville Bell gave his son that aroused in Alexander Graham Bell a deep interest in the subject of sound and equipped him for his work in telephony. Melville Bell had been well known in Scotland and England for his public readings; in fact, he told us that he had been rebuked by the elders of his church for reading aloud the works of so "ungodly" a man as Dickens, and that this had so disgusted him with the church that he had never again "darkened its doors." He was an authority on elocution and the author of many books on the subject, and was the inventor of Visible Speech. This is a system of phonetic writing which uses as its symbols abbreviated diagrams of the mouth positions required for each elementary sound. It is of great help in correcting defects of speech, in teaching a foreign language, and in teaching speech to deaf children.

One of the main reasons for Alexander Graham Bell's coming to this country was to introduce his father's system of Visible Speech. This led to his meeting with Mr. Hubbard, whose unremitting efforts were bent toward giving to all deaf children the same chance his own little deaf daughter had had to learn to speak and read lips. When Mabel Hubbard had lost her hearing, the schools for the deaf taught only the sign language and the manual alphabet, giving them no equipment to encourage association with normal people. Mr. and Mrs. Hubbard refused to have their daughter doomed to the isolation and remoteness from the hearing world which this would mean. Aided by a gifted and imaginative young teacher, Mary H. True, they built upon the slender store of speech which the five-year-old child already had, and educated Mabel with her hearing sisters.

Mr. Hubbard was largely responsible for inducing the Legislature of Massachusetts to include oral schools for the deaf in its educational program, and he was one of the founders of the Clarke School at Northampton which is outstanding among schools for the deaf in the world today.

After Mr. Melville Bell's death in July, 1905, the family went immediately to their home in Nova Scotia, where Mr. Bell's laboratories were busily turning out thousands of fragile wooden frames covered with silk for his tetrahedral kites. Marian and I remained in Washington at Twin Oaks.

While these things were happening outside the Office, matters of con-

Alexander Graham Bell as we often saw him in his study, busy over his records and statistical tables on the inheritance of longevity.

314A

Frank N. Meyer was an ideal agricultural explorer for China, as he was a great walker and thus could travel in a land of no roads.

314B

siderable moment were taking place there too. A young Hollander had come to America. His name was Frank N. Meyer, and he had been head gardener under Hugo De Vries in Amsterdam. It seemed possible that Meyer might prove to be the man we had been looking for to send to China.

"He's a strange fellow," said Pieters. "A bit erratic perhaps, for he doesn't seem to care about staying in one place. He had a letter to Erwin Smith, and Erwin gave him a job in the greenhouses, but Meyer spent all his spare time tramping around the country. He walked down to Mt. Vernon through the fields along the river, and on the way back spent the night in an old barn. Meyer told me that he heard noises around the barn and thought it must be the Indians. In fact, Meyer was surprised that he had not seen any redskins during the whole trip. Like so many European boys, he has been fascinated by James Fenimore Cooper's novels and thought the Indians still occupied the country. From Washington he went to Cuba, California and Mexico. I understand that he walked hundreds of miles in Mexico. Recently he has been at the Shaw School of Botany in St. Louis."

I was much impressed with the fact that Meyer was a great walker, for I knew that there were no roads in China, and a man must either be carried in a sedan-chair or walk if he is to get anywhere throughout the interior. Pieters wired to Meyer to come to Washington, and I remember our first interview as plainly as though it were yesterday. It was a boiling hot day and Meyer was one of those full-blooded men who had spent his life out-of-doors and perspired freely. He cared nothing about his dress. Somewhere he had picked up a striped shirt, and when he came to see me it was wringing wet and the stripes had run. But he sat on the edge of the chair with an eagerness and quick intelligence that won me in an instant. His lack of pose, his willingness to work for any reasonable sum, and his evident passion for plants, all were evident in that first interview. Meyer told me that some of the bamboos which Mr. Lathrop and I had sent to California had been planted by a stubborn plant pathologist who did not know enough to mulch them and would not let Meyer do it either. They had died in consequence, and, as Meyer told me about it, his eyes filled with tears. From that moment Meyer and I were friends, and for thirteen years I travelled with him, in spirit if not in body, through the farms, gardens, forests and deserts of Asia.

The World Was My Garden

I was anxious to introduce Meyer to Marian's family, but he was so unconventionally dressed that I tried to spruce him up a bit before taking him there to dinner. I even presented him with a tuxedo thinking that he would need it in the Orient, but he brought it back three years afterwards and dragged it out of his trunk, green with mold. He fascinated Mr. and Mrs. Bell and all who met him by his keen interest in everything he saw, and by his eagerness to learn.

As I have remarked in previous chapters, securing an authorization to travel in a foreign country has always been a major operation in the Department. Sending a man to China for three years required almost more red tape and engineering than Pieters and I were capable of. We finally managed to start Meyer off, telling him to get a receipt for every expenditure he made amounting to more than fifty cents in American money. His first accounts were loaded with piles of yellow slips like those the Chinese laundryman gives you with Chinese characters scribbled all over them.

It was early autumn before the final papers were drawn and Marian and I really owned our forty acres in the woods of Maryland. We immediately carried some plants out to start a little nursery in a bend of the brook.

The autumn leaves were turning as we made our way up through the scrub pines and dogwoods to the large white oaks which crowned the hill. The squirrels were gathering nuts from the shagbark hickories overlooking Rock Creek and a row of tulip poplars gleamed a rich gold in the autumn sunlight. Like children, we waded in the brook, chased the squirrels and built a fire of twigs beneath the oaks. When night came, we could not bear to leave, and, as we had sent out a hammock and a comfortable steamer chair, we spent our first night there under the oaks, and watched the dawn break through our own forest in Maryland.

I was appalled by my abysmal ignorance of practical agriculture when I started to create my own home. I soon discovered that through all my years of association with professors of horticulture, agricultural colleges, and experiment stations, I had developed a blind spot towards the practical formulæ of such an immensely simple thing as the making of a flower or vegetable garden. Marian, of course, depended upon me for all this knowledge which I had only theoretically, and in gardening you may as well not know anything at all as to know it in a theoretical way. If you

316

are not actually sure exactly how to do a thing in gardening, you do not know how to do it, and that's that!

It was soon evident that Marian loved every tree on the place, even the scrub pines which covered the hill on which we proposed to build our house. She flatly refused to have any of them cut down, and, not knowing where to put the house, we compromised by building a garage in a space among the pines just large enough to squeeze it in.

So afraid was Marian that the wrong trees would be cut, that long after the garage was built and we had turned it into a living room and added bedrooms, a sleeping porch, and a kitchen, a hickory tree still stood so close that the door could scarcely be opened, and spindling scrub pines obstructed the view from every window.

The struggle between us kept up for a year or so, and then I made the great discovery that, if I cut down a tree and did not tell her, nine times out of ten Marian failed to miss it. This subterfuge eventually worked both ways and reached a climax when Marian retaliated by having one of the large tulip trees cut down and every vestige removed. When I returned, I could not tell what was missing, although I had an uneasy feeling that there had been a change.

We combed the catalogues and ordered a comprehensive shipment of European trees and shrubs for our small nursery, and I wrote to my old friend Suzuki and ordered a hundred and twenty-five Japanese flowering cherry trees of twenty-five choice varieties. The English ornamental trees were disappointing almost without exception. I think perhaps the fact that they represented species already common in America led me to take less interest in them than I did in those imported from Japan.

There was no lasting pleasure in a Colorado blue spruce either. The light blue foliage refused to harmonize with the native trees, even though I gave it a prominent position near the house. I argued in its favor with Waller Page one day, but, pointing his long finger at it, he said,

"You'll move that, you'll get sick of it."

And I found that already I was tired of it. I moved it three times, until I got it out of sight behind the barn. However, it taught me a lesson that "specimen" trees belong in arboretums and not conspicuously dotted about one's home.

By the time neighbor Welch had built a road for us, and the carpenter had erected the framework of the little cottage among the pines, our shipment arrived from Japan, beautifully baled with bamboo staves.

On a perfect spring day, Marian and I motored out to open the crates of cherry trees, which had cost us, landed there, about ten cents apiece. The afternoon passed so quickly that it seemed a shame to leave our trees and go back to town. Marian suggested that we lean the loose boards up against the framework of the garage for a shelter, and spend the night. Of course I consented. We arranged our makeshift tent and were cooking our supper when a storm came up. Rivulets trickled through our board shelter but we managed to get some sleep, and the morning broke warm and bright as April mornings have a way of doing in Maryland. When my secretary arrived, we were busy planting the cherry trees. Fastened on each one was a little cloth tag, on which was painted in Japanese characters the Japanese name with its English equivalent. The planting of these trees was the real beginning of our experiences on the place which Mr. Lathrop named for us "In the Woods."

CHAPTER XXIII

Washington, Madeira, and "In the Woods"

I T WOULD REQUIRE a separate volume to give an adequate idea of the varied activities of the winter of 1905–1906—and every year, in fact. In my desk-book is the record of visitors from China, the Argentine, Persia, Mexico, the Philippines, Majorca, Bogota, Manchuria, Hawaii and of course from every state in the Union. The problems and questions were as varied as the countries from which my callers came. There was never time to be bored.

With the arrival of Doctor Albert Mann, who applied the latest microscopic technique to the barley kernel, and the visits of Max Wallerstein, a remarkable brewing chemist from New York, we felt that we were accomplishing something substantial in the improvement of American barleys. Wallerstein soon discovered and patented a process of clarifying beers by the use of a small quantity of the dried juice of the papaya plant, importing this dried juice or so-called papaïn from Ceylon. Later, I helped him establish the first large plantation of papayas in South Florida. But that is another story.

In December, 1905, Doctor W. C. Gorgas called at my office with a letter of introduction from Doctor W. W. Keen of Philadelphia. When Doctor Keen had been in the Orient, he had developed a taste for the mangosteen and had written to the Secretary of Agriculture asking why the trees could not be grown in Panama. The letter had been referred to me and I had replied that I thought that they would grow if arrangements could be made to establish an orchard there. Doctor Keen had been a noted surgeon in the Civil War and, having faith in the military, he turned to Doctor Gorgas to make something happen. In the note he wrote me he said, "I am sending my friend, Doctor Gorgas, to you, as he is going as the Chief Sanitary Engineer to the Canal Zone and perhaps can help you to establish the mangosteen there."

The World Was My Garden

I was delighted by the friendly, interested attitude of my caller. He was sailing from New York in three days and was rushed to death with official calls in Washington, but he had taken the time to see me.

"I don't know what a mangosteen is or anything about it, but I'll do anything in the world for Doctor Keen," he said.

Doctor Gorgas told me that as chief sanitary officer he would have the appointment of a great many field men. I told him that, although work on the Canal had scarcely begun, it would some day be one of the wonders of the world, and it was an unusual opportunity for a good landscape gardener to beautify the zone with gorgeous flowering trees and palms. Since it would require a great deal of time to assemble the plants, I hoped that it might be feasible for him to appoint a landscape gardener who could start a plant introduction station in Panama, where mangosteen trees and other valuable fruit and ornamental trees could be cared for as we sent them from time to time.

Doctor Gorgas agreed to this idea, and we decided such details as salary and the man's living quarters. This led to the appointment of Mr. Schultz, and in less than a year's time we had delivered into his hands twenty mangosteen plants together with other tropical fruit trees. He wrote that he could handle five hundred more trees as soon as we could get them to him. The complete abandonment of this experiment I have always charged to a lack of vision on the part of General Goethals, whose marvellous engineering mind was not interested in things which could not be blue-printed. The Plant Introduction Garden went uncared for and seventeen years' time was lost. It was not until I made a trip to Panama in 1922 and interested Governor Morrow in the idea, that a second start was made which has ultimately resulted in an orchard of mangosteens, and the splendid Experiment Garden at Summit in the Canal Zone which delights visitors today.

The summer was enlivened by the arrival of our first-born, who was named after his Grandfather Bell. He was a lively little youngster who persistently sucked his thumb.

The germ theory of disease was still in that unbalanced form which encouraged the idea that everything about the baby's basket should be sterilized, and we went the limit to prevent any possible infection. At the same time, we began our pioneering of the outdoor sleeping idea. We built probably the first sleeping porch in Washington under our big white oaks; not only was it open on all sides but the sashes of the glass roof could be opened in fair weather so that we actually slept under the open sky, while in rainy weather, with the sash closed, we could watch the drops patter down through the branches onto the glass roof.

Our winter was spent with Mr. and Mrs. Bell, but, as soon as spring came, we took the child to our little house in the woods. There we would put him in a wicker clothes basket out under the trees.

I had a theory that the most impressionable period of a human being's life is the period before he or she learns to talk—before the child enters the world of words. To utilize this theory, I filled a test tube of tough glass, which the baby could not break, with living ants and bees and fastened the cork down securely, and for many days it was the child's only toy. By coincidence, and I am far from wishing to suggest that it was anything more, my son has become an entomologist.

While I do not suppose that his plaything in any way determined Graham's bent for natural history, I do believe that the many months he spent alone in the forest gave him an opportunity to develop those powers of observation which I believe he brought with him into the world as his hereditary set-up.

Before he was three years old he startled me by saying,

"Daddy, you know some trees have gum and some don't."

He also pleased me by pointing out and naming little seedlings of pine, cedar and tulip-tree before they were three inches high. I realized that he had what many people seem to lack, a memory for form, without which I doubt if any one can be a first-rate naturalist.

E. H. Wilson had returned from China, and my attempts to arrange with Professor Sargent to combine forces for future exploration of China were unsuccessful. Cooperation between two such entirely different personalities as those of Meyer and Wilson presented difficulties aside from Sargent's and my own different temperaments. Due to my personal experiences as a plant explorer, I did not insist on any collaboration between these men, preferring to retain the loyalty and devotion of Frank Meyer, who was developing rapidly into a first-class agricultural explorer.

The entire charge of the Office had again been assigned to me, and it was growing rapidly and expanding its force of investigators in all branches of agriculture. The stream of callers increased almost daily; the details of the established plant gardens demanded constant attention; and the foreign correspondence had become formidable.

The strain was exhausting, and during the winter of 1906, my bronchial tubes rebelled against the climate of Washington. We therefore decided on a trip to Madeira and, leaving the baby with his grandmother, we set sail.

Our steamer entered the harbor of Funchal at sunrise, as a rainbow arched over the harbor. Great masses of red bougainvillea hung over the cliffs, interspersed with the silvery-green aloes growing in crevices in the rock.

We were greeted as old friends by Mr. and Mrs. Blandy, whom I had known before, and enjoyed the charming blue garden Mrs. Blandy had created, using these borages, those loveliest of all the plants of the Canary Islands. One day, when we were lunching with them, Mr. Blandy suggested our studying Funchal harbor through the large telescope on the terrace.

"It was through this glass," he said, "that my sister and Lord Kelvin carried on their courtship when he was on the boat sent out to locate the broken ends of the Atlantic cable."

When Marian said that Lord Kelvin was a great friend of Mr. Bell, Mr. Blandy told us how Lord Kelvin had decided where to look for the break in the Atlantic cable. Kelvin had gone to a dinner party in London with some young people, but his mind was on this problem and he became oblivious of his surroundings. Enjoying his abstraction, the young people had turned back the rug and moved the tables and chairs, leaving him marooned in the middle of the room, as they danced around him. For a long time Lord Kelvin sat there; then suddenly he slapped his thigh, cried "I've got it!" and strode out the door. He had worked out a scheme for measuring the electrical resistance of a definite length of cable, and this enabled him to compute the length of cable to the break.

Lord Kelvin had no faith in flying machines and was grieved that Mr. Bell was wasting time in his efforts to accomplish aerial flight. He wrote Mrs. Bell to this effect in answer to a letter from her defending the work which her husband was doing and the support which Mr. Bell was giving to Professor Langley. The letter has been published in Miss McKenzie's life of Mr. Bell, and it is interesting to recall that as great a mind

as Lord Kelvin's completely failed to foresee the coming of man's naviga-
tion of the "universal highway overhead," as Langley called it.

The weather was cold and damp in Funchal, and we huddled before
our little fireplace and shivered. Clouds hung over the peaks, and the
valleys were dark and mysterious. We made attempts to penetrate them
in hammocks carried on the shoulders of mountaineers, but the mist was
too gray and dreary. It was so evident that we were there in the wrong
season that we soon retreated and remained in the lowlands, while I
contented myself with the study of the cultivated plants on the terraces
around Funchal.

I found a number of interesting things. It was the season for the
cherimoya (*Annona Cherimola*) and great heaping baskets of them were
peddled on the streets. I have never seen finer examples of these delicious
but curiously neglected fruits. Cherimoyas rank in delicacy of flavor and
texture with the best fruits that grow, and yet so little have they found
their way into the markets of the world that I doubt whether one person
out of ten thousand in the United States has ever seen one. The fruit
grows best in an oceanic climate at altitudes of one thousand feet or so,
but nevertheless I now grow hybrids of it and every year enjoy its fragrance and delicate flavor. Some day perhaps when methods of transportation are even better than today, a cherimoya industry will develop on
the island of Madeira, and the markets of Europe be supplied with such
delicious specimens as I saw there in 1907. It thrives in California.

One morning as we were strolling through the center of the town,
Marian called my attention to a tree loaded with pendant clusters of
flowers like the snowball (Viburnum), but the flowers were a beautiful
shell-pink. The gardener called it *Dombeya Wallichii,* but it is now
known as *Assonia Wallichii*. It is a native of Madagascar. I gathered
cuttings and sent them in to South Florida, where they were propagated
and widely distributed. I always like to think that those I see growing in
Florida came from that introduction, although in Doctor Nehrling's
Plant World in Florida, he mentions having it in his Gotha garden two or
three years before this. I presume that his was one of Pliny Reasoner's
gifts to Florida.

Probably the most important crop of the islands was the sweet potato,
which was planted on the terraces far up the mountain slopes, and even
on the mountainsides, which one would think were too steep to offer foot-

hold for anything. It is curious how this American plant has travelled throughout the world and represents today one of the cheapest and most nourishing foods of mankind.

On the lowlands near Camacho I found patches of taro (*Colocasia esculentum*) and the tubers were sold in the markets after being boiled so that they were ready for immediate consumption. We had become much interested in this plant, known as the dasheen in Florida and the West Indies, but it never occurred to me that dasheen exports from Madeira to the Portuguese colony in Boston would interfere with the development of the vegetable in Florida!

A month's stay in Funchal restored my health, and when we heard that a boat of the Amazon-Lisbon Royal Mail Line would touch at Funchal en route to Portugal, we engaged passage, not realizing how dirty the boat would be. It was so filthy, in fact, that we spent the trip disinfecting everything we had to touch.

One of the most noted private estates in Europe at that time was Monserrat near Cintra, some distance from Lisbon, situated on a high pinnacle of rock overlooking the mouth of the Tagus. The garden represented the intelligent collecting of Sir Francis Cook, an English amateur who had spent large sums of money in landscaping the hills and valleys surrounding his palatial residence. With notebook in hand, I studied the exotic conifers, water plants, vines, palms, and border plants. Climatic conditions closely resembled southern California, and most of the species growing at Cintra later found their way into the beautiful private gardens of Santa Barbara, Riverside, and Pasadena.

As we heard that there was a large cork factory in Seville, we stopped there on our way south. At the factory I learned something of the cork business. The manager was very proud of a new machine which had been invented in Baltimore and cut 20,000 corks a day. It reduced the cost from thirty-five cents to one cent a thousand, as the old method had been essentially a hand method in which the operators used a sharp, broad-bladed knife.

We were not in season to see the "mascalage" which takes place between the months of May and August. But I did see trees only six years old from which the rough, outer bark had been removed. It is amazing that a great oak tree, two feet in diameter, will permit the removal of bands of bark three feet high extending clear around the tree, and that

Closely resembling its relative the Cherimoya, the Ilama (*Annona diversifolia*) of Guatemala ranks with the best fruits that grow.

The spectacular *Echium fastuosum* of Madeira has giant blue spikes taller than a man's head.

An old Cork Oak in California. The "mascalage"—removal of the outer layer of bark—may be done every seven years.

324B

Marian gave two Berber women their first taste of ice.

Buried under enormous loads of bark, burros ambled down the dusty highway from
the cork oak forests on the hillsides of southeastern Spain.

We acquired an unusual gardener, a delightful Japanese peasant boy named Mori.
He is seen here in his patch of giant radishes.

324D

this "mascalage," as it is called, is done every seven years. During the intervening seven years, the cambium, which remains attached to the trunk when the outer bark is peeled off, forms a layer of new cork over two inches thick. To girdle a tree ordinarily means to kill it, but in the method of mascalage the growing cambium (which is the living inner layer of cells near the wood) is not disturbed. As with most trees, a ring of bark of considerable width can be removed without seriously injuring the tree, as long as the cut is not too deep, for the wound soon heals. The ascent of water in a tree is through its young or sap wood and not through its bark; the descending sap, however, does flow through the inner bark—the outer part of the cambium.

After a day in picturesque Ronda, we hired a couple of mules and were off on the six hours' trip over the mountains to Grazalema.

The mountain road wound through cork forests where the gnarled, rough trunks and spreading branches stretched over great, irregular blocks of gray limestone, making an unforgettable picture. As we threaded the narrow path, we looked down into the valley below and listened to the calls and songs of the peasants at their work.

The little walled town is like a swallow's nest against the cliffs, one's dream of a mediæval fortified village. Our visit there was one of the most beautiful experiences which Marian and I have had. We approached the town by a steep ramp, and entered the city gate under the portcullis, imagining ourselves some knight and his lady in the Middle Ages. We rode directly into the little hostelry, dismounted, and saw the mules led away to stalls on the ground floor. We then ascended to the dining room and sleeping quarters. The granary was in the attic and a chute ran through our room to the stable below. The hotelkeeper and his wife had never entertained Americans before and did their best to make us comfortable.

Marian became as fond as I of hard Spanish bread and highly flavored Andalusian foods colored with saffron. She also liked the red Spanish wine and the fresh almonds which were gathered on the hillside nearby. Grazalema was inconceivably quaint and delightfully quiet except for the ringing of the church bells, to which we eventually became accustomed.

A crowd of children followed us everywhere and showed us all the sights, and we became very attached to them. Marian proposed that we

325

buy some candy for them as they were evidently so poor that they had only the roughest food. When we began to give away the candies, the children gathered as when the Pied Piper blew the first notes on his flute. They came from every doorway until I felt that I had never seen so many children in my life before. The narrow streets were soon blocked as we continued to throw candies to those near enough. Soon larger boys appeared and elbowed their way through the crowd. In less time than it takes to tell the tale the little tots were being trampled underfoot and in danger of being seriously hurt. We escaped through the candy shop and made our way back to the hotel. But the affair had roused the town, and the following morning an armed guard appeared, a gendarme with a long sabre at his side. He had come, he told us, to protect us against the children and from that time on we never left our quarters without an armed escort.

When next I visited Ronda, twenty years later, a motor road had been built to Grazalema, but I preferred to keep my memories of the place when it was a truly mediæval village.

After our return to Washington that spring, we acquired an unusual servant, a delightful Japanese peasant boy named Mori.

Mori proved ideally suited both to us and the place, for everything he did was done in a picturesque Japanese way. As we had Japanese flowering cherry trees and udo growing beside them, as well as Japanese maples, dogwoods, bamboos and the wasabi, his personality and skill fitted most perfectly into the surroundings.

Planting "common garden" trees and shrubs is easy compared to growing new varieties and species of which you want to keep track. I had replaced the cloth tags on the cherry trees with wooden ones, but the copper wire was cutting into the trees, strangling them. There was definite danger of their girdling and killing the trees; yet changing the labels whenever the wire became too tight was a chore.

Mr. Bell loved the woods and the quiet, and spent much time with us in a little house, called his "Retreat," which we built for him in the forest overlooking Rock Creek. In cool weather he would spend his days in front of the fire with his big notebook, his "desk book" he called it, before him. A sofa, a table, a fireplace or stove, a box of shot in which he could stick his pens and pencils, a few pipe cleaners, and a rug of some kind to throw over him, were all the material comforts which Mr. Bell ever required. But the most important thing of all was quiet. Above everything he loved to meditate, to think, to dream in the inventor's sense, and to be free from interruptions. As he expressed it, he had "a yearning for something deeper than the bare facts." I wrote these words down as he said them with the date, February 3, 1908.

One day, as he lay on the sofa, I asked him if he could suggest a way to tag my cherry trees. He listened attentively and then said,

"Why don't you put the wire through your trees?"

I had never thought of it or seen it done. It seemed cruel to run a wire through the trunk of a little tree and leave it there. But I knew that barbed wire often became embedded in trees without damaging them, so I tried the experiment with different kinds of wire, everything I could think of from silver to lead. The cherry trees had a gum which oozed out around the wire and, in rainy weather, formed a jellylike mass. I thought perhaps the wire irritated the plant tissues and produced the gum, but a noted entomologist told me that the cherry gum is produced by a bacterium. What looks like gum oozing from the tree is really being formed by bacterial action near the surface of the wound.

The tag experiment proved a decided success; the wires were several inches long, and my tags remained attached for years until sometimes the tree trunk engulfed the wire and the copper tag too!

I became interested in the gardening work among children in the public schools and sent for trees of the drooping cherry to give to each school in the District of Columbia. This shipment arrived from Yokohama in May, 1907, and the officials of the Chevy Chase car line, for whom I also ordered three hundred trees to plant along the line, provided a special car to bring out a boy from each school. The Superintendent of Schools accompanied them and, with Mori to help me, I showed every boy how to dig and ball and cut back a tree for planting.

I gave Mori a big ledger in which he pasted the labels of all the "S. P. I." seeds and plants which we tried in our garden, and for nearly twenty years I ran small scale experiments with seeds and plants selected from the stream of introductions pouring into the Office. Mori grew the Chinese cabbage, which has become a rather popular vegetable in this country, and experimented in growing and forcing sea kale. We grew soybeans as a vegetable and had to endure the criticism that they took too long to shell. We also had a field of the wax melons which the Chinese use in their soups, and even sold some of these wax-covered fruits to a Chinese restaurant on Ninth Street.

Mori grew the Sakurajima, the giant Japanese radishes, to perfection, and we kept a patch of udo going until the plants grew eight or ten feet high and were destroyed by a miserable fungus, a Sclerotium which attacked their roots. One year we had a patch of the Cape gooseberry, *Physalis peruviana,* which fruited enormously, and our cook took a fancy to it and put up over a hundred jars of what we called Inca Conserve, as the plant is a Peruvian one and has nothing to do with the Cape of Good Hope. Inca Conserve met with such universal favor that we had great plans for setting Emelie up in business.

In Japan, the incredibly long stems and masses of big, soft leaves of the Kudzu vine (*Pueraria Thunbergiana*) are fed to cattle. I tried to establish it on our place, but the expensive roots I imported all failed. Finding a nursery which offered seeds for sale, I grew a lot of seedlings and scattered them about rather recklessly, having in mind the plant's stubborn refusal to grow for me before. The seedlings all took root with a vengeance, grew over the bushes and climbed the pines, smothering them with masses of vegetation which bent them to the ground and became an awful, tangled nuisance. I spent over two hundred dollars in the years which followed trying to get rid of it, but when we sold the place there was still some Kudzu behind the house and the new owner pastured his cow on it all one summer. I see by the paper today that the Kudzu vine is now being used by the Soil Conservation Service to control gullying and erosion throughout the South.

One spring we put up a trellis which became covered with the South African pipe gourd, and we had a fascinating time making long pipes out of their crooked necks. The pipes attracted the attention of the American Tobacco Company and led to the publication of a bulletin on the subject which was so in demand that the American Tobacco Company distributed thousands through their stores throughout the country. Momentarily we started a craze for South African pipe gourd growing among children. It was amazing how much technique those gourds required in order to produce good pipes, for most of them had necks so crooked that they were difficult to adapt.

When John Tull went to Japan to study the matting rushes, he sent us a collection of Japanese morning-glory seeds, including some rare forms which cost a dollar or so a seed. They grew well and our trellises were gay with the enormous blossoms, some of them seven inches across.

In Japan, I had crossed Tokyo at dawn to see a morning-glory exhibition when the flowers were at their best. Some were double and of unusual size and color, but most of them were the fancy, potted morning glories, exhibited because of their strange, unusual forms which are not like the conventional morning-glory trumpets, but have flowers composed of separate petals. The technique involved in growing these Japanese potted morning glories was too intricate for me; I never mastered it.

In a nursery behind the house, where I could care for them personally, I grew each year about a hundred species of new plant introductions. If they grew well, and passed the winter safely, I set them out around the house or among the trees. That little garden of new plants brought me such pleasure that I cannot understand why most people are contented to grow the same things their neighbors grow, instead of trying something which has never been grown before in the region where they live.

Among our cherry trees, in sheltered positions, I planted some Japanese apricots or "mumes," although I had never seen them in bloom and they were supposed to flower too early to make them worth growing in America. A few did prove difficult to keep alive, but other varieties grew like weeds, and flowered in late March or early April, even though they frequently were caught by the late snow storms. Of course the least attractive ones with small white flowers proved the hardiest, but when there is nothing else flowering in one's yard, a gnarled Japanese apricot with delicate-scented blossoms scattered picturesquely over it is as fascinating a harbinger of spring as it is possible to imagine. Some of the rarer forms, moreover, with their great shell-pink blossoms, are among the most delicately beautiful of early flowering shrubs, and I lost my·heart to them when they occasionally flowered in all their perfection.

CHAPTER XXIV

Baddeck, Nova Scotia

IN AUGUST, 1907, we were at last able to go to the beloved estate called "Beinn Bhreagh" at Baddeck, Nova Scotia. Laden with cases of Walker-Gordon milk and much paraphernalia for the baby, we arrived one afternoon at the Grand Narrows which separate the Big and Little Bras d'Or Lakes. There Mrs. Bell met us and we went by steamer to Baddeck. The sun was setting as we came in sight of the big house on the point and passed the red stone cliffs which gave the name of "Red Head" to the promontory until Mr. Bell changed it to "Beinn Bhreagh" which is the Gaelic for "beautiful mountain."

Mr. Bell, in knickerbockers, Norfolk jacket, and with a beret from the Pyrenees on his head, was waiting for us on the pier with the Grosvenors, Mr. and Mrs. George Kennan, and all the men of the place. What a welcome! Bonfires had been lighted on the mountainside and strings of lanterns decorated the half mile of road from the wharf to the house, while rafts illuminated by burning fagots were anchored on the lake. This homecoming was so spontaneous and gay that it left a glow in one's heart.

Marian had told me that I would find Beinn Bhreagh a busy, interesting place, but until I reached there I did not realize the extent of the many, whirlwind activities of the inventor of the telephone. The next morning Mr. Bell showed me the shops where men were making tetrahedral cells of wood and red silk for the giant kites with which he was conducting aeronautic experiments. We watched the latest kite make its first flight, and I met young F. W. Baldwin of Toronto, nicknamed "Casey," and renewed my acquaintance with young Douglas McCurdy, who was to be the first person in the British Empire to fly a heavier-than-air machine and also would be the first flyer to attempt to cross by plane from Florida to Cuba.

Baddeck, Nova Scotia

The kite field was a sloping hillside, stretching upwards from the frame structure in which the kites were built. The kites were of all sizes and shapes, some of them enormous affairs over forty feet in length, shaped like the roof of a house, and composed of hundreds of tetrahedral cells. Hanging down from the rafters of the shop were models representing ideas which Mr. Bell had tried and cast aside.

All the tests were carefully recorded, and after each experiment Mr. Bell, Douglas, and Casey discussed every detail while Miss MacKenzie recorded the data in one of the large notebooks which are inevitably associated in my mind with Mr. Bell. He always had a notebook with him and, with a meticulousness which never slackened, he dated and initialled each observation and the account of every experiment. His untiring capacity for detail was something which I have never seen surpassed.

In the house, Mr. Bell had a study lined with books. Here he worked every night from the time the house quieted down at eleven or twelve o'clock, until dawn reminded him that his own bedtime had arrived. Then, wrapping a bath towel around his head to protect his eyes from light, he would drop into a sleep so profound that it was difficult to awaken him, even in emergencies. He usually slept until ten o'clock, breakfasted in his study, and drove in his little phaeton drawn by an old white horse, to the field where the kite trials began.

Sometimes the experiments took place on the Bras d'Or Lake, and the kites were towed behind the motor boat until they soared into the air. At these trials Mrs. Bell also would be present. The waterfront by the laboratory was very beautiful in those days. The lovely harbor was enclosed by a spit of land and there was a little pond from which the water came and went, according to the tides. A sailboat or two and a couple of canoes would be drawn up on the shelving beach, and there were no ugly buildings such as now disfigure the place.

Gilbert and Elsie Grosvenor and their young family lived in the Lodge down by the harbor where their children went wading on the smooth beach. Gilbert had set up an editorial tent there among the spruces and firs and was working over one of the fall numbers of *The National Geographic Magazine*.

Mr. Bell had been interested in the possibilities of aerial flight ever since his boyhood. He used to say that the two things which most annoyed him, as a scientific man, were to watch the flight of birds and realize that man was incapable of flying too, and to read accounts of men dying of thirst in open boats at sea with "Water, water everywhere and not a drop to drink."

Since 1888 when Professor Langley, Secretary of the Smithsonian Institution, had begun his researches on what he called "the internal work of the wind," Mr. Bell had been his staunch supporter. However, Mr. Bell's experiments had been largely with kites which he believed had elements of safety which Professor Langley's machine with outspread wings had not. The War Department had furnished Langley $50,000 to construct his machine; but on December 8, 1903, the machine had taken an unceremonious plunge into the muddy waters of the Potomac with Charley Manley on board, because the "release mechanism" failed to work and properly release the machine from its metal ways on top of a house-boat anchored in the river. At the time of which I write, no more money was available from the War Department to continue experiments, largely due to the unsympathetic ridicule of the newspapers throughout the country, who dubbed the machine "Langley's Buzzard." Professor Langley had previously constructed a model of his machine in which the motive power was a tiny little steam engine. Mr. Bell had seen the model fly and had made the only photograph, I believe, that was ever taken of it.

Professor Langley's discouragement, after the plunge which wrecked his only large-scale machine, is reflected in an account given me by Doctor John Brashear, a noted astronomer who was for many years the best-loved citizen of Pittsburgh. Doctor Brashear said that he was in Washington shortly after the disastrous plunge into the Potomac, and Professor Langley sent word that he would like to see him. They had been very warm friends when Professor Langley was at the Allegheny Observatory in Pittsburgh.

"As I entered his office in the Smithsonian," Doctor Brashear said, "Professor Langley met me and grasped me by both hands and said, 'Brashear, I'm ruined, my life is a failure.' He took me to the desk and showed me two bits of steel. They were the two triggers which had failed to release his machine as it slid down the ways to get its start in the air. Instead they had caught and held it, with the result that the machine dived into the water.

Every Saturday, Alexander Graham Bell drove to an old houseboat beached on the quiet Nova Scotian shore and disappeared from our sight until Monday with his thoughts and note-books to keep him company. To test his ingenuity, he occasionally imagined himself shipwrecked here, and made clothing from moss and spruce roots, and a porridge from the seeds of a wild sedge.

Marian looking across the Little Bras d'Or Lake from the deck of Mr. Bell's retreat.

Langley's flying machine, financed by the War Department, ready for flight from the roof of a houseboat at Widewater, Va.

"Casey" Baldwin seated in one of Mr. Bell's giant tetrahedral kites.

332B

© *National Geographic Society.*
At the trial flight, October 3, 1903, the release mechanism failed to function, and Langley's plane plunged into the Potomac. Public ridicule discouraged further appropriations.

A branch of the Tamopan Persimmon in an orchard near the Ming Tombs of China.
This is the most important commercial fruit of North China.

In China, the Tamopan or Grindstone Persimmons are ripened under bamboo mats.
Freezing eliminates the astringency.   (Page 337)

Baddeck, Nova Scotia

"He cried like a child," continued Doctor Brashear. "You may not know it, but he was a very emotional man. I tried to comfort him. I reminded him of his discoveries in astrophysics, but he would not be consoled. I believe," said Doctor Brashear, "that he died a bitterly disappointed man."

Some years after these events, I met an American naval officer in the smoking room of a Pullman on which I was bound for Nova Scotia. He asked me where I was going, and told me that he was Rear-Admiral C. H. Davis and that in 1898 he, then a Commander, and Lieutenant-Colonel George W. Davis—and he remarked on the curious coincidence of names—had been detailed by the War Department to investigate and report on Professor Langley's request for an appropriation to make a practical flying machine. He said that Colonel Davis and he, after listening attentively to Professor Langley's description, were not impressed and looked at each other across the table knowingly. But Professor Langley had asked Mr. Bell to be there to assist him in presenting his case. Mr. Bell made such a forceful speech, lending the prestige of his personality and his past accomplishments, that both Admiral and Colonel Davis sent in favorable reports, and the fifty-thousand-dollar grant was made.

I sensed vaguely in 1907 that I had arrived at Baddeck at a critical period in the development of the heavier-than-air flying machines, and, while I shared Mr. Bell's enthusiasm for his tetrahedral kites as having a peculiar degree of stability, I caught the interesting currents of opinion with regard to the kites which were developing in the minds of his young assistants, Baldwin and McCurdy.

I had dined with Mr. Bell in Washington the past winter when Professor Simon Newcombe, the astronomer, Professor Octave Chanute, the noted engineer, and Professor Langley had been present.

The newspapers had already carried accounts of the Wright brothers' secret flights at Dayton and Kittyhawk, but these stories, not being illustrated by photographs, had failed to convince the public, and every one seemed inclined to doubt their authenticity. Mr. Bell turned to Professor Chanute; I can see him now as he posed the question.

"What evidence have we, Professor Chanute, that the Wrights have flown?"

To this Professor Chanute replied:

"I have seen them do it."

The World Was My Garden

These words, from a man who had been identified with many experiments in the use of gliders, produced a tremendous impression on those around the table. Today, when our ears hear the constant drone of aeroplanes, it is impossible to realize how revolutionary was the vista opening before us in the days when flying verged on the supernatural.

I had thrilled when Professor Strasburger announced that his friend Röntgen could photograph the skeletons within us, but to hear from Professor Chanute that man had actually flown made Aladdin's wonderful lamp seem simple child's play. There are few moments in my life which have compared to that in interest and excitement.

By the summer of 1907, therefore, we had the knowledge of successful flights by the Wright brothers and knew that their flying machine was a biplane, entirely different from the tetrahedral kites. It was evident that Mr. Bell was disappointed to realize that his two assistants were losing interest in the principle of the kites, and were becoming enthusiastic about the biplane principle, but he was too big a person not to recognize that perhaps the biplane might hasten the solution of the problem of aerial flight.

Lieutenant Thomas Selfridge of the United States Army had obtained permission from President Theodore Roosevelt to observe Mr. Bell's kite experiments, and was already in Baddeck when we arrived. Glenn Curtiss of Hammondsport, New York, a maker of motorcycle engines, who had won the world's record for speed in the races held the year before at Daytona Beach, arrived a few days after we did. He had been invited there by Mr. Bell to give his advice about the construction of a light engine with which Mr. Bell proposed to equip a tetrahedral kite.

Mr. Bell and the four brilliant young men were such a happy combination of personalities that Mrs. Bell could not bear to have the association end with the summer. At her suggestion, they formed themselves into an Aerial Experiment Association to work together to "get into the air," and Mrs. Bell agreed to finance their experiments. All the associates were to have an equal interest in the Association and each was to contribute any discoveries he made. The younger men were to assist Mr. Bell in carrying out his ideas, and he in turn would assist them in carrying out theirs. As a beginning, the four went to work on gliders, and later in the fall transferred their work to Hammondsport, New York, where Curtiss was manufacturing the lightest possible engines, and where they proposed to build a biplane and try one of his engines in it.

Both Mr. and Mrs. Bell were interested in my experiments with plants and thoroughly tested everything I sent them. The udo in particular did remarkably well in Mrs. Bell's vegetable garden on the hillside, and when Davidson, the gardener, mounded up the plants in order to blanch the young shoots as they came up through the soil, Mr. Bell, who loved a quiet joke, marked the mounds with wooden crosses calling them "graves of David's Japanese immigrants."

Until that summer, I had not appreciated Mr. Bell's activities as a scientist. He was at work on a paper on eugenics which he published in February the following year, and every Saturday he used to leave the house with his little black satchel containing his notebooks, and be driven to an old house-boat which was drawn up on the shore between a tiny fresh-water lake and the gravel beach. There he disappeared from our sight until Monday morning.

No one dared to disturb him, and no noises reached his ears in the "Mabel of Beinn Bhreagh" save the throbbing of the surf on the pebbly beach, the cries of the gulls and the sighing of the wind through the fir trees. It was the most idyllic spot in all the world to one who loves the forest and the sea and solitude, and there Mr. Bell thought out the broad theories which gave him his position as a leader in the science of eugenics.

All too soon, the eventful days of my first vacation in Nova Scotia drew to a close. A volume of interesting work was piling up in Washington. Two remarkable men joined us at this time, Homer C. Skeels and Robert A. Young. Although quite different personalities, their devotion, persistence and capacity for detail fitted them both peculiarly well to become as they did two of the most important "work horses" of the whole organization.

Skeels was an excellent botanist interested, above everything else in life, in building up a carefully named collection of seeds. In the Office of Plant Introduction he saw his opportunity, as more than two thousand different kinds of seeds from all over the world were pouring in there every year. The definiteness of Skeels' choice of life work and the joy which he took in the correct classification of each sample of seed were delightful to see. In the thousands of times that I consulted him, I never failed to get from him the same kind of inspiration which had come to me in the biological station in Naples from dear old Paul Mayer, whose meticulous habits and passion for accuracy were so stimulating. In his discussions with me, Skeels almost always held in his hands a bottle of seeds, the actual objects we were talking about, and there was none of the aridity about him which one associates with those dealing with so-called "facts and figures."

So keenly did he feel the responsibility of an accurate identification and unquestionable verification of each specimen, that only once do I recall a slip on his part. It was amusing, and not really serious, and I am sure that if he were alive today he would pardon me for telling about it, particularly since it illustrates the important rôle which his seed collection, now a tangible monument to his memory, played in the work of the organization.

Ilex paraguariensis, from the leaves of which the drink Maté is made, is similar to tea in some respects, containing the same alkaloid and having less tannin. We were much interested in establishing it in this country if possible. Several attempts had failed, when one day a collection of seeds arrived from a little botanical garden at El Saff, Egypt, including one packet labelled *Ilex paraguariensis.* Coming from a "botanical garden," the presumption was that they would be correctly named, and Skeels checked them rather perfunctorily. Finding that they looked like those already in his collection, he gave them his O.K. The seeds were propagated in our Chico garden and grew well. No one there had ever seen maté growing, and the little plants passed easily for *I. paraguariensis* and were so distributed. Several years elapsed, and people who had received the so-called maté tree wrote in enthusiastically about it and some declared that the dried leaves made an excellent tea.

George Mitchell, the tea expert of the Department, had begun his investigations of a native American holly (*Ilex Cassine*) from which he extracted the alkaloid theine, and, hearing of the success of our maté trees in Florida, he wrote for leaves, dried them and discovered that they had no theine in them at all. Our suspicions were immediately aroused and we found that the so-called maté from the botanic station at El Saff was an entirely different thing, *Ehretia buxifolia,* a plant of the borage family, the leaves of which by a curious coincidence were sometimes used for tea in the Philippines. Skeels was deeply mortified, but, when he brought me the two vials, the true Ilex seeds and the tiny pods of the Ehretia appeared so strikingly alike that I did not wonder in the least that he had made the mistake.

I have often told this story to illustrate how essential it was to maintain a collection of samples of seeds of every species of plant which was introduced into the country through our organization. During the many years when the Office was housed in one flimsy building after another, one of our chief concerns was lest some fire destroy this collection. I once waged a successful fight in its defense against no less a politician than Senator Smoot of Utah, who wanted me to move the collection from a place of comparative safety into one of the temporary war shacks on the Mall. The fight was futile after all, as such fights against politicians usually are, for later the seeds were housed in a shack so liable to fire that every now and then a fire-drill gong was rung and every clerk in the building was driven out of it onto the street.

By the grace of God, the Skeels collection, to which he devoted the best years of his life, has finally been stored in a fireproof building.

The other new member of the force, R. A. Young of Ohio State University, was of a peace-loving, retiring disposition, with a quiet, persistent curiosity and a retentive memory. Although rather frail in body, Young has never been distracted by the multiplicity of things around him, and has pushed his work from year to year, carefully recording his results. His experiments with the West Indian yams, the chayote, and the dasheen, together with a mass of other work during these thirty years, and his studies of the bamboos, entitle him to special mention in this narrative.

In the spring of 1908, Doctor W. A. Taylor, and Doctor H. C. Gore, a young chemist, joined me in an investigation of the ripening process of the Oriental persimmon. I had heard that in California the Japanese ripened their persimmons in tubs in which the rice wine or sake had been kept, and I secured a tub from a Japanese importer. Gore discovered that the importance of the old tub was that it contained a large amount of carbon dioxide and therefore, when it was tightly closed, the persimmons were surrounded by CO_2 which excluded the air. This was the beginning of a long and interesting investigation which Dorsett continued and expanded during his explorations in Manchuria and northern China. Dorsett discovered that the persimmons in the Peiping markets have the "pucker" taken out of them, so that they can be eaten when still hard without shrivelling up one's mouth, by storing them in great piles under mats where in early fall they are subjected to low temperatures but are not actually frozen.

During his travels through the persimmon-growing regions of China, Frank Meyer had sent us several seedless forms of this remarkable fruit, particularly of the famous Tamopan or Grindstone persimmon grown in the province of Chihli. These Chinese persimmons differed materially from the Japanese forms previously introduced, and engaged the interest of our Office for over twenty years. Every autumn when the great, flat Grindstone persimmons come on the market, I eat them with sugar and cream as a breakfast dish and think of dear old Meyer and Dorsett and the persimmon orchard which we established in Chico, as well as the other orchards we fostered in Virginia, Florida, and elsewhere. These trees are a gorgeous sight in October when loaded down with fruit. Unlike other orchard trees, the persimmon drops its foliage before the fruit is ripe, and the brilliant orange fruits hang on bare branches silhouetted against the blue sky.

By this time, Mr. Bell's interest in new foods had been aroused and he suggested that at his "Wednesday Evenings" he would like to serve any new dishes which we wanted tested by an intelligent audience. As our plantings had developed enough to furnish us material, I served new introductions one by one, generally with success. The suppers at the "Wednesday Evenings" were simple, but Mr. Bell's guests were pleasantly hungry when the party broke up somewhere about midnight.

These practical experiments in serving our new introductions were a great encouragement to all of us in the Office, and, as the National Geographic Society then held an annual banquet of its Washington members, Gilbert Grosvenor came to me with the suggestion that I serve a new vegetable at their next banquet. The idea appealed to me, and at four consecutive banquets I served the following introductions: American grown Deglet Noor dates from the new plantations established by Swingle near Indio, California; West Indian dasheens from Florida; chayotes from our arbors at Brooksville, Florida; and preserved Chinese jujubes or T'saos from Chico, California.

Considerable advertisement was given these new foods in this way and, had the growers of these crops been in a position to supply an increasing demand, much could have been accomplished. I came to appreciate through this free advertising the utter inadequacy of the machinery at my disposal for commercializing a new food. I was told that the Postum Company expended a million dollars a year in the restricted area around Chicago alone, which was much more than double the amount our Office had spent in all the years since its inception.

CHAPTER XXV

Mostly Aviation

A BOUT THIS TIME it seemed the conventional thing to have our baby christened, but as we had no church affiliations we did not know quite what to do. Finally we decided to ask Edward Everett Hale, who was a dear friend of Mrs. Hubbard and was interested in her grandson. Doctor Hale was beloved for his story, *The Man Without a Country* as well as for his philosophy of life which was embodied in his words,

"Look up and not down; look out, not in; look forward, not backward; and lend a hand."

We called on Doctor Hale and asked him if he would christen little Alexander Graham Bell Fairchild.

"What do you call him?" he said. "What is his nickname?"

I replied that we called the boy Sandy, the name which Mrs. Bell often used for her husband.

Doctor Hale seemed curiously delighted and announced the fact to Mrs. Hale who was in an adjoining room.

"And do you let him play in the mud?" he said. "I don't think much of boys who don't get dirty."

I had to admit that we could not keep Sandy out of it.

The ceremony at Twin Oaks was the simple service of the Unitarian Church, except that I provided a photographer, which was something of an innovation, and I have a charming picture of Doctor Hale with Sandy on his lap and Mrs. Hubbard and Mrs. Bell by his side.

One day late in June we received a telegram from Mr. Bell who was in Baddeck, asking us to go to Hammondsport, New York, to witness the flight of the new biplane *June Bug*. It was going to try for the trophy which *The Scientific American* was offering to the first aeroplane to make a public flight of a measured kilometer. This was the first trophy ever offered for a flight of a heavier-than-air flying machine, and we were so excited that we could think of nothing else. I had a little camera, a fixed focus affair, and I bought a lot of films for it.

Marian and I took a Lackawanna train from New York which brought us into the little town of Bath early in the morning of July 4, 1908. We were met by two of the grimiest, dirtiest boys I have ever seen, Casey Baldwin and Douglas McCurdy, who escorted us to breakfast in the little hotel. I say escorted us, but I must modify that statement. Since they had spent the night in the machine shop and were in their working clothes, the head waiter of the tiny hotel mistook them for mechanics and refused them entrance to the dining room. After many remonstrances from us, these two pioneers of aviation were finally permitted to have breakfast, and later drove us in a car to Hammondsport. There we found Glenn Curtiss, his young wife, and Kathleen, Casey's bride, whom we had not seen before. Curtiss was simply "Glenn" to the people of the town, who had known him since childhood, and had brought their bicycles to his little shop on Main Street for repair. His establishment had now grown, and consisted of a collection of hastily constructed wooden buildings, scattered up the hillside.

The sky was overcast, the day was windy, and it had rained the night before. Every one wondered what the weather would be like in the afternoon. We visited the engine shops and were shown the way the wooden propellers were built up of thin layers of wood and were tested by being run for hours attached to one of Curtiss' special engines. After enjoying our amazement at the current of air which the propeller created, Curtiss suggested that we go and see the *June Bug,* which had been taken to the field where she was to be flown.

Curtiss got out his automobile and we were given the seats of honor, while fourteen other men and boys climbed onto it too. I had never seen so many people in a car before. Glenn drove to the canvas tent which housed the flying machine. The field was an old abandoned race track some distance out of town and was a level stretch of land surrounded by hills on which were growing the grapes from which the Hammondsport wine is manufactured. The little half-mile track, partly overgrown with grass, had no grandstand, while a potato patch and vineyard crowded it uncomfortably close on both sides.

The *June Bug* looked frail but trim, with its struts, wires and white

341

canvas surfaces mounted on bicycle wheels. It bore as little resemblance to the great, forty-two-passenger clipper ships which fly over my Florida study today as Stevenson's first steam carriage bore to the giant modern express trains. But the principles of flight were there.

The *June Bug* was a biplane; it had upper and lower wings made of canvas, stretched on the lightest, strongest framework which could be made at that time. Both wings were curved, and the space between them was greater in the center than at the tips. Curtiss' engine had been mounted in the middle, between the wings, with the propeller behind it. Out in front of the motor, on a fragile framework, was the driver's seat; and in front of it was the steering wheel which controlled the vertical and horizontal rudders. A U-shaped arrangement of pipe gripped the pilot's shoulders and was attached by wires to the movable tips of the wings so that by leaning his body to one side or the other the pilot could raise one set of tips and depress the other simultaneously. These movable wing tips furnished lateral stability and prevented the sidewise tipping of the plane. In fact, they have proved essential and are now to be seen in modified form on all aeroplanes. Today they are called ailerons.

I photographed the machine from every possible angle, and even took a picture of Marian in the driver's seat with her hands on the wheel. As we stood there looking at the machine, others who had come to see the flight began arriving.

There was Manley, who had built the engine for Professor Langley's *Buzzard* and had been rescued by Langley when it plunged into the Potomac. The representative of *The Scientific American* was there with an enormous Graflex camera and a general air of skepticism which I did not like. (The boys had told me that when *The Scientific American* announced its intention of offering a trophy, they had written to the editor and received a reply which angered them so much that they went to New York and convinced the editor that they had a machine which could actually fly.) Professor A. F. Zahm of Washington, who had watched the Langley experiments with great interest, was also there, and that picturesque enthusiast, Augustus Post, who later had the distinction of falling in a balloon through the roof of a house and landing in a lady's boudoir. But these various spectators were overshadowed for the moment in all our minds by a man by the name of Herring who posed as having a marvellous new machine up his sleeve, an aeroplane so small and so efficient that he could pack it in a suitcase. I took a violent dislike to Herring and was not surprised to learn as time went on that he was nothing but a bluff. Later, through his relations with Glenn Curtiss, he caused an amazing lot of trouble.

The judges measured off a course of exactly one kilometer. It led over a potato patch and along the side of a vineyard and I recall a feeling of distinct apprehension as I saw them lay it out, for I pictured Curtiss being impaled on one of the posts or becoming entangled in the wires of the vineyard. I had read of the death of the German pioneer, Otto Lilienthal, in one of his gliders, and it haunted me as I studied the situation.

As the course was being measured, a sharp thunder storm came up and the lightning struck something on the hillside. I had taken shelter under a large oak tree out in the field, from which the representative of *The Scientific American* soon drove me by his argument that lightning had a way of striking trees which stood alone, and that cattle and men were often found dead under them. It was the first time that I had heard this stated.

After a visit to see one of the wine cellars on the hillside overlooking Keuka Lake and a simple luncheon, we returned to the track at the appointed hour. An audience had gathered and was scattered along the railroad embankment and over the field, a few hundred people, I suppose, but not a large crowd. The sky was still overcast and gusts of wind were blowing, enough to make a flight out of the question. We waited anxiously, cursing the weather and becoming more and more discouraged as the hours passed. Late in the afternoon, however, the sky cleared and the boys brought out the *June Bug,* pushing it along on its own wheels to the position determined upon for the start.

Marian took up a position in the potato patch with Douglas McCurdy, and Mrs. Curtiss and I stood not far from them. My attention was focused on my camera, for I had determined to secure a picture of the first flight.

Suddenly we heard the roar of the propeller, saw the dust cloud which it raised behind it, and then, against the pale gray of the evening sky, there came towards us this strange, white, flying apparition, with the long, slender form of Curtiss out in front, and a whirling thing behind him. With his hair streaming in the wind he flew almost over us, thirty feet or so above the ground. As he passed, I felt Mrs. Curtiss grip my arm as she cried out in alarm,

"Why does he fly so high?"

His first flight landed him some distance short of the mark but not, as I had feared would be the case, in the vineyard itself.

We all of us rushed pell-mell to see if Curtiss were hurt, and found him standing beside the plane uninjured and ready for another attempt. He had looked the machine over and found that it was also uninjured.

There was barely time for another trial before dark, so together we pushed the machine from its position half a mile or so away, back onto the race track.

Again we stood and watched as he flew over us; this time it was so late that I could not see him clearly as he passed. Nor did I know until cries came from the end of the course, that his second trial had carried him over the mark and he had won the trophy.

That brief afternoon at Hammondsport had changed my vision of the world as it was to be. There was no longer the shadow of a doubt in my mind that the sky would be full of aeroplanes, and that the time would come when people would travel through the air faster and more safely than they did then on the surface of the earth.

Marian and I returned to Washington the next morning and on the train I wrote down my impressions of the flight for Mr. Bell.

When we reached home we intended to write a fuller account of the flight but, since the newspapers had published accounts of the event, the magazines looked upon the incident with comparative indifference. In fact, many of my own friends were quite apathetic towards the whole matter, seeing no particular significance in this trial flight. It was really not until Lindbergh's flight across the Atlantic in 1927 that the general public and newspapers really awoke to full realization of the fact that what Langley called "the universal highway overhead" had been opened.

This skepticism was even shared by General Henry T. Allen, head of the Signal Corps of the United States Army. He was so completely indifferent to the idea advocated by Captain George Owen Squier, that the Army should purchase a heavier-than-air flying machine, that he told Squier to go ahead and write the specifications himself. General Allen was not even present when the Wright brothers assisted Captain Squier in drawing up the first specifications for a flying machine. This was told me many years later by General Squier himself.

General Allen maintained his stubborn skepticism regarding heavier-

Glenn Curtiss at the wheel of the "June Bug," July 4th, 1908.

In the tent before the flight, Marian at the steering wheel, her shoulders gripped by the fork made of gas pipe which controlled the movable wing tips. The wings of cotton cloth were "doped" with parawax dissolved in gasoline to make them air-tight.

Glenn Curtiss, at the wheel, before starting on the first public flight of a
heavier-than-air flying machine over a measured course.

The "June Bug" thirty feet above the ground at Hammondsport, N. Y., July 4th, 1908.

Orville Wright (with light cap) on September 9, 1908, manœuvring his machine just before the first public flight given by either of the Wright brothers. The Wright machine had long, wooden skids and required a complicated launching device.

Frank N. Meyer's caravan approaching the Wu Tai Shan mountains. This is one of the photographs with which Theodore Roosevelt illustrated his message to Congress.

Meyer's outfit under a Siberian Larch—a tree now in demand for the great plains of the United States and Canada.

than-air flying machines even after the first mail plane had carried its historic load of letters from Washington to New York. I happened to be lunching in the Army and Navy Club with Admiral Albert Ross the day of this flight, and seeing the General at a table across the dining room, I went over to congratulate him on the event. To my amazement he was completely pessimistic about the whole idea of air-mail service, saying that he did not believe it would ever amount to anything. I would not conceal my disgust and turned abruptly away.

Shortly after Glenn Curtiss' flight, Frank Meyer returned from his first three years in China, bringing back to San Francisco seventy-seven different species and varieties of plants.

His collection included eighteen named varieties of the soy bean, with descriptions of their characteristics and the uses to which they were put; several varieties of a small bean (*Phaseolus angularis*) used by the Chinese as a vegetable when sprouted, and a bean from the warm region around Tangsi which has grown well in southern Florida. Its flat pods sliced and boiled make an agreeable vegetable, and its dried beans are used in soups. There were also varieties of clover, oats, millet, hemp, watermelons and cotton as well as a large collection of useful bamboos.

Meyer's shipments showed that he knew how to pack and ship plants and could recognize the species which he saw. Furthermore, his accounts were so meticulously kept that they passed the Disbursing Office with their attached yellow Chinese laundry slips intact. Their uniqueness even aroused the interest of the clerks. Meyer was a most economical traveller; he walked instead of being carried in a sedan chair as are most travellers in the interior of China, and he lived in Chinese inns and ate native food.

Meyer's arrival in Washington was of course an event to be celebrated by the Office, for everybody had read his letters filled with delightful glimpses of the Chinese people and customs. Marian and I gave a bonfire party at "In the Woods" one moonlight night to which the entire Office staff was invited. We were none of us orators, but, with such ceremony and speeches as we were capable of, we complimented the explorer on the success of his three years of strenuous, even dangerous plant hunting in China. Meyer had been attacked by ruffians in Harbin and had experienced many hardships and discomforts while travelling on foot through the Korean forests, the plains of Manchuria and the Wu Tai Shan Mountains.

In deference to the teetotalers among us, only cider was served. But some of the boys doctored the cider a bit, and every one thought it the best which they had ever tasted. Meyer appreciated the occasion and did his share; as he lay stretched out on the ground before the dying embers of the bonfire, he sang us his favorite Dutch songs and gave us an account of the uncomfortable brick beds, the vegetable gardens and the family lives of the poor agricultural class of China. His descriptions seemed more intimate and real than any book or travelogue because they were associated in our minds with the actual plant material which he sent in and which we were growing.

Meyer and I became very close friends that summer, for he would come out to "In the Woods" to escape the excessive heat of Washington, which he felt keenly. He told me about his life in Holland. His parents were people of modest means and simple tastes and he had early become captivated by a Utopian colony which was started by the poet Van Eden. Meyer had been employed as gardener for this idealistic enterprise, the tenets of which caused a good deal of comment in Holland because of the radical ideas they contained. I asked him why it finally broke up, and he said the intellectuals in the colony wanted to stay up at night and read books and educate themselves, but the non-intellectuals objected to the cost of kerosene for their lamps, and insisted that the whole colony should not be made to bear this expense. The intellectuals, they said, should get up at daybreak as the others did and not stay up and read at night. Arguments over this issue broke up the colony.

I expected that, after a summer in America, Meyer would be ready to go back to China, and I was much disappointed when he declared he would not return and begged to be sent to the Caucasus and Chinese Turkestan. It was with a good deal of reluctance that I acceded to his request, for I did not think those regions likely to yield as much valuable material as the far interior of Western China which he had not yet visited. But Meyer was essentially a roamer and wanted to be pushing on into new fields.

Some of the nurserymen of the United States began to take an interest in Meyer's introductions, and Henry H. Hicks of Long Island begged me to let Meyer visit him as he wanted to take him to Oyster Bay to see Theodore Roosevelt. At my request, Meyer reluctantly accepted. He had some remarkable photographs of the deforestation in the Wu Tai Shan Mountains, west of Peking, which showed clearly the ghastly effects of erosion following in the wake of deforestation, and, as President Roosevelt was then in the midst of his program of conservation, he was much impressed by these photographs, and asked Meyer for some of them to be used in his forthcoming message to Congress on the subject of conservation. Meyer came back excited enough. He had travelled fast and far! From washing the windows of Smith's laboratory in 1903, he had risen by 1908 to a position of such prominence that his photographs were going to appear in the President's message to Congress, the only time that a Presidential message was ever illustrated.

At the end of the summer, our attention was turned toward aviation again. The War Department had called for bids for a heavier-than-air flying machine, and for a dirigible as well, and the trial flights of both were to take place in September, 1908, at Fort Myer, Virginia. Glenn Curtiss had supplied one of his light engines to Captain Thomas Baldwin who was entering a dirigible which he had constructed; so we had a personal interest in that. But of course the great event was the prospect of seeing the Wrights' machine.

Tom Selfridge came down from Baddeck to be present, and Marian and I found ourselves occupying front seats, so to speak, at the historical occurrences which took place on the parade ground of Fort Myer.

Wilbur Wright had taken one of their machines to France, and Orville had come to Washington to make the flight for the War Department with a similar machine. The secrecy with which the Wright brothers had surrounded their early trials at Kittyhawk, North Carolina, and Dayton, Ohio, had been so complete that it could truly be said that they had not yet flown in public. There were hundreds of people in Washington who wanted to witness the event, and the street-cars to Fort Myer were filled with spectators, but there was no such crowd or enthusiasm as now attends the daily landing of the clipper ships at the Miami airport all winter.

Tom Selfridge was frequently at our house, and brought Orville Wright out with him the Sunday afternoon before his first flight. Wright seemed delighted with our woods and we wandered through them to the brook

347

which led into Rock Creek, while Tom and his dog romped happily with Graham.

"There are no such woods with virgin timber around my home in Dayton," Wright remarked.

I was appalled to learn that the great forests of Ohio, in which my father had shot squirrels and of which I had heard so much in my boyhood, had all been destroyed, and that there was nothing left now but scattered "wood lots."

The day of the flight, September 9, 1908, Glenn Curtiss and I went to Fort Myer together and there at last I saw the Wright machine. Behind it was the launching device, a tripod about twenty feet in height with a heavy weight suspended from it. From the weight, a rope ran over a pulley at the top of the tripod, descended to another pulley at the base, and from there went between the runways on which the airplane was resting, over still another pulley in front of the machine, returning to the bow of the airplane itself, where it was attached with a release mechanism.

Having seen the *June Bug* run along the ground on wheels and rise into the air carrying the wheels with it, I was not as impressed with the Wright machine as I would otherwise have been. Its long skids, shaped like skis, and the tripod and weight, seemed amateurish compared to the rubber-tired bicycle wheels of the machine which had been manufactured by Mrs. Bell's Aerial Experiment Association. As I stood there with Glenn Curtiss, he also criticised this feature of the machine. The wings, unlike those of the *June Bug,* were squared at the ends and there were no wing tips or ailerons attached. The control mechanism of the Wright machine involved a flexing of the planes themselves, which were not rigid as were the slightly bowed planes of the *June Bug.*

Wright took his place at the controls; two assistants whirled the pair of propellers located behind the engine in front of which Wright sat. The weight in the tripod was dropped, giving a quick jerk to the rope attached to the front of the machine. The machine slid rapidly down the ways, rose into the air and flew around the field a number of times at a height of forty or fifty feet above the ground, landing easily and without accident at the far end of the parade ground. The crowd was of course thrilled and excited and gathered around the plane, but there was no great crush of people to mark this historic event; no such crowd as gathers at a football game or around a race track. Most of the people in Washington were too busy to bother coming to see the first public flight of one of the Wright brothers, who were the first men in history really to fly.

While this flight was taking place at Fort Myer, Wilbur Wright was making preparations to fly their other machine in France. Orville had told me that he and Wilbur had agreed that Wilbur would not make his first public flight in France until Orville had flown in America. They wished the first public demonstration of their invention to be made in their own country, not abroad.

As Curtiss and I returned from Fort Myer in the street-car, he said to me, as though the idea had just come to him:

"I believe that I could fly the *June Bug* upside down—could turn over in it," and I remember telling him that he was crazy, the idea seemed utterly fantastic.

I do not remember how many flights Orville Wright made alone before he took a passenger. The specifications of the War Department demanded that the machine should be able to carry two persons and remain in the air for one hour, and also that it should be designed to have a speed of at least forty miles an hour in still air. The speed was to be determined by taking average of the time over a course of more than five miles, with a flying start, passing the starting point at full speed at both ends of the course.

I saw Orville one morning in the Cosmos Club talking to Professor Chanute, and asked him why Wilbur and he had kept their flights secret from the public for nearly five years. Chanute spoke up:

"I think that you made a mistake, Orville," he said.

In reply, Orville called attention to the French contracts which they would not have secured had they made their invention public.

Chanute waved his hand and said:

"What do they amount to? A few hundred thousand dollars. Before you convinced the public that you really flew, other inventors have made public flights." He had in mind the "hops" and short flights of Santos Dumont, Farman, Delagrange, and Curtiss; none of them equalling the Wrights' secret flight in the fall of 1905, at Dayton, when Wilbur Wright flew for thirty-eight minutes and covered a distance of twenty-four miles.

The great pioneer Chanute had been interested in the Wright brothers' experiments from the very beginning, and his sentiments, so frankly expressed, made a great impression on me. I have always felt that he was right in his contention that, had the Wright brothers made their flights at Kittyhawk public in 1903, their fame as inventors of the flying machine and as the first human beings actually to make sustained free flights in a heavier-than-air machine would have stood out more dramatically than they do today. They had made four completed, though short, flights nine days after Langley's failure, but they were made in secret.

I last saw Chanute in Chicago when Mr. Bell and I visited him in the hospital. He was an intellectual giant and a truly scientific man who cared nothing for publicity and deserves a high place of honor among the pioneers of aerial flight. Verily, aviation has had many fathers.

We had become so "air-minded" that Marian and I had postponed our vacation to Nova Scotia in order to see these flights, but we finally arranged to leave on September 18. Mr. Lathrop had arrived and was staying at the Willard, accompanied by his young friend, Drummond McGavin, who was also an intimate friend of Tom Selfridge. On September 17, Tom was to be Orville Wright's first passenger, and he wanted Drummond and Mr. Lathrop to see the flight. Tom had been up in the air in December, 1907, in one of Mr. Bell's tetrahedral kites, the *Cygnet,* when he had been towed in it a short distance over the Bras d'Or Lakes at Baddeck. I think that he had also made a short hop at Hammondsport with Glenn Curtiss.

As Marian and I were leaving the next day, we were too occupied with last-minute details to go to Fort Myer, but I agreed to dine with Mr. Lathrop, Tom, and Drummond that evening. We were staying at Twin Oaks, and were dressing for dinner when a cousin, Mr. C. J. Bell, returned from Fort Myer and we ran to the door to hear about the flight.

"They had an accident," he said. "The plane crashed and I'm afraid Lieutenant Selfridge is fatally injured."

I hurried into my clothes and hastened to the Willard Hotel as quickly as possible. Mr. Lathrop and Drummond had just come in and were terribly upset.

When they had arrived at Fort Myer, Tom had taken them to a spot near where he expected to circle in the flight, some distance from the crowd on the parade ground. The plane took off all right, and, as it passed almost over their heads, Tom waved his hand to them. It circled a second time and Tom waved once more. Then came the crash. The crowd rushed in and there was considerable delay, as apparently no preparations had been made in case of an accident. Finally both Tom and Orville Wright were taken to the hospital, and, as Mr. Lathrop found that there was nothing he could do to help in any way, he and Drummond returned to the hotel.

Mr. Lathrop's nerves were thoroughly shaken and for years afterwards he never cared to see a flight. He called the aeroplane "a thing of rags and tatters."

We sat down to dinner in very low spirits indeed. We had expected Tom to be with us. Instead, towards the close of the gloomy meal, a message came that Tom had passed away at the hospital at Fort Myer, but that Wright would soon recover, although his ankle was badly smashed.

There was much difference of opinion with regard to the cause of this accident, the first aeroplane "crash" in history. I went out to Fort Myer the next day and saw with my own eyes the broken propeller which had evidently struck or been struck by one of the wire stays. The wire had apparently snapped, making a gash of an inch or more in the propeller and breaking it. Whether this was the cause of the accident remains, I understand, a mooted point.

W. S. Clime, the expert photographer of the Department, was a very reliable observer and had acquired great proficiency in photographing objects in motion. He had seen this fatal flight and came to me the morning after the accident, and I have in my notebook his "deposition" on the subject. As this was the first crash, and I was so close to the early history of aviation, I include the details as I wrote them down on September 18.

The left propeller broke with a pistol shot explosion. Clime thought that it was one. Propeller flew south, spinning, machine tilted front up and fell backwards almost to ground, then tilted forward again. Hard to tell, but thinks it hit the ground on the forward edge of the lower plane and turned completely over so that the skis were in the air. Two cavalrymen there when Clime arrived at the wreck. They lifted up the bottom plane, seemed to Clime that the engine had fallen off the plane to the ground. Gasoline spilled on everything. He lifted the plane and saw Wright was on his stomach across a bar or wire. Selfridge was on right and just underneath Wright. Wright said merely, "Help me," and tried to move. The propeller was going the fastest Clime ever saw it go when it broke. He said, "I'd go into court and swear that the machine turned up its fore part and fell backwards for half the distance to the earth." Doesn't know whether the other propeller was buzzing when the machine fell backwards. Clime was there ten to fifteen seconds from the time the machine fell. He suggests the possibility of the guy wire having snapped and caught on the propeller and thus cut the propeller in two. (An idea that turned out to be approximately correct, but that does not explain why the wire broke.)

Until recently I always believed that the snapping of the guy wire and its impact on the propeller were the causes of the accident. Last summer (1936) Douglas McCurdy told me that many pilots classed this crash as a "tail spin," believing that the wire broke after the spin started.

Tom's tragic death and Wright's accident, which resulted in a permanently stiff ankle, cast a cloud over the whole flying program for a time, and the repercussions were world-wide. No less a person than the German Kaiser sent a cablegram of condolence to Tom's mother.

CHAPTER XXVI

Lacquer and Wild Wheat

W HEN WE RETURNED to Washington that fall, we settled ourselves in the annex to the Bell house at 1331 Connecticut Avenue, the apartment built by Mr. Bell for his father.

It was a hectic winter in the Office, filled with conferences with all sorts of people. What we were doing was becoming known, and visitors appeared from all over the world.

I used to see Mark Carleton now and then and always discussed the progress of durum wheats. I have recorded in my notes that the crop of 1907 had been forty-five million bushels, with an average yield per acre of twelve to eighteen bushels as compared with an eight- or ten-bushel average of ordinary wheat. Carleton was then in the midst of his struggle with the millers, trying to make them remodel their milling machinery so that it would grind the harder-kernelled durum wheats.

When we distributed the Japanese paper plant, Mitsumata, from the Chico garden in California, some plants were sent to W. T. Ashford of Atlanta, Georgia. The Mitsumata grew beautifully on the hillside along Peach Tree Drive and I was more enthusiastic about its possibilities. Like the daphne shrub, the bark of the Mitsumata is so tough and so loosely attached to the wood that, if it is done skillfully, every bit of bark can be stripped from the plant by simply pulling it off as a lady strips off a tight-fitting glove; in other words, by turning it inside out. This bark is then macerated in vats until its fibers come apart and, with the juice of a certain species of hibiscus as a binder, beautiful paper pulp is formed. The test of the Atlanta plantation was a qualitative test only, that is, it proved that the trees would grow there, while the quantitative test to prove whether Mitsumata paper could be produced at a profit had yet to be made. It never was. It has never seemed sufficiently promising to any one to induce him to put money into the experiment.

The World Was My Garden

Like every visitor to Japan, I had lost my heart to the exquisite lacquered ware. It was not yet common in America, but I had read about the manufacture of lacquer and knew that it was made from the sap of a species of sumac (*Rhus vernicifera*) which is so poisonous that only those not susceptible could work with it. A childhood experience with poison sumac (*Rhus vernix*) made me unenthusiastic about the introduction of this tree. However, the lacquer industry in Japan was an important one, and at least some of the Japanese had apparently learned how to handle the acrid sap. So it seemed foolish not to attempt to grow this sumac in America, where so many of its relatives occur.

Unlike shellac or any of the common varnishes with which our painters were familiar, this lacquer hardens better in a moist atmosphere than in a dry one. In other words, lacquered articles are put in a room saturated with moisture to "dry." This is the chemical process known as polymerization, which takes place more rapidly in the presence of moisture and is quite different from the phenomenon of drying as commonly understood.

Since the sumac seeds were not supposed to be acrid, we imported some and planted them in one of the nurseries near Washington. Before the seedlings were a year old, requests came from the propagators asking permission to throw them out; they reported that several of the men had been poisoned. I hated to abandon the experiment so soon, and a clerk named Bliss offered to grow some of the trees in his own yard as he said that he was never poisoned by ivy or sumac. During the first and second years his trees flourished and he boasted of their growth, but in the spring of the third year Bliss appeared at the office one day with a face swollen almost beyond recognition. His case of poisoning would have done justice to the swamp sumac of Michigan. I was not surprised to learn that he had already burned every Japanese sumac tree in his yard! Still, it seems a pity that American ingenuity did not have a chance to experiment with so unique a substance.

A later experience finally convinced me that I should have to leave the investigation of lacquer to others.

Doctor Kin, who headed a hospital in Nanking, repeatedly referred to the fact that their tables were covered with a white lacquer which was marvellous in its resistance to alcohol and acids. The workmen applied it like paint. At the time, I was fitting up my private laboratory, and I asked Doctor Kin to send some of this lacquer to me. When it arrived, in

354

the package was one of those ingenious Chinese brushes which are enclosed in wood like a pencil. I examined the brush, but refrained from opening the cans of lacquer. However, the following day my face was covered with a dreadful rash. I hastened to my doctor, and he gave me a heroic treatment, breaking the blisters and painting my face with permanganate of potash until I assumed the color and appearance of a Chinese god. Even my intimate friends were frightened at the sight of me.

While this experience was still fresh in my memory, I received a letter from my sister Agnes in Kansas City, saying that her husband had a strange skin trouble which the doctors believed to be something obscure and frightening. Later, word came that he had recovered, followed by a frantic wire saying that the eruption had appeared again and that he had been taken back to the hospital. Eventually his trouble was traced to a new Mah Jong set given him at Christmas. In fact, eighty cases of "Mah Jong poisoning" were recorded that year in Kansas City alone. Some Chinese firm had hurried the sets to market when the lacquer was still green.

These incidents serve to prove the complexity of any attempt to introduce a plant industry from a foreign country.

"In the Woods" was left almost deserted during the winter of 1909. Mori had charge of the place and lived alone in a little house under the oaks. The winter was severe, with heavy snows, and whenever I visited Mori, he seemed very lonesome. I remember vividly going out to see him one moonlight night when the crust on the snow was thick enough to bear my weight. The light from his lamp shone through the trees and he seemed somewhat alarmed when I called him. We tramped around and looked at the place in the moonlight and I would gladly have taken him back with me to Washington. But Mori did not like town life and preferred working by himself among the Japanese plants. I am afraid this winter alone in the woods had something to do with his mental breakdown later, for Mori was never the same after 1909.

In the early spring a little daughter was born to us and we called her Barbara Lathrop after my dear old friend. In consequence of this event, the cherry trees at "In the Woods" bloomed without our appreciative presence there.

I have referred to my meeting with the African explorer, Professor

The World Was My Garden

Schweinfurth. Some time during the winter he sent me a clipping from a Munich paper, describing the discovery by Aaron Aaronsohn of a wild wheat in Palestine. Aside from the botanical interest attached to such a discovery, there was the possibility that this progenitor of our cultivated wheat might provide characters of resistance to disease or drought. I wrote Schweinfurth suggesting that if Aaronsohn had any idea of exploiting the wild wheat, he should come over to America and give our wheat breeders a description of it, bringing specimens to show them.

Early in June, a short, light-complexioned Jew walked into my office and introduced himself in broken English as Aaron Aaronsohn from Palestine. He spoke little English, but we resorted to German and I soon discovered that I was in the presence of an extraordinary man. Although Aaronsohn had never been there, his knowledge of California almost equalled his knowledge of Palestine. No foreigner had ever been in my office who had so keen an understanding of the soils, climates, and adaptability of plants to their environment as had this friend of Schweinfurth. He had studied at the School of Agriculture at Montpellier, France, but was largely self-educated, having collected a comprehensive library at his home in Haifa. His grasp of dry-land agricultural problems was astonishing.

Aaronsohn's discovery of the wild wheat had a twofold interest. The origin of wheat, like that of almost all of our cultivated plants, was shrouded in mystery and every botanist was curious to know from what wild species of grass it originated and whether the progenitor of our cultivated wheat was still in existence. This long-bearded, large-headed wheat of Aaronsohn's had certain primitive characters which distinguished it from the common wheat and made it seem likely that it was the ancestor of that great cereal.

If this were so, then the question arose as to whether it might not have qualities that had been lost in the long period of wheat cultivation, and whether, through breeding, these might not be brought back into our cultivated wheats to "rejuvenate" them; in order to give them a greater resistance to rust and drouth, for example.

I introduced Aaronsohn to my friends, particularly those interested in wheat, and Cook, after a number of years' investigation, named the new species *Triticum hermoni* because Aaronsohn had discovered it on Mount Hermon.

Lacquer and Wild Wheat

The speed with which Aaronsohn picked up English was amazing. In a week's time I heard him carrying on technical conversations, comparing the flora of Palestine and California. His store of information proved so unusual that he was requested to write a bulletin on the cultivated plants of Palestine. His was one of those rare pioneer minds which quickly leap to the essentials, and he sat down and in short order drafted an article covering a wide range of useful plants which, in his opinion, should find a congenial home in America. I remember that his opening paragraph contained the expression "ex Oriente lux"—light comes from the Orient.

Inasmuch as he had not seen the Pacific Coast, we arranged a trip for him that summer, and put him in touch with the agricultural investigators of our western States. This trip proved a great success, and Aaronsohn even made public addresses in English before a Dry Farming Congress in Wyoming, a Foresters' convention, and a special meeting of Botanists in Washington.

When Aaronsohn left for California, he told us that he was writing to Palestine to have seeds of his wild wheat collected for us so that we could test it as a grain crop. Also, he promised us other wild and cultivated plants which he considered worthy of introduction into California. Just as he stepped on the train for the West, he said to me,

"I have a special request to make of you, but I will save it until I see you in the fall."

The office, always a busy place, seemed to reach fever heat in the summer of 1909, for Meyer was preparing for his trip through the Caucasus into Chinese Turkestan, and every spare moment I had was devoted to ransacking the library for information regarding desirable plants from these regions.

The Brazilian Government had suddenly requested a plant expert who could establish a garden of economic plants at the mouth of the Amazon near Para, where C. F. Baker, the entomologist, was stationed. Fisher was sent down from the Office and constantly forwarded us material and interesting accounts of the flora of Brazil. His shipments always gave me a nostalgia for the tropics as I pored over them before forwarding them to the little garden in Miami.

When the suffocating summer days of July began, Marian and the children departed for Baddeck. But Mrs. Hubbard had made us both promise to meet her in England later in August. Mrs. Bell was delighted to care for the grandchildren, so Marian and I met in Montreal and sailed for Liverpool. We reached Windermere on August 9 in time for afternoon tea with Mrs. Hubbard, and I well remember how green and inviting the English Lake District seemed after a bleak week at sea.

Washington had suffered from drought that summer and it had been so hot that it was difficult to work except in the early morning. The glare of noon resembled the sunlight on the equator, and altogether the charms of gardening had largely vanished. Here in England, the weather was as cool as April is with us, and the sun's rays were tempered by the moisture in the air. Until I experienced this quick contrast between the summer climate of our eastern States and that of England, I did not fully understand why England is filled with luxurious border plants and charming cottage gardens and why so many English women take an interest in their gardens. Summer gardening in Great Britain, from the Highlands through to Devon, is a delightful pastime, and the cool, moist air inhibits the development of many of those insects and fungus diseases which harass a gardener in the essentially tropical summers prevailing practically everywhere in the central and eastern United States. Even house flies breed much more slowly in England than they do with us.

Perhaps more than anything else, we were amazed by the collections of conifers, many of them of American species which had been gathered around the castles and country places of Scotland. Our experiences reached a climax when we visited Murthly Castle with its magnificent Douglas firs, and its long, century-old avenue of yews. It was a perfect day and I felt that I had only then appreciated how superbly beautiful conifers could be, with their shades of green, deep shadows, perfect forms, picturesque cones, and the carpet of soft needles underneath them dotted with bright-colored mushrooms.

We returned to our hotel intoxicated by the beauty we had seen. I am normally a good sleeper, but the pictures of that place, and the strange, sacred stillness of it, kept me awake throughout the night. There may be other collections of conifers more complete, but I never expect to see a more enchanting place than Murthly near Dunkeld.

At Perth we ran into what the inhabitants considered unbearably hot weather. We found it warm enough to sit out-of-doors in overcoats, but Katie, Mrs. Hubbard's maid, came back from church with the report of a woman who had had sunstroke. I think that the temperature was slightly over seventy.

For some time I had been corresponding with Mr. A. K. Bulley, a cotton broker at Ness Neston, who had one of the largest collections of Chinese ornamental plants in Great Britain, having sent his own plant collector, George Forrest, to Western China. I spent a day with Mr. Bulley and met Forrest, who greatly impressed me by his knowledge of plants. I arranged with Bert Grosvenor to publish an article by Forrest entitled "The Land of the Cross Bow," an interesting account of his thrilling adventures on the divide between Burma and Western China.

This visit to England resulted later in the introduction of interesting plants into America through correspondence with the gardeners on the various estates which we visited.

There is something about a rose petal in quality and color that is seldom equalled by any other flower, and it is little wonder that hybrid roses have become so much a part of the life of human beings. In fact, the words "garden" and "roses" seem indissolubly connected in the mind.

I once said as much to Louis C. Tiffany as we were standing one brilliant morning in Miami, gazing at the gorgeous red flowers of the *Bombax malabaricum*. I complained somewhat peevishly that Northern people who settled in South Florida insisted on spending time and money growing roses, and I could not wean them from the idea, even when they saw gorgeous trees like the Bombax.

"Don't you know," he said, "that the vast majority of people are influenced more by their memory of flower gardens, than by their appreciation of the beautiful? To most people a garden means roses, and they will always try to grow them even in Florida, where they may have to plant them every year."

Marian and I bore out the truth of this statement, for we brought back from England many of the choicest varieties of roses then grown in the nurseries there. We planted them in the approved fashion in a little glade in our woods, and had a wonderful time with them, revelling in their gorgeous colors and perfumes.

At Mr. Bulley's nursery I noticed the gardeners using "cloches," thin glass bell-jars made in Belgium, under which to grow their cuttings. These seemed such simple devices that I imported some, but I soon found them unsuited for the severe winter and spring climate of Maryland, and had little success with them. I believe that their use requires vastly different climatic conditions from those we have in America, and less expensive labor.

It was on this trip to England that I learned to use that greatest of all books on gardening, W. Robinson's *English Flower Garden,* which has ever seemed to grow in my estimation, for it represents the actual experiences of a man who has grown all the plants which he describes, telling about them as a mother might speak of the children whom she has raised from babyhood. I heard recently that this remarkable English gardener has just passed away. He leaves behind him one of the most enviable monuments a man could have, something which touches the finest, most æsthetic qualities in human personality, for he has added to the appreciation of natural beauty among the things closely surrounding men's lives.

Of course I went out to Kew Gardens and renewed my acquaintance with Sir Thistelton Dyer and Sereno Watson, and filled a notebook with names and suggestions for new Chinese plants to introduce.

A rose espaliered against a south wall greatly took my fancy as it was covered with great, single white blooms. It was *Rosa bracteata,* the Macartney rose from South China. Its thick, dark green leaves and densely tomentose stems form a striking background for the brilliant white flowers. I took a plant home with me and was delighted when it bloomed on a trellis at "In the Woods." After it had grown there for several years and I had proudly shown it to many of my friends, I happened to be in Louisiana visiting Ned McIlhenny. He was driving me across the low, flat delta region of the Mississippi when I saw in the distance a long, hedgelike mass of bushes, a quarter of a mile long and perhaps forty feet across. I could not imagine what it was.

"Oh, that," Mr. McIlhenny said, "is the worst weed we have down here. That's the Chickasaw or Macartney rose, introduced by the early settlers some time in the seventies I believe. Since its introduction it has grown wild and covers large areas of land and has been impossible to eradicate."

I gasped, for I realized that I had been tenderly nursing the worst weed of Louisiana. This rose, introduced into England in 1753 and named after Lord Macartney, behaves in a ladylike manner in more northern climates.

Mr. McIlhenny had many interests. He was the manufacturer of Tabasco sauce, for which he grew many acres of bird peppers; he operated

Chains of pollen-grains hang from the anthers of the honeysuckle and the sticky
stigma protrudes beyond. The flower is awaiting some insect visitor to pollinate the
stigma and carry pollen to the next flower visited.

360A

The brilliant red flowers of the *Bombax malabaricum* glisten in the Florida sunlight, gorgeous and tropical; yet people struggle to cultivate the northern roses which constitute their idea of garden flowers.

360B

The McCartney rose (*Rosa bracteata*) is a treasured climber in England, but has become a troublesome weed in Louisiana. Mr. McIlhenny standing beside a thorny thicket on Avery Island.

360c

One of Frank Meyer's gifts to the country was the Chinese Elm — thousands are now growing throughout the Middle West

a large salt mine; and he had created a wild fowl refuge, forty acres in extent, where thousands of white herons came to nest each spring. At the time of my visit, he was setting out a new plantation with young bamboos which he had propagated from those we sent in, and, in the main grove, the young shoots were coming up so fast that each morning there were scores where there had been but one the afternoon before. These shoots were delicious eating, and I rejoiced that at least one southern plantation had begun the cultivation of Mr. Lathrop's favorite introduction.

When we returned home that fall, the dogwood leaves were turning and to my surprise the leaves of the Japanese cherry trees were turning too. Each variety had its own particular coloring—browns and purples and greens so beautiful that we decided some varieties of the Japanese flowering cherry are quite as lovely in the autumn as when laden with flowers in the spring. But the loveliest of all the introduced species around the house was *Acer nikoense,* a maple from Japan and Central China. Its delicate leaflets turned to a more beautiful and brilliant scarlet than the autumn foliage of any other maple I had ever seen.

Almost before we were settled and while we were eagerly looking forward to the autumn and perhaps even winter in our little cottages in the woods, we were crushed by the tragedy of Mrs. Hubbard's death. Nelson, the chauffeur who had motored us through England and Scotland, brought Mrs. Hubbard out to see our Barbara, her youngest great-granddaughter, one afternoon, and she remained until nearly dusk. The Chevy Chase car-line was, as usual, undergoing changes and repairs, and it was necessary for automobiles to run on the street-car track for several blocks.

As they were returning to Washington, a trolley car loaded with workmen, run by a raw recruit of a boy, got out of control coming down the hill. Before Nelson could speed up the automobile to keep ahead of it, he was overtaken. The trolley smashed into them, and Mrs. Hubbard was knocked unconscious. The accident happened within a stone's throw of Twin Oaks, within sight of the oaks themselves.

The tragedy came so swiftly that we were stunned. It seemed that we had scarcely entered the house after bidding Mrs. Hubbard farewell, when the telephone rang and Peter Bisset's frightened voice was telling us that she had been taken to the Garfield Hospital and was not expected to live.

361

Our ride into town on the street-car, with its agonizing delays, can be imagined. Automobile accidents seem nowadays to invade nearly every household, but in those days they were rare and seemed even more terrible. We came too late; we never saw her lovely face again. We spent the winter that followed with Mr. and Mrs. Bell at Twin Oaks and it was a sad time for us all.

CHAPTER XXVII

Aaron Aaronsohn and Joseph Rock

THE EXPERIMENTAL GARDENS were filling up with interesting newcomers, and I was anxious to strengthen our cooperation with the other branches of the Bureau of Plant Industry by calling their attention to the economic plants which were coming in, and encouraging them to use these plants in their breeding and other experiments. I therefore inaugurated a typewritten bulletin to which, at Mr. Bell's suggestion, I gave the title of "Plant Immigrants." The first number appeared on August 19, 1908, and from being merely an inter-departmental affair, its readers soon included a continually widening circle of amateurs (designated as "experimenters") who seemed to take a lively interest in anything which they could grow in their own particular localities. The bulletin was an amateurish affair, but we derived a great deal of satisfaction from it, and in the spring of 1909 I began to illustrate it with photographs.

Two hundred and ten numbers of *Plant Immigrants* were published, containing 340 full-page photographs taken by members of the Office staff. It continued until 1924, when a controversy arose with the Government Printing Office while I was on an expedition with Mr. Allison Armour to the Orient. Issuance of the bulletin was stopped, and this vital contact with the growing number of experimenters was ended. I cannot write of this with equanimity, convinced as I still am that this bulletin served a most useful purpose, and gave an opportunity for others to see some of our thousands of photographs showing the behavior of introduced plants from all over the world. These pictures have continued to accumulate in the files of the Department at Washington, but few people except the group that is handling them ever see them now.

While Howard Dorsett was creating the plant introduction garden at

363

Chico, California, one tragedy after another overtook him. Finally, broken-hearted from the death of his wife and a daughter, he returned to Washington, settled in Alexandria, and took up commercial floriculture. I missed him sadly and begged him to come back to the Office, but in vain. He had not been long in the East when another tragedy occurred, the death of his youngest daughter. After the funeral of this child, one of those terrible experiences made more grim by the bitter winter weather, I succeeded in making him promise to come back to us. Since then his untiring energy, loyalty and faith in the cause of plant introduction have made him the cornerstone of the whole organization.

That same year, after Mrs. Hubbard's death, I suggested that her gardener, Peter Bisset, join our force. He brought with him a practical knowledge of the art of growing plants, and a wide acquaintance with ornamental species, to supplement Dorsett's and my knowledge of economic plants.

The Aerial Experiment Association had been definitely organized for one year's work, and in that time it had fulfilled its avowed purpose of "getting a flying machine into the air," four of them in fact, had been successfully flown. The existence of the Society, therefore, came to an end. Mr. Bell felt that the era of the commercial development of aerial locomotion had arrived, and not being a business man, he was content to leave this to others. All the A. E. A. patents were later sold to the Curtiss Airplane Corporation.

Mr. and Mrs. Bell had always wanted to make a trip to Australia, where Mr. Bell had a number of relatives, and this seemed an opportune time to go. Taking Casey and Kathleen Baldwin with them, they sailed from San Francisco, and were gone for a year.

By this time our interest in our place in Maryland had become so great, and the city seemed so unsuitable a place in which to bring up a family, that we determined to build a permanent home at "In the Woods" and spend our winters as well as summers in the country. We loved the little wooden houses we had built and had done our best to make them comfortable, but had not succeeded very well. So a real house seemed the only alternative.

Our life in the little one-story frame cottages, close to the ground, surrounded by vines, trees and shrubs, had given us fascinating pictures from our windows, impossible to reproduce from a second story. We therefore determined that the new house must also be a rambling affair. It seems strange today to think that in 1910, throughout the whole region around Washington, we did not know of a single country house from which the owner could step directly into the garden from his drawing-room or bedroom. When we talked with various architects and told them of the charming homes we had seen in England, they replied that a house without a basement was unsanitary; that a building must be high enough above the ground to ventilate the timbers under it; also, that we must have a basement kitchen-laundry, and how could it be lighted if the house were flush with the ground?

At last, when we were almost desperate, some one suggested Edward Clarence Dean, who had just returned from his architectural studies in Europe. We took Clarence to see our little wooden houses under the oaks and they captivated him, as they had us. He set to work and drew us charming plans with French doorways from every room leading down only one step into the garden.

I also had insisted that there should be trellises for vines on every available wall space, and that they should be so made that they could be taken down with ease, yet strong enough to hold the most rampant creeper we might want to plant on them. Nothing in the way of devices ever gave me more pleasure than those trellises where I exhibited forty different species of vines at one time, most of them new introductions from China or Japan. The idea so prevalent among most people that vines make a house damp did not prove to be the case, for the trellises were set out from the wall and there was a circulation of air between them and the house. Nothing annoys me more, even to this day, than to see a new house go up with no provision whatsoever for vines. I know that sooner or later the owner will tack up a piece of chicken wire in a vain attempt to train a flowering creeper against the wall.

Not only did our house fit into its surroundings, but its architect, Clarence Dean, and his charming wife came into our lives and are among our most cherished friends. In 1927, Clarence built for us our Florida house—an equally satisfying home. For it he designed special trellises much stronger than those employed at "In the Woods," which the tropical lianas have not been able to destroy.

When Aaron Aaronsohn returned from his successful western trip, the bulletin he had written was in the page-proof stage and he brought it to me one morning and reminded me of my promise to do something for him.

"I want you," he said, "to introduce me to some of the wealthy Jews in this country."

I protested that I did not know any financial leaders, as the only prominent Jew of my acquaintance was Cyrus Adler, who had been librarian of the Smithsonian Institution. However, the next Sunday I took him to Philadelphia, where Adler was President of Dropsie College. Doctor Adler was very glad to see me and introduced us to his house guest, Mr. Oscar Straus, who had recently been appointed Ambassador to Turkey.

I was curious to know why Aaronsohn wanted these introductions, and he explained to me that if he went himself with his story of the wild wheat, they would not believe him, but would immediately ask what the Department of Agriculture thought of his discovery. I was, of course, much interested in this wild wheat, but when Doctor Adler and Mr. Straus inquired whether its introduction would revolutionize wheat-growing (which was rather the inference to be made from some of Aaronsohn's claims) I had to hedge, and throw the burden of proof onto the shoulders of the wheat breeders and let it rest there.

However, as Swingle pointed out in our discussions with Aaronsohn, aside from the matter of the wheat, it was highly desirable to have an up-to-date experiment station and plant-breeding garden somewhere at the eastern end of the Mediterranean. We could keep in touch with it, and learn much about the agricultural crops of that region and their suitability for cultivation in California. We therefore hoped that Aaronsohn would make friends among the prominent Jews and secure their support for such a garden in Palestine. These were the early days of the Zionist movement, about which little was known as yet in America.

Aaronsohn's charm and forceful personality soon attracted the attention of a wide circle of influential men in New York and Chicago. Not long after I left him with Doctor Adler, he telephoned me from New York, asking me to lunch with Julius Rosenwald and himself at Delmonico's. Mr. Rosenwald was a delightful person who disclaimed any knowledge of agriculture, but seemed much interested in young Aaronsohn. I could not assure Mr. Rosenwald that the wild wheat would profoundly affect wheat-growing throughout the world, but I was able to interest him in the idea of an experiment station at Haifa and he and his friends agreed to underwrite a fund for its establishment.

The bulletin on Palestine plants came out and attracted widespread attention, serving to interest many Jews in Aaronsohn's projects. He returned to Haifa, built some laboratories and rapidly increased his agricultural library. Soon he was not only introducing new plants into Palestine but was making a study of malaria and the methods of its eradication, as well as investigating the diseases of crops. We kept in constant communication with him and were proud to think that he was building up such a useful institution.

After the outbreak of the World War, another phase of Aaronsohn's career began. Our first intimation of trouble was the receipt of a cable begging for help to prevent his whole establishment from being destroyed by the Turks. I consulted Paul Warburg of the Federal Reserve Board, and he in turn persuaded the State Department to send a cable to Admiral Decker in command of the American Fleet cruising in eastern Mediterranean waters. The message suggested that, if possible, the Admiral rescue Aaron Aaronsohn, a collaborator of the United States Department of Agriculture, and his valuable collections.

After all our trouble in sending this cablegram, Aaronsohn refused to come on board Admiral Decker's man-of-war—apparently refusing the assistance he himself had requested. We were all much mystified and were still more amazed when a second cablegram arrived from Denmark several months later, requesting that we cable the British Minister in Copenhagen stating that Aaronsohn's presence in America was desired. I took this cablegram to Secretary Houston. As Aaronsohn's refusal of assistance from Admiral Decker was still unexplained, the Secretary was not inclined to do anything more, but finally he decided that, as Aaronsohn really was a collaborator of the Office, he would send a carefully worded cable of recommendation. Some time passed, I do not recall how many weeks, when in the morning paper a dispatch appeared stating that Aaronsohn had been arrested on landing in England on his way to America, and interned as a Turkish spy. We were then more puzzled than ever.

Late one afternoon, the telephone bell rang and Aaronsohn's voice came over the wire.

"I am telephoning from Mr. Warburg's office in the Treasury Building," he said. "I am coming around to see you in Justice Brandeis' car. Wait for me."

When he appeared, I demanded an explanation of his mysterious behavior, and he told me this strange story:

Acting in accordance with our cablegram, Admiral Decker had made a request which was interpreted by the Turkish authorities as indicating that Aaronsohn was not a Turkish subject. Aaronsohn realized that if he did not manage to repudiate this, his family, who had been living in Haifa for years as Turkish subjects, would be massacred. So he sent word that he did not want assistance, and began to think of other ways of escape, for he was determined to get through to America.

Among the plants he described in the bulletin which we had published, was the sesame, the important oil-producing crop of Palestine. It happened that he had found a mutation in one of the fields, a variety of sesame with extremely large, double pods, which he thought might, by selection, be made to yield a double quantity of the oily grain. As vegetable oils were becoming scarce in Palestine on account of the blockade, Aaronsohn convinced the authorities that it was very important that he should get botanists in Berlin to help him in his attempt to create a heavy-yielding sesame which might prove of inestimable value for the country.

When he reached Berlin, he found no geneticist there and persuaded the authorities to let him visit Professor W. Johannsen, the noted geneticist in Copenhagen. Although under suspicion, and watched closely in Copenhagen as a possible spy, his assiduous application to his botanical studies threw his watchers off their guard, and he was able to make contact with the British Minister and impress him with the fact that his institute in Haifa was really an American affair and that he was a member of the staff of the Office of Plant Introduction in Washington. The Minister required some confirmation, and, when Secretary Houston's cablegram reached him, he provided Aaronsohn with papers permitting him to pass through England.

As soon as he boarded the steamer, Aaronsohn tore up his passport and papers and threw them into the sea. He felt sure that he would be arrested on trying to land in England without papers, and he confided as much to a passenger who was going to America and made the man promise to see that the story of his arrest was in the newspapers. His object in doing this was to protect his family in Palestine, for the news that he had been interned as a spy would be cabled back to the Turkish authorities and they would then believe that he was still a loyal Turkish subject.

When questioned by the authorities in Southampton, Aaronsohn told them of his connections in America, particularly of his acquaintance with Ambassador James Bryce, whom he had met at one of Mr. Bell's Wednesday evenings. Viscount Bryce identified him and the British allowed him to leave for New York. But, as he desired, the story of his internment had been printed.

Aaronsohn was dining with Justice Brandeis that evening, and I remember driving up Connecticut Avenue with him as he explained Bolshevism to me, for it was a new term then.

"It is a kind of faith," he said; "not a theory. Its effect on the individual is that of an unlocking; a disintegrating force rather than a keying-up process. It is this which people over here cannot seem to understand. It lets things down instead of tightening them up."

Aaronsohn stayed on for several days and we had many conversations. I learned a fact which he had never told me before: that he was a protégé of Baron Rothschild. He told me that the Jewish colonists in Palestine were speaking and writing Hebrew and that he was doing all that he could to revive the ancient language of his race, as he believed it capable of expressing their ideas better than any alien language could. He was full of schemes for making Palestine a homeland for the Jews, and told me that he was going to see Balfour on his return to England. I believe that he did, and I have always wondered what part this brilliant botanist played in the Palestine negotiations. Aaronsohn was definitely a seer of visions rather than a dreamer of dreams, for he was always active; constructive, never passive.

I heard rumors through mutual acquaintances that Aaronsohn had something to do with the withdrawal of Bulgaria from the War, but I never saw him again to ask him. The aeroplane in which he left London for Paris in June, 1919, crashed near Boulogne. His body was never found, so far as I know. For years we wondered if the office door might not open some day and Aaronsohn walk in. He had come through so many tight places that I felt he could surmount any difficulty.

Of Aaronsohn's discovery—the wild wheat—nothing spectacular has come, and the exaggerated claims made about it are strange reading now. Yet his death was a real loss. He was a positive and dynamic force which the world could ill afford to lose. He believed in himself; he was never hesitant, never afraid to trust his judgment. His activity in the field of politics may have overshadowed his scientific accomplishments; even so, he was first and foremost a scientific man. Even when his convictions led him to give his energies to developing the Zionist movement, he longed for his books and the quiet of his laboratory. Had he lived, interesting and useful things would have come from the Experimental Station at Haifa.

Our Introduction Gardens began to attract a good deal of attention among Congressmen, and Senator P. J. Macumber of North Dakota wanted one in his State. He presented a bill for the establishment of such a garden at Mandan. The stations we already had were more or less starved, and we were loath to take on the responsibility of another one. However, we realized the value which it might be to the great northern plains region and were not unsympathetic, provided the bill carried adequate funds for its permanent maintenance. Senator Macumber had many friends and the bill finally passed.

When advocating the bill, Senator Macumber had stressed the need for plant introduction, so I went to see him to find out just what he really wanted. He explained that his State was almost treeless although he understood that trees could be grown there; in fact, he had noticed some along the river banks.

'I'd like to see the state covered with trees," he said. "There's one that I hear makes an amazingly rapid growth, but I can't remember the name of the damned thing."

Thinking he might have heard of the eucalyptus of California I suggested its name.

"That's it," he said. "The eucalyptus. I want to see eucalyptus trees planted all over North Dakota."

"Mr. Senator," I replied, "the only trouble with your idea is that the eucalyptus can't stand cold weather. It is semitropical."

I never quite recovered from this revelation of ignorance. Apparently Senator Macumber had secured funds for a North Dakota garden in which to grow eucalyptus without securing any information about them. And yet it was possible that the place might turn out to be just as valuable as if it had not been based on a fallacious idea.

Aaron Aaronsohn and Joseph Rock

Only the other day I was reminded of this episode when a doctor, who was born not far from Mandan, told me of making a visit to the garden as a boy and begging a handful of little seedlings of a dry-land elm from North China.

"These tiny elms weren't any larger than a darning needle," the Doctor said, "but you should have seen them grow. They made trees in no time and we have them all over the place there now. They are magnificent trees."

This elm, *Ulmus pumila,* is one of Meyer's introductions and now has a wide-spread popularity throughout the Mississippi Valley from Canada to the Gulf. My doubts about the garden at Mandan were groundless. No matter what its beginning, it has proved of great value.

It was during that summer of 1910 that my attention was attracted by a statement regarding the threatened extinction of *Kokia drynarioides,* a Hawaiian tree related to the cotton plant. The statement was made by Joseph F. Rock, a botanist in the University of Hawaii, who believed that there was only one tree in existence. This one tree was growing where it had been found and described by Nelson, a companion of Captain James Cook, the discoverer of the islands. Rock later found six trees of a closely related species on the slopes of the volcano Hualalai, and Lewton named them for him, *Kokia Rockii.*

"If not properly protected both these species will soon belong to the past," Rock wrote. "The natives strip the bark off the trunks to obtain the rich, reddish-brown sap which they use for dyeing fish nets. The cattle are also very fond of the leaves."

The threatened disappearance of two cotton relatives at a time when the cotton-breeding program was still at its height so aroused me that I framed a letter to the Governor of the Islands and requested Secretary Wilson to sign it, asking that, if possible, precautions be taken to prevent the extinction of the original tree, at least until seeds of it could be secured for propagation. This correspondence brought me into touch with Rock himself, and for several years he collected all the seeds available for us to plant and preserve in foreign botanic gardens.

While this correspondence was going on, Rock made an expedition to Burma and Siam, and wrote me from there that he was returning through Europe to America. His letters had impressed me as those of a discontented man, and when we met I was much pleased to find him a delight-

371

ful personality. His emotions had been aroused by the tragedy of the lepers on Molokai, and he was stirred by the possibility of relieving their sufferings through the use of Powers' newly discovered ethyl ester of chaulmoogric acid, an acid derived from the oil in the seeds of certain forest trees of Burma.

During his travels in Burma, he had reached the conclusion that a plantation of the best of these trees should be started, as it was not safe to depend on the wild forest trees for supplies of oil. He wanted to go back to Burma, make a thorough study of the chaulmoogra oil-producing trees, and bring in quantities of seed for cultivation in the American tropics. Tears came into his eyes as he pleaded his case, even proposing to establish a plantation of his own. Although we were short of funds, I could not resist this appeal, and money was found for his trip into Burma.

It was during this journey that Rock had the novel experience—novel as far as explorers of the Office were concerned—of being tracked through the forests by a tiger which eventually killed the wife of one of his native companions. The little village in which he stopped was aroused by the tragedy, built a tiger trap, and baited it with the remains of the victim. Rock shot the tiger in the trap the next day.

Rock's researches in the forest of Burma cleared up the confusion which had existed regarding the various species which yield chaulmoogra oil. He took innumerable photographs, and the seeds which he brought back grew and were distributed to numerous locations where they are growing today. However, so far, the supply of oil from the wild trees appears to cover the demand, and I believe that no extensive commercial plantations have been required as yet.

Rock's wide botanical knowledge, and his extraordinary acquaintance with Oriental languages, including Chinese which he both wrote and spoke easily, as well as his unusual skill as a collector and photographer, contributed to his great success as an explorer. He tramped the mountains of Siam and Burma and made a most difficult journey on foot across the mountains into Yunnan, over trails never before travelled by botanists. He brought back a species of Castanopsis (tall trees related to the chestnut), new cherries, rare lilies from snow ranges in the province of Yunnan, distinct species of wild apples, and wild as well as cultivated peaches, raspberries, plums and grapes, also gorgeous forms of primroses and wild peonies, not to mention rhododendrons, azaleas and spruces with which the fanciers of plants have now been experimenting for years. *Castanopsis Delavayi* has shown itself adapted to our southern States, where it has been widely disseminated.

I cannot do justice here to Rock's expeditions. They became too extensive and expensive for the Office to support, and the National Geographic Society engaged his services for a number of years. Readers of the *National Geographic Magazine* have had the pleasure of seeing his splendid photographs and following him along the China-Tibet frontier among the Lamas and to the Holy Mountain of the outlaws.

In 1933 I visited a leper settlement in British Guiana to which some of Rock's chaulmoogra trees were sent, and saw the doctor in charge extracting the oil from the seeds for his patients. It was pleasant to realize that I had had even a slight part in bringing the chaulmoogra tree from Burma.

CHAPTER XXVIII

More Plants, Introduction Gardens and Mr. Bell

DURING 1910, plant immigrants came in at the rate of about ten a day, many of them from the most out-of-the-way parts of the world. The provisions for their care were totally inadequate, both the greenhouse facilities in Washington and the arrangements to care for tropical material in the South. We did the best that we could to see that nothing valuable was lost, although our task was made even more difficult by the entomologists and pathologists who insisted that every packet of seeds, cuttings or plants must be thoroughly inspected or held in the crowded quarantine greenhouses.

The interest in mango growing had become so great in Florida that we made six separate introductions that year from the principal mango regions of India, and landed fifty-five varieties of this tropical fruit in Wardian cases. By the close of the season, we were testing one hundred varieties.

Swingle had established a grove of hybrid citrus fruits on Little River, five miles north of Miami, then considered a long way out among the sticks. By 1910 this had become an important garden containing hundreds of hybrids which Swingle and Webber had made. An enormous amount of record-making and care was involved, as each tree was carefully tagged and its ancestry scrupulously recorded. This place was the nearest approach, in extent and number of seedlings tested, to Burbank's work in California. Out of it have come the delicious tangelos which are slowly winning their way into public recognition. It is a great pity that more plant-breeding gardens were not established and maintained over a

374

period of years, for from them would certainly have come varieties of great value to the horticulture of this country. Unfortunately, the land where Swingle grew his citrus fruits was not owned by the government, and all too soon the tentacles of the growing city reached it.

When Swingle's tangelos, particularly the one he named the Sampson (a hybrid between the Dancey tangerine and the ordinary Florida grapefruit), proved to be of suprisingly good quality, he turned over to us quantities of budwood which we distributed widely. As I write these lines, my own tangelos are ripening again, and I have come to prefer tangelo juice as a drink, or the fruit eaten whole, to either the orange or grapefruit, for the tangelo combines the characteristics of both.

We had been trying for some time to get a drug plant known as khat, (*Catha edulis*), the young shoots of which are chewed by the Arabs in the bazaars of Yemen on the coast of the Red Sea. These shoots are so highly prized that they are heavily taxed by the government, and special means are taken to prevent their being bootlegged into the towns from the plantations where they are grown. Consul C. K. Moser of Aden secured this plant for us, and wrote a glowing account of the effects of khat-chewing. However, none of us has been able to experience similar effects from the American-grown khat, although it grows successfully in South Florida, forming rather attractive evergreen shrubs in our gardens.

Fom the Caucasus, Frank Meyer sent us cuttings of the wild apple, *Pyrus Malus,* var. *paradisiaca.* This is the dwarfing or "paradise" stock commonly used by French nurserymen and we wanted to compare it with those from France. From there he travelled eastward through Tashkent, visited the oasis of Merv, and pushed on into Chinese Turkestan. From these regions he sent in the bush almond, *Amygdalus Fenzliana;* eleven varieties of sweet-kernelled apricots grown in Turkestan not only for their fruits but also for their kernels, which are eaten like almonds; a number of sand-binding plants from Chartchui; the Afghanistan pistache which it was thought might be useful as a stock for the pistache nut; and a number of wild roses and many other promising things.

By this time, we had five Introduction Gardens. In addition to those which I have already mentioned, we had a special garden for bamboos at Brooksville, Florida. These Gardens constituted the core of our whole program, and deserve a separate volume, but I fear that such a book will never be written. The general inclination of the splendid men who have been in charge of them has been to find their pleasure in working with the plants instead of writing about them.

It is sad that some of these Gardens have passed away instead of having the permanency of old-world botanic gardens, but, in the early days, the Department had to contend with a strong feeling against the purchase of land by a Federal department, such as the Department of Agriculture. Congress had been apparently loath to appropriate monies for such a purpose and the only way of securing any land was by interesting the people of a community and inspiring them to offer land as a gift to the government. I have described how the first offer of land, that on Brickell Avenue, Miami, had to be refused by the Department, and that this forced us to establish our introduced plants on ground leased for ten years. Consequently this Garden was forced to move twice before reaching a haven at Chapman Field where today it is again threatened by a projected Bombing Field of the War Department.

The Brooksville location was given to us by a land company headed by a public-spirited citizen of the village. The Chico, California, land was presented by a Citizens' Association, formed for the purpose. Strangely enough, there was no government land in the neighborhood of Washington suitable for our purposes, although we badly needed a site for an experimental nursery. Leigh Hunt owned a large tract outside the District of Columbia; so I laid my difficulties before him.

"You can have part of the place on the Rockville Pike," he said, "and you can have the use of the land for a dollar a year. If my plans go through, I'll give it to you."

The place was not ideal, but at least it was a foothold, and at Yarrow we grew a large assortment of newly introduced plants, propagating them there and sending them out all over the country. The soil conditions embarrassed us as time went on, and it also became apparent that Leigh Hunt's affairs had become involved, and that he would not be able to make good his plan of giving the land to the government. By this time it was less difficult to secure money from Congress, and Doctor Galloway succeeded in getting an appropriation with which we purchased land near a trolley-stop called Bell, Maryland, to which were transferred all the plants at Yarrow. This is now the permanent plant-introduction garden near Washington, and known as the Bell Garden.

Left: The Guatemalan Avocado, introduced by G. N. Collins, from which the "Collinson" originated. *Right:* The Sampson Tangelo, a hybrid produced by Swingle, who crossed the Dancey tangerine with the ordinary Florida grapefruit, producing a new and most delicious fruit now found on some of our northern markets.

Frank N. Meyer, Agricultural Explorer, among the wild quinces of the Caucasus, in 1910.

Clearing the land for the garden at Brooksville, Florida, was a major operation.

376B

More Plants, Introduction Gardens and Mr. Bell

The story of Brooksville, Florida, differs in that we knew definitely what we wanted and were able to secure it without political interference. Our main purpose was to find a suitable location for the bamboo collection which Meyer had assembled in China. Meyer and Clarke made an extensive trip through the South, for Meyer had very definite ideas regarding the requirements of the bamboos as he had seen them growing in China. He had been impressed by the fact that the groves there were underlaid with clay, and that the bamboos required a stiff subsoil in which the long rhizomes, characteristic of all the running or forest types of bamboo, could spread and be firmly fixed in order to support the stems which in the Chinese species attain a height of fifty feet.

I had not realized how difficult it would be to find land which was suitable and which some one would donate to the government. When, however, they settled upon a tract in western Florida, near Brooksville, and sent in samples of the soil and accounts of the magnificent live-oak and other deciduous trees which grew on it, we all felt that they had made an excellent selection. However, at first we regretted that Brooksville was so inaccessible, being at the end of a little jerkwater branch off the main railway line from Jacksonville to Tampa. This isolation, which we deplored, turned out to be a blessing in disguise. Although the Pullman porter threw us out at the junction at five o'clock in the morning and we had to wait several hours for the little train for Brooksville, once we reached there we could spend our time *with the plants themselves,* and not with a lot of people who came to talk to us about them, or with newspaper men who wanted some story which they could improve and use as land-advertising propaganda.

In order to make room for our introduced plants, we had to make a clearing in the dense forest, cutting down some magnificent live-oaks. There was always a conflict between our desire to have a trim, business-like looking experimental garden, and our desire to preserve the natural beauty of the place. I was in frequent difficulty with the more practical men like Dorsett and Bisset, when I attempted to save some of the huge old trees from the axe. As things have turned out, I am sorry that I did not save more of them.

In this garden, we established not only bamboos but also the collection of taros—or dasheens, as we had decided to call them. This collection, started in Puerto Rico by O. W. Barrett, was transferred to Brooksville in the autumn of 1911 from the garden of Doctor Henry Nehrling at Gotha, Florida.

Brooksville proved a very suitable place to test those newly introduced plants whose hardiness we wanted to determine, for nearly every winter it was visited by killing frosts. These frosts generally came in January and February, and we kept careful photographic records of the behavior of the plants after each freeze. I doubt very much whether there exist outside our files more detailed records of frost injury than those taken at Brooksville.

I think that my associates will agree with me that the days we spent in that forest clearing in Florida were filled with more of the real romance of plant introduction than almost any other days of our lives. I used to spend much time in the groves of Chinese and Japanese bamboo, studying their characters, trying to find out why they grew so slowly, and what their insect and fungus enemies were. The arbors of chayote vines from Guatemala were fascinating. Their pendant fruits of jade and ivory were beautiful and I never tired of watching their development. I could not understand why so delightful a vegetable was not grown throughout the southern States. The fields of dasheens with their great, broad, peltate leaves looked very tropical. The harvesting of the tubers, although more difficult than the harvesting of potatoes, was an interesting performance, made still more so by the quiet enthusiasm of Robert Young, who for a number of years spent every spring and autumn planting and harvesting and selecting the best varieties from the long list of those he was testing.

Curiously enough, those very soil conditions which had appealed to Meyer as best suited to the bamboo were ill-adapted for its culture. The fact that land supports giant forest trees is not an indication of its suitability for every other type of plant. The blue clay subsoil, overlain with almost pure sand, became a swamp in rainy weather and a frying pan in dry seasons. Also, the bamboo rhizomes were utterly unable to penetrate the subsoil, and therefore made an unsatisfactory, disappointing growth. This and other considerations subsequently led to the abandonment of the Brooksville garden and the transfer of the bamboo collection to a location on the Ogeechee River south of Savannah, Georgia. But that is another story!

For over a year, Mr. and Mrs. Bell explored the other side of the globe, spending some months in India where they suffered much from the heat. As usual, this experience stirred Mr. Bell's keen intellect, and upon their return home he commented to me on the curious stupidity of man, who had learned to heat his houses but had not thought of cooling them. The next year, when he was forced by business to spend part of the summer in Washington, he set to work to see what could be done about this problem.

Knowing cold air to be heavier than hot, he moved his desk into the empty swimming pool in the basement, and arranged electric fans so that they blew over ice placed above him at the edge of the pool. The chilled air of course descended and filled the pool, and Mr. Bell worked contentedly in a temperature twenty to thirty degrees below that of the rest of Washington.

It was typical of Mr. Bell that thoughts should be followed by action and, as his thoughts came fast, life near him was constantly invigorated by new and interesting experiments in every field of science. Many of his ideas have been developed and are now in common use everywhere, just as his air-conditioning is being made practical and possible for every home owner today.

The spring of 1911 found us in our new house with its forty trellises. The little wooden houses had been moved, and the smallest one was being equipped as a laboratory for me with Nathan Cobb's new microscopic equipment and one of his revolving tables. This equipment was the most complete thing of its kind which I had ever seen and was adaptable for all kinds of photographic as well as microscopic work. Influenced by Doctor Cobb's enthusiasm, my interest in the microscopic world revived and I began to study subjects which I found around me in the woods.

Also, I became more and more drawn into Mr. Bell's activities. He never lost interest in his profession as a teacher of speech to the deaf, and devoted much time to the building up of the Volta Bureau for the Increase and Diffusion of Knowledge Relating to the Deaf. Mr. Bell had created and liberally endowed this institution and, through his efforts and those of his father, it had become the center of information with regard to deafness. All deaf children were particularly dear to Mr. Bell's heart, and his efforts were untiring to make it possible that every deaf child in America should have the opportunity of learning to speak and to read the lips.

The Volta Bureau was in Georgetown and took its name from the

Volta Prize which had been awarded to Mr. Bell by the French Government for his invention of the telephone. Mr. Bell had utilized the original prize money in establishing the laboratory, the Volta Laboratory
as he called it, in a house his father owned in Georgetown, and there, with
his cousin Chichester Bell and Sumner Tainter, he hoped to develop some
invention which would in time support the Laboratory and enable each
man to go on with his own hobby.

They selected Edison's phonograph, then little more than a scientific
toy, as having commercial possibilities, and went to work to perfect it.
They invented the flat disk and patented a method of engraving its wax
surface. A gouge-like instrument attached to the diaphragm cut out a
small shaving in the hard wax, making a groove in which the needle of
the transmitting diaphragm could travel and, in travelling, vibrate the
diaphragm and reproduce the sound. When Edison's associates discovered
what the Volta Laboratory was doing, they revived Edison's interest in
the phonograph and induced him, so I have been reliably informed, to
purchase the new patents which Mr. Bell and his associates had taken out.
Mr. Bell's share of this money was added to the endowment of the Volta
Bureau.

I was naturally very proud when Mr. Bell asked me to join the American Association for the Promotion of Teaching of Speech to the Deaf, the
"A. A. P. T. S. D.," as it was called in the family. As an officer of this
Association I came in touch with the teachers, and worked with Mr. Bell
in building up a magazine called the *Volta Review* which was designed to
interest not only the deaf but the general public as well in all the problems
of the deaf, particularly of those who had lost their hearing in childhood.

That January I gave two lectures on the subject of plant introduction
before quite different audiences. The one in Trenton, New Jersey, was
only as mildly interested in plants as most people are who spend the whole
of their lives surrounded by buildings; but the other audience was the
Massachusetts Horticultural Society, then next to the oldest, and the best
known in America. I spent a great deal of time on this address, and the
Society gave me a royal welcome, and some years later honored me with
the George Robert White Gold Medal.

That midwinter trip to Boston gave me an opportunity to become better
acquainted with E. H. Wilson. Professor Sargent as usual invited me to

lunch and spent the whole day with Wilson and me going over the most interesting of the Chinese introductions and discussing a possible expedition to Kamchatka. I was astonished to learn that Wilson had never attempted to send in cuttings or scions of plants for budding purposes, but had confined his shipments to seeds and bulbs. Both he and Sargent were incredulous when I told them how many of Meyer's most successful introductions had come as living budwood or cuttings.

I had learned to enjoy the Professor's sarcasm and the pitfalls he set for my ignorance about some species with which he was familiar, even though his remarks often seemed rather severe and almost brutal. Sargent's points were usually well taken when pertaining to ornamental plants, but both he and Wilson showed an indifference to economic species which was often annoying to me. Wilson was even guilty of sending us some naked barley marked "wheat." This lack of interest may have been due in part to Professor Sargent's belief that it was impossible to get people to eat new foods. In fact, he once told me that it was useless to try to introduce any new vegetable as it simply could not be done. But as this was in the face of a certain degree of success which we had already had, I discounted his opinion in regard to crop plants. It seemed difficult for him to appreciate that our activities in the Department must necessarily be directed along economic lines. This is indicated by a sarcastic remark of his, which I never forgot, to the effect that we botanists in Washington were interested only in things to eat.

When in Boston, I always spent as much time as I could with Jackson Dawson, for his skill as a propagator aroused my profound admiration. Any man who can successfully propagate a wide range of plants, and who knows from experience the various methods of budding, grafting and growing plants from cuttings, deserves a high degree of respect. This type of horticulture has attached to it a mysterious charm, the charm of making plants perform, of creating many individuals out of one. In Dawson's little greenhouse, where every cutting he stuck into the ground seemed to take root instantly, I found a number of pots of *Actinidia polygama.*

"Do you know," he said, "the cats have somehow discovered that this new Chinese plant is good to eat, and they give me any amount of trouble in its propagation. I keep an old cat here in the greenhouse to catch the rats, and she has bitten off the potted plants nearly to the ground. For a long time I couldn't find out what was cutting back my actinidia plants, but one day I saw some cat's hairs on the plant and later saw the cat at work. The large vine which we set out in the arboretum was clawed to pieces, until we put netting around it."

The fact that the Boston cats in two years had learned to eat a newly introduced vegetable, whereas Professor Sargent had just declared that I could not induce the people of the country to touch one, struck me as so amusing that I published a note on the subject in *Science,* and was rewarded by a letter from Edward Everett Hale, who had also had a similar experience with an actinidia vine.

CHAPTER XXIX

Florida in 1912

I T WAS NOW FOURTEEN YEARS since my visit to Florida in 1898. During this time we had sent many thousands of plants there, and I felt that it was high time for me to see them. Scarcely a week passed, summer or winter, without a call from some one who had a new home in that great southern State. These visitors always came begging for plants, and generally had but a vague idea of the minimum temperatures prevailing in their particular localities. This ignorance, combined with our own inexperience regarding the frost-resistance of the plants we were introducing, necessarily caused considerable wastage of plant material.

Marian had never been in South Florida and was keen to see our experiments, so it was with high spirits that we left Washington on the seventeenth of February, 1912, headed south. Upon our return to Washington, I wrote up the experiences of this trip in an illustrated report which I have before me, and which I find extremely interesting after the lapse of a quarter of a century. Naturally some of the predictions which I indulged in have not come true. This I realized with a touch of sadness when I stood the other day and looked at the wreck of the Florida Overseas Railroad, which in 1912 had recently been completed and now is a mass of ruins, great stretches of track having been swept away by the hurricane of September 2, 1935. This overseas line was expected to do wonderful things for Key West and Cuba.

Farther on in the report I wrote with considerable temerity: "I would not be surprised to see the day when one hundred thousand people are living along the coast between Palm Beach and Homestead. During my visit in 1898, when people talked of the East Coast as the Riviera of America, I thought that they were dreaming. Now, in 1912, I believe that this is likely to come about. The drift of all classes of people from the cold north southward is one of the most interesting spectacles in American life. I believe that it is a psychological phenomenon which will grow immensely during the next few years, and affect the development of subtropical agriculture more rapidly than any other existing factor."

Those were the early days of the grapefruit industry, and it was exciting to hear from Mr. Ingraham that "at Kendall, sixteen miles south of Miami, a grove of twenty acres, containing 2000 grapefruit trees belonging to the Florida East Coast Railroad, when nine years old netted $11,430, or $567 per acre." This grove, now thirty-five years old, deserves a place in the annals of Florida because it represents one of the early, successful groves of a new fruit which until 1880 was practically unknown to the markets of the world. According to Harold Hume in his excellent work *The Cultivation of Citrus Fruits,* "somewhere between 1880 and 1885 the first pomelos (grapefruit) were shipped from the state of Florida, sold in New York and Philadelphia and netted the shippers about fifty cents a barrel." 1880 is not so long ago, and today grapefruit not only is sold in every market in the United States but has invaded the principal markets of the old world as well.

My report continued:

"We drove to the Brickell Avenue garden at about six-thirty in the morning. Simmonds showed us through Swingle's jungle of citrus hybrids, and Mrs. Simmonds insisted on our living in the house at the garden. This stay was altogether a memorable experience; seldom have I enjoyed so close an acquaintance with plants as during our week there. I am convinced of the fallacy of large laboratory buildings. A shanty in the midst of plants is better. I believe the Bureau is wasting thousands of dollars and infinitely more in lost inspiration, by not housing its men in the midst of the plants."

As I was supposed to "cover" Florida in this trip, we spent only a week in the Miami garden, but it converted me to the possibilities of South Florida as a place into which interesting tropical plants could be introduced and established. The years which followed have only served to strengthen my conviction regardless of the fact that real-estate booms, hurricanes, freezes, and depressions have held back the development of horticulture much more than any human being could have suspected was possible.

At Brooksville, Florida, a field of Dasheens (Taro)—a vegetable with the mealy texture of a potato and a flavor of chestnut. (Page 378)

The huge, melon-like fruits of the Papaya grow on what is probably the most rapid-growing but short-lived tree in the world.

Florida in 1912

Edward Simmonds was in charge of the little garden, and was busy testing the avocado and mango varieties which the Office had been gathering from different parts of the world. He had mastered the art of propagating tropical trees, and his slat-houses were filled with neat rows of wooden boxes in which were young plant immigrants ready to be distributed as rapidly as possible to experimenters throughout the region.

The *Carica Papaya* was practically unknown in the Miami region and I had taken particular pleasure in securing all the varieties I could from the West Indies, Java, the Philippines, South America, East Africa, Mexico and Panama. There were some eighty or more introductions, and it was a delight to see many of them growing side by side in a well-regulated testing garden. The enormous melon-like fruits of the papaya hang on what is probably the most rapid-growing but short-lived tree in the world, and are rather startling when one sees them for the first time. The great peltate leaves alternate with the fruits and spread a curious pattern of shade on the ground.

The problem which worried dear old Simmonds about the papaya was the fact that it is a diœcious organism, and you cannot tell by looking at the seedlings which ones will turn out to be male and which female. Therefore, half the trees in a plantation might be male and unproductive. He had taken a particular fancy to a variety from Panama which he always spoke of as "28533" and for many years he propagated seedlings from this papaya in preference to all others. Unlike the flowers of many others, its flowers had, at the base of the ovary, small, poorly developed stamens which produced good pollen, and these were apparently responsible for maintaining a certain uniformity in the strain. Plants grown from the seed of self-fertilized flowers proved somewhat less variable than those produced by the crossing of male and female flowers from different trees.

It was my old favorite the mango, however, which attracted my especial attention in the little garden. There I studied the collection which I had gone to great pains to assemble. The trees were coming into bloom, and the long sprays with a thousand flowers on each are an attractive sight somewhat resembling the bloom of the American chestnut.

A serious disease had already appeared which loomed as a very real cloud upon the horizon of the mango industry. This disease I would have called in my younger days "an interesting, fascinating fungus," similar to one which I had studied on the apple and called *Colletotrichum glœosporioides.* It was believed to be responsible for the black spots on the mangos which marred their appearance and often made them unsalable. Knowing something about the virulence of this genus of fungi, I was glad to see that the men were busy spraying the trees with Bordeaux in an attempt to circumvent it. I suspected that the dead flower stems which were allowed to remain on the trees carried over the spores of this fungus from season to season, and I soon proved to my satisfaction that this was the case. Later, a young entomologist named McMuran, working on the diseases of tropical fruits at the Miami laboratory, successfully sprayed mango trees when in bloom, preventing the disease without injuring the flowers or affecting the setting of the fruit.

The great freezes which had visited Florida, particularly those of 1894 and 1895, had made us cautious about planting young mango trees in the open. Reasoner Brothers at Oneco had built a tall, expensive slathouse for those which they were growing, and had maintained this even after the trees came into bearing. Not wishing to have our mango collection wiped out, we also planted our newly introduced varieties under a slathouse. But, as the years passed, these trees grew up through the roof, and we discovered that, at least along the East Coast from Palm Beach southward, young mango trees in the open, while they might be killed back to some extent by an unusual freeze, soon recovered and needed no slathouse protection.

In 1912 the possibilities of the avocado (*Persea americana*) becoming an industry appeared more promising than those of the mango. We had gathered together about thirty varieties to test before distribution through the State. Among them was a Guatemalan variety which G. N. Collins of the Department had collected in 1906 in Guatemala, and which later was named for him. It originated a line of valuable seedlings.

While we were in Miami, Mr. Ingraham wired me that Henry M. Flagler would like to see me if I would come to Palm Beach. I had always wanted to thank him for giving the little laboratory to the Department, and was glad of this opportunity to meet him. This visit was towards the close of his remarkable career, and he was in poor health. We were shown into the large drawing-room at Whitehall which had been darkened because Mr. Flagler was having trouble with his eyes. We had to wait some time as he had two stenographers with him, being still quite active in business notwithstanding the state of his health. When he came into the room, he wore a green shade to protect his eyes from the light and spoke pathetically of the way in which, as he expressed it, life was closing in on him.

Like most of the business men with whom I came in contact, the obvious crop plants appeared to Mr. Flagler the most important. He was strongly of the opinion that there was a great future for sugar-cane in the Everglades, and said that a big company was planning to plant a thousand acres. I am afraid that I was not very successful in interesting him in the avocado and mango, for his mind had been too long focused on "larger game," and our little beginnings seemed of small importance to him.

The interview was brief and rather formal, but I was glad to assure him that the acre of land and the thousand dollars which he had contributed towards the establishment of the laboratory had been well spent. From the view-point of 1938, I believe that one could easily prove that this thousand dollars was better invested than some of the millions which went into the railroad development.

I paid a visit to the old Mulgoba mango in West Palm Beach. Professor Gale's son George told me that the tree had been blown down twice and put back into place with a block and tackle. It had been badly neglected, had not fruited for seven years, and he had ceased to fertilize it. I tried later to persuade him to give this tree to the Department so that it might be preserved, but he took no interest in my proposal and refused to answer my letters.

W. J. Krome, the young, active engineer whom Mr. Flagler had put in charge of building the Overseas Railroad, was at heart almost more interested in commercial horticulture than in that great engineering work. He and his associates had homesteaded land around what is now the town of Homestead, and his own well-cared-for place was run on a paying basis. It was there that I first saw a tangelo in fruit. Krome had two bearing trees, an acre of young plants, and was planning to put out ten additional acres the following year.

The idea of draining the great Everglades and converting them into farm land had captured the imagination of the public, and a series of canals had been planned, leading from Lake Okeechobee to the coast. The question as to whether draining the Everglades has been a mistake is too controversial to discuss here. It is true that almost everybody except expert soil chemists had an exaggerated idea of the fertility of the Everglade soils. Also, the hazards from frosts and flooding have proved greater than any one considered possible. This is, I believe, admitted today by all whose memories go back far enough to give them a perspective on the situation.

Some business men of Miami who were interested in the sale of Everglade lands had established an experiment station at a little settlement called Davie, west of Dania. The water table was only two feet below the surface. I see from my report that the situation impressed me as somewhat similar to conditions in the Sacramento Valley, for the soil was a mixture of peat and muck and would not, in my opinion, be true land for many years. Two feet below the surface, the roots of the trees would strike water, as they had in Sacramento, and would refuse to bear fruit of good quality. Although there seemed to be a fair flow of water in the drainage canals, I thought that keeping them open would prove expensive. The temperatures on the Glades would be lower than on the rocky soils about Miami, and altogether the outlook impressed me as precarious.

The most practical way to reach the West Coast in 1912 was via Key West. This appealed to me, as I wanted to see Mr. Krome, who had not been at home when we visited his grove, but was at Marathon, the headquarters for the Overseas Railroad construction gangs. I was surprised to find him a slender, wiry young man who looked far too youthful to hold so important a position. We were told that Flagler's chief engineer had died, and Krome, his assistant, had been promoted when he was still in his twenties and put in charge of a construction program involving millions of dollars and requiring ingenuity and originality of a very high order.

One of the difficulties in building the gigantic causeway to Key West was the possibility of hurricanes during the autumn months. This was before the days of radioed weather reports, and Krome told us that he carried a barometer and looked at it every hour during the hurricane season. At any decided drop in the pressure, he was prepared to jump on an engine which he kept always ready, and rush down over the line. Great barges loaded with the expensive machinery were anchored at night inside a framework of piles. At the first signs of danger, the plugs in their bottoms were taken out and the barges were scuttled to prevent their being blown away. Before Krome had learned to do this, a group of his men living on one of the barges had been blown out to sea in a storm. They were picked up by passing vessels, and one man was taken to Australia and did not turn up again for six months.

Krome told us that on No Name Key an old Russian was growing some very large sapodillas (*Acras Sapota*). Krome said that the fruits were the size of saucers and of excellent quality, and he proposed that we go in his launch to see them. We chugged over to a rickety old wharf which led up to Nicholas Mateovitch's house.

Mateovitch was a picturesque old fellow, with a great shaggy beard, large head and perfectly enormous hands. After fighting in our Civil War, he had homesteaded a hundred and sixty acres on the Key in 1868, and, although he had a wife and son in Key West, he had lived as a hermit on this little island for forty-three years. In a haphazard way, he had planted patches of fruit trees and vegetables, and as he was particularly fond of sapodillas, he had many trees of this delicious fruit. More than once he had nearly starved to death and doubtless would have done so had it not been for Mr. Krome's kindness in sending him food from time to time. He was morbidly suspicious and set gun traps in the brush, stretching invisible wires about, so that visitors to the Key were in danger of being shot. As we were with Mr. Krome, we were received with great cordiality and even allowed a peep into his shack, which was incredibly disorderly and unkempt.

Unfortunately his sapodillas were not in fruit. The difficulties of propagating this tropical tree have prevented its becoming an orchard crop of South Florida. Large fruited seedlings exist, but even today there are but few budded trees of them.

It was so late when we started back that we missed the down train to Key West, and Mr. Krome had to flag a freight for us. It was a glorious moonlight night, and we rumbled over the viaducts with the side of the car open, enjoying the silver light on the water. Even the meager vegetation on the Keys created mysterious shadows on the white sand.

We stopped off at Sugar Loaf Key to see Mr. Charles W. Chase, who was conducting a sponge farm in the lagoon there. He had advertised his Key as an ideal spot for the cultivation of tropical crops such as sea-island cotton and sisal, both of which had been sent him by the Department. As there was salt water two or three feet under the surface all over the Key,

The World Was My Garden

I made up my mind that the possibilities for either cotton or fruit trees were very limited. In fact, I felt that there was little promise in the agriculture of nearly all of these Keys and that they would be a long time developing.

Mr. Chase's sponge farm was the first undersea agriculture that I had ever seen, and I was fascinated by the technique which had been developed. Men and women on floats anchored in the shallow lagoon were attaching bits of living sponge to flat disks of cement a foot in diameter, and then throwing them as far as they could in every direction. Mr. Chase said that these bits of living sponge grew rapidly on the cement disks, and in two years' time an excellent bath sponge was formed. Bath sponges were then bringing a fair price on the market, and a good many people had bought stock in his company.

I noted in my report that there appeared to be two unsettled questions. Would the sponges grow wherever the disks were thrown, regardless of the condition of the lagoon bottom? And what rôle would disease, for there seemed to be diseases, play in this artificial propagation of the sponge? I think the disease factor was involved in the failure which finally overcame the sponge farm, but also the radical change in the technique of the American bathroom made the market for sponges less profitable year by year.

Key West was quite a busy place with its fishing industry, naval stores and railroad shops. Many people considered the possibilities of Cuban development to be phenomenally promising. Americans were not only visiting Cuba in greater numbers, but were acquiring land in Cuba, particularly in the Isle of Pines.

Marian and I spent a night in the principal hotel at Key West, and I noticed that the bathtub equipment was imperfect, and the overflow pipe had not been attached. It seemed a quaint method of conserving the limited water supply while, at the same time, offering the guests the semblance of a bathroom.

A year or so later, this same bathtub got Douglas McCurdy into serious difficulties. He was in Key West waiting for favorable weather to attempt the first flight ever made to Cuba, and stayed at this same hotel. Paying no attention to the sign saying that guests were responsible for any damage caused by water from the bathtub, he let the tiny trickle of water run all night, hoping to have enough for a bath next morning. Instead of a bath, the manager presented Douglas with an exorbitant bill for damages, which Douglas refused to pay on the ground that if you are charged for a bathroom you are entitled to water and plumbing.

In the altercation which ensued, Douglas suddenly realized that the hotelkeeper was going to attach the plane. He hastily appealed to an officer on the United States cruiser which had been detailed to follow his flight, and the officer agreed to buy the plane. So the two young men jumped into a taxi and just managed to beat the infuriated hotelkeeper to the magistrate's office, arriving there in time to make the legal transfer before his very eyes.

I do not know how many people there were in Key West at the time of our visit, but I am confident that no human intelligence could have foreseen the plight in which the town finds itself today, with the Overseas Railroad destroyed and 10,000 people isolated there, forced by circumstances over which they have no control either to migrate to other regions or to grow poorer with each passing season. Man boasts of his abilities to control the economic forces which surround him, but nature proves his words only a boast. Still, maybe the new motor road will solve this problem.

From Key West we went by boat to Tampa and made our way to Oneco, which is near Bradenton. I shall never forget my first visit to the great Reasoner nursery. Our train brought us into the station at about six o'clock and, after a cup of coffee, we drove out to the nursery where Mr. Egbert N. Reasoner met us. As we stood there in the early morning sunlight, he began talking about his plants, and I felt that I had never met a more charming plantsman in all my travels. He spoke of his brother Pliny and explained that he had kept the firm name "Reasoner Brothers" in his honor, although Pliny had been dead twenty-four years. It was natural that he should hold him in high esteem, for aside from being a lovable character, Pliny Reasoner was without doubt one of the first and greatest of the horticultural pioneers of Florida. His passion for the new plants which he introduced from all over the world, and his practical abilities as a grower and disseminator, won for him an enviable reputation, although he died of yellow fever at the early age of twenty-five, after only six years in Florida.

We were much impressed by the Reasoner collection of palms and keenly felt the desirability of procuring additional species for trial in the

State. Also, the whole nursery inspired me with a desire to broaden our field in order to include other than the obviously economic species of plants.

I was delighted to see plants of the Chinese litchi. Augustine Henry had told me of the wonderful varieties in the island of Hainan and said that this fruit had saved his life at one time when he lived on litchis almost exclusively during a severe illness contracted there. With great optimism, I believed that litchi-growing would rapidly develop into an industry in southern Florida, but unfortunately the tree is not easily established. Enough, however, have been grown to make it apparent that commercial orchards will come into existence as soon as certain points in the technique of establishing the young trees have been worked out.

My stay with Mr. Reasoner was altogether too short, and I sincerely wished that I had not undertaken to see all of Florida in so short a time. Twenty-four days in which to form an idea of the plant industries of a vast commonwealth was only long enough to form snap judgments and prejudices. However, in my case the prejudices were in favor of the State and not against it, and after this lapse of years I find little to revise with regard to my general impression of the possibilities of horticulture which await development in Florida.

From Tampa we continued to Brooksville, where we found W. H. F. Gomme, the superintendent of the garden, awaiting us. The major problem at the moment was how to propagate the bamboo. In Japan, where bamboos are abundant, they merely dig up an offshoot of a considerable size and transplant it. But we had a limited number of clumps, and we wanted not scores, but thousands of young plants. In the available literature, a method of propagation by means of rhizome cuttings was recorded, but we had never seen it done. Gomme and I therefore set ourselves to find out how pieces of rhizomes could be rooted satisfactorily.

There was real beauty in those yellow rhizomes, and we found that their buds, from which the shoots grow, were almost as brittle as glass, and had to be handled with great care. Once broken, they either died or only made a rosette of leaves and "croaked," as nurserymen are wont to say, perhaps remaining alive for years while producing only roots and a few leaves without a single shoot.

The dasheens had grown amazingly in the rich soil between the stumps, and Gomme had learned to bake their tubers in the ashes of the burning stumps as he cleared the land. No food ever tasted more delicious than those small dasheens which we cooked and ate out in the field.

From Brooksville we went to Gainesville, the seat of the University of Florida, to check on the new plants we had sent to the State Experiment Station for trial. I was pleased to find Chinese persimmons doing well, and to see that the dasheens had come through a rather severe winter. To my surprise my pet Kaffir orange, *Strychnos spinosa,* had withstood a temperature of 24°.

The interesting experiments at both the Miami and Brooksville gardens convinced me that some diary of their doings should be kept. On my return to Washington I had impressive books made with neat gilt headings and gave instructions that all the important events should be entered, with the observations of the man in charge of the Garden about the behavior of the newly introduced plants. Alas, it is much easier to have such a book prepared than it is to make a gardener write in it. The events came thick and fast in these gardens, the stream of callers was a continual interruption, and the men were more at home with trowel than pen.

CHAPTER XXX

Monsters of the Backyard

W E RETURNED to Washington March 13, in time to see our cherry trees come into bloom and to harvest our udo and try it out on various of our distinguished friends.

The summer was as busy as ever, but by this time the machinery of the office had become organized so that it ran more smoothly, even though the volume of work was gratifyingly large. I was quite proud of the totals. During the year we had distributed over forty thousand plants from the introduction gardens, ten thousand copies of the little bulletin called *Plant Immigrants,* and four numbers of the *Plant Inventories.* The total budget in 1912 was over $80,000 and I had a staff of sixty-eight employees.

After the long hours at my desk and endless conferences, my holidays and Sundays were spent pottering around my garden or in my little study where, besides my microscope, I also had a dark room and a special camera capable of photographing microscopic objects. The world under my microscope was a world of quiet, where no insistent emotional personalities could enter, and, although I belonged to the Chevy Chase Golf Club and passed it every day, I could never find recreation in playing there. The club and links were always full of people, and I could not get away from the sound of voices and those problems and prejudices which form so large a part of casual human conversation.

One Sunday afternoon, using my microscopic camera, I succeeded in enlarging the head of a large milleped into striking proportions, and I then turned my attention to a grasshopper which I had picked up near the door. I was amazed by the result, for it made me realize that I did not know what insects looked like. I had always viewed them from above, as one sees them pinned in boxes by the entomologists.

Monsters of the Backyard

About this time E. L. Crandall began experimenting with long focus lenses and I used to visit his studio when he was photographing our new plants as they fruited in our gardens. He showed me a large lens, a twenty-five-inch focus, Cooke's anastigmat. As I held it in my hand, I unconsciously pointed it at an iris on the table, and there flashed before my eyes a greatly enlarged picture of a fragment of the flower, which grew larger as I moved my head away from the lens. The magnification was less than that which my small Zeiss six-inch lens gave me, but the field was so much larger that I decided to borrow this lens and try it on my grasshopper. In my little carpenter shop, I improvised a camera twelve feet long out of thin boards, blackened on the inside, and fitted the contraption with a piece of ground glass at the farther end from the lens.

The first opportunity I had to try it out was the Fourth of July, a boiling hot day with a temperature of 102°. For hours we tried to make the thing work, shoving the box around on the grass and propping it up on stones and turf as I lay prone with my head under a suffocating cloth. Marian kept putting grasshoppers in front of the lens but I could not get the focus, until suddenly, late in the afternoon, when I was pretty well worn out, there appeared on the ground glass a king grasshopper so marvellous, so enormous, that it startled me. This opened up a new world, the world of familiar objects magnified three, four or ten times, as contrasted with the microscopic world with which I was familiar.

I determined to make full-sized photographs, and ultimately elongated my improvised camera to twenty-five feet. I also evolved outdoor methods of lighting the under side of the insects; but for some time we were completely defeated when it came to mounting the insects so that they would look lifelike, for the exposures were often a minute in duration.

When we killed the insects, their legs stiffened and curled under them and all the pictures were obviously the photographs of corpses. I tried to fasten their legs down with glue or sticky flypaper, but with no result. One Sunday afternoon, as I racked my brains for some way to mount the insects as a taxidermist mounts a bird, my eye fell upon the butt of a candle. It seemed an answer to prayer, and I dripped some candle wax on a block of wood and let it harden; then, with a needle heated in the candle flame, I softened a tiny spot of wax under each leg of the grasshopper in turn and, on cooling, this held the creature's leg in place. The white surface made the insects look as though they were on a block of

The World Was My Garden

ice, but I improved my technique by laying a leaf on the wax and mounting the insect on it by puncturing it with the needle and allowing a little drop of wax to rise through the leaf and harden into a tiny, almost microscopic mass beneath the insect's foot. Marian enjoyed this part of the work more than I did—except for the time when she mounted a praying mantis in such a life-like pose that it haunted her in her dreams. The poor creature had simply refused to die. After being left for hours in a lethal chamber, it would slowly turn its head and look reproachfully at her.

The photographing of these "monsters," as we called them, became a fascinating pastime. We would rise at dawn to take advantage of the stillness, for a breath of wind would make a fly's antennæ quiver and blur the picture, and there was no way to fix antennæ. Holidays, early mornings, late afternoons, and even my month's summer vacation were all too short in which to photograph even a fraction of the insects all about us.

Thinking that Mr. Bell would be interested in the pictures, I sent him some. Characteristically, he immediately wired back:

"Your pictures are magnificent. Go ahead and publish them."

However, my entomological friends did not like the photographs. As L. O. Howard, Chief of the Bureau of Entomology, expressed it, they made his familiar acquaintances look like hideous monsters.

Doubleday, Page and Company were bringing out illustrated books on insects and I wrote to Walter Hines Page, asking if I might show him mine. He invited me to luncheon at Garden City and was delighted with the pictures, declaring that I had opened up a new field of illustration. Mr. Doubleday was not so enthusiastic, and they finally suggested that they might publish the best one as a frontispiece or something of the kind. But I was not prepared to give up my project of publishing the whole collection, and, closing my albums in disappointment, I brought them back to Washington.

Gilbert Grosvenor took a different view. Offering me the largest honorarium *The National Geographic Magazine* had ever paid for a contribution, he published thirty-eight of my insect photographs in the May, 1913, *Geographic,* with an article I wrote entitled "Monsters of Our Back Yards." This attracted wide attention, and determined me to go ahead and make more photographs to publish in book form.

The Book of Monsters, over which Marian and I spent many nights

396

The Author's improvised camera, affectionately called "Long Tom."

This member of the Horsefly family had a ferocious look when enlarged.

396A

The King Grasshopper eyed me through the lens with a rather saucy stare.

In Florida we devised a caterpillar-like variety of our "Long Tom" camera to photograph the tropical fruits of the Kampong.

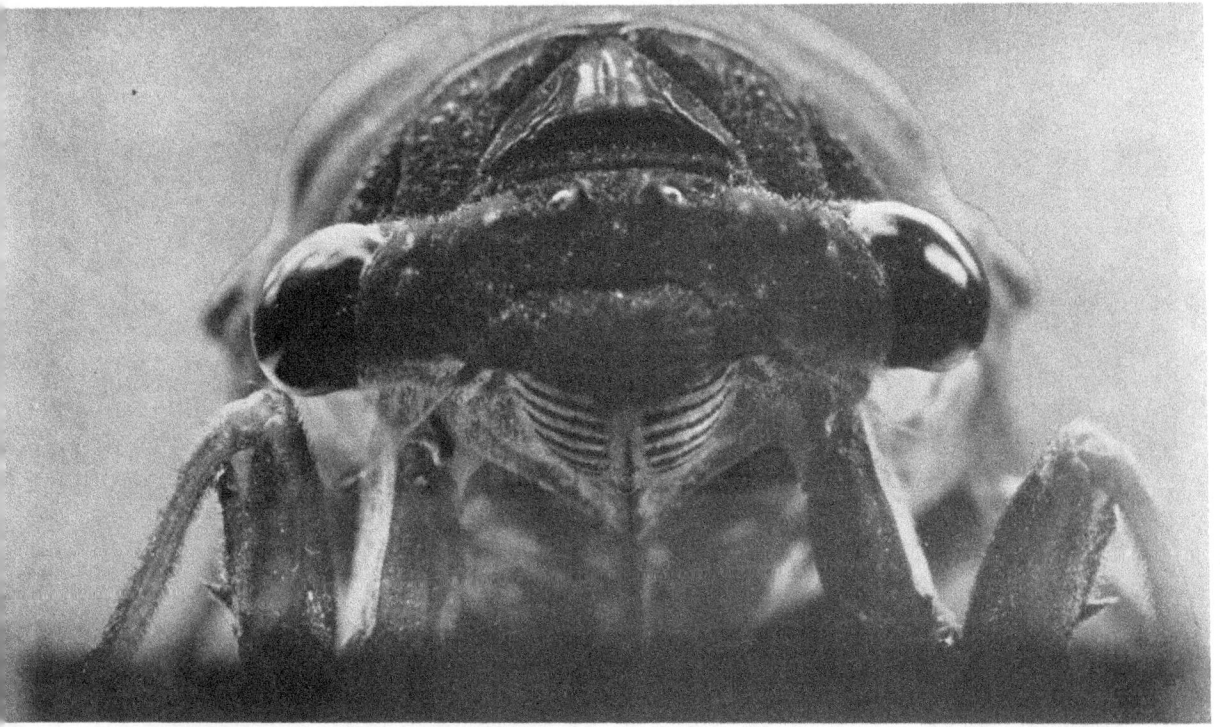

The protruding eyes of the cicada.

A turret-making spider.

396c

Frank Meyer in April, 1911, camping in Chinese Turkestan at an elevation of 7800 feet.

A White-Barked Pine in the courtyard of the Jade Buddha Temple of the Winter Palace in Peking.

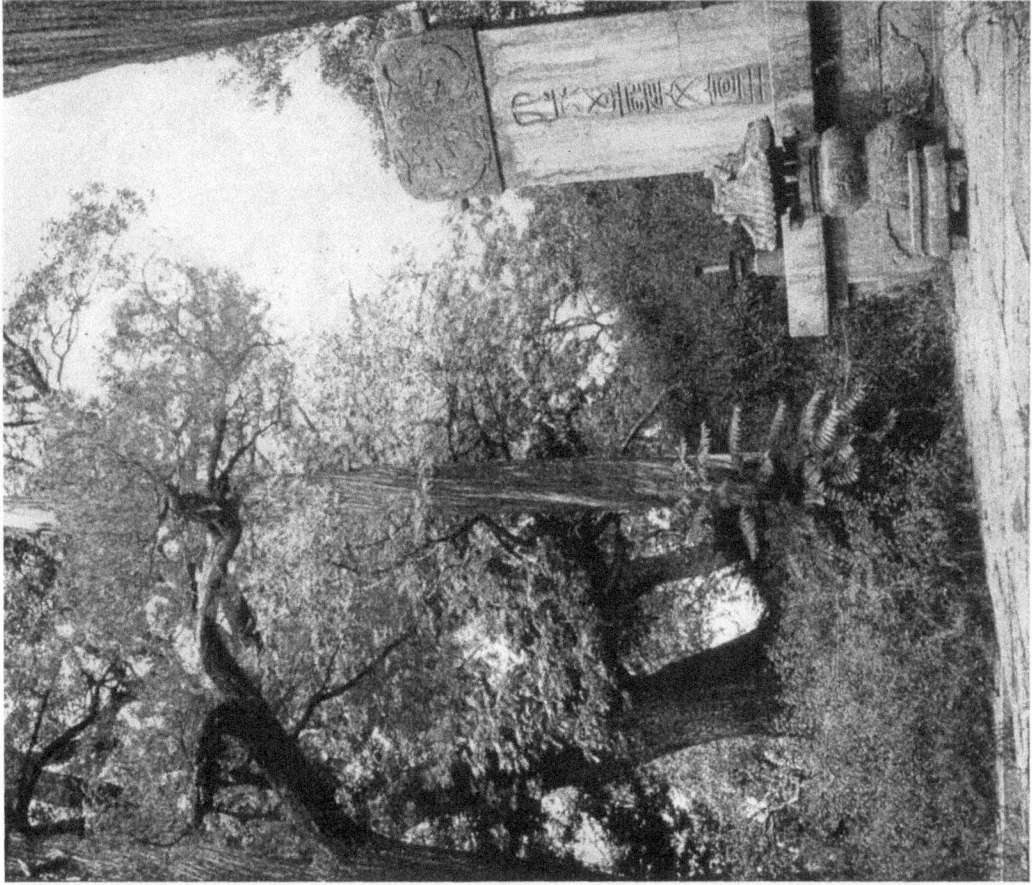

Frank Meyer sent a photograph of the tomb of Confucius, shaded by a venerable Chinese Pistache tree.

One of the best fruiters of Meyer's Chinese Chestnuts (*Castanea mollissima*) is still living on Howard Dorsett's place at Glenn Dale, Md., where it has borne many crops of fruit.

396F

Monsters of the Backyard

and holidays, was finally published early in July, 1914, by the National Geographic Society. On July 28, 1914, the day War was declared, Gilbert Grosvenor showed me a cablegram from Sweden asking for the rights to translate the book into Swedish, and also a letter from *The Illustrated London News* requesting the rights to reproduce a certain number of the illustrations. Of course, the War interfered with the success of the book; only five thousand copies were sold, and the Swedish publisher withdrew his offer. But from many letters received we had the satisfaction of knowing that we had interested a number of people who heretofore had not known the difference between a bug and a beetle. And *The Illustrated London News* published half a dozen of the monster pictures, enlarged until they made half-page illustrations.

But to return to the summer of 1912. Swingle was still trying to solve the riddle of the Wahi date, and we had commissioned Aaronsohn to go from Palestine to the Sudan to find it. Aaronsohn sent in thirteen other varieties and believed that he saw the Wahi date, but he did not secure suckers of it.

The arrival of this shipment of dates was written up by an amusing reporter on *The Washington Star*—a man badly crippled with rheumatism who had discovered in the happenings of the Office of Plant Introduction a field for his humorous stories. (He had written one with regard to the udo which ended, "You may think u do know the udo but you don't.") This shipment from Aaronsohn came in the early days of plant quarantine and the cases were deposited in the quarantine house. Every possible precaution was to be taken in order that no microscopic insect on the palms might escape to infect the young date orchards in California and Arizona. When the cases were opened, a stowaway in the form of a large rat ran out and hid among the various sacks and boxes of the receiving room. The reporter immediately raised the question, "which division of the Department of Agriculture could claim the rat?" Did it belong to the Bureau of Plant Industry which had brought it in, or the Bureau of Animal Industry, or the Bureau of Public Health, or the Biological Survey? He pictured all of these various bureau chiefs running post-haste to the quarantine house to put in their claims, while, in the meantime, the workers there had struck, on the grounds that the Government did not employ them to kill rats.

397

The World Was My Garden

Professor Sargent turned over to us a collection of E. H. Wilson's introductions which were unsuited to the climate of Boston, as the Arnold Arboretum had no adequate machinery for distributing them for trial elsewhere.

In this collection was an Oriental gum tree, *Liquidambar formosana,* belonging to a small genus of plants of which only four species are known, one being the gum tree of our southern States. The fossil remains and leaf impressions of the genus indicate its wide distribution in Tertiary times. It was not surprising, therefore, that, like the American tulip tree and hickory, which also are ancient forms of vegetation found both in eastern America and in China, this closely related species of our southern gum tree should occur in China and Formosa. The foliage of this Oriental species turns crimson in the autumn, like our native gum tree, and we were glad to plant it at Brooksville, Savannah and other places in the South. To our great disappointment, destructive cankers appeared on its branches. They were caused by a fungus, believed to be of American origin, which is found on the native species but does little damage to it. This canker fungus has played havoc with the introduced Liquidambar tree, and has threatened to exterminate it.

This case illustrates an interesting fact. It is well known that an introduced species of plant may bring into a new country a fungus (like that of the chestnut bark disease) which may wipe out a species of plant closely related to the one on which the disease has come in. However, because of its higher resistance, the carrier itself is little bothered by the disease. In the instance of the Liquidambar tree, the newly introduced plant fell a prey to a parasite on our native species which does only minor damage to our gum trees. The subject of immunity in plants, as in animals, is a complicated one, and much remains to be done in the way of explaining its chemical background.

During that summer, Meyer was back again and we saw a good deal of him at "In the Woods." His second expedition had covered more territory than his first one, taking him through Europe, the Crimea, the Caucasus, both Russian and Chinese Turkestan and into Siberia. He was full of interesting and valuable information and I was most anxious that he put his observations into permanent shape in the form of a bulletin. His letters, properly translated into correct English, would have made a fascinating narrative, but would not have been prosaic enough for a government publication. As he found it difficult to write a technical bulletin, I used to get him into my little laboratory, where he liked to lie prone on the floor and talk while I jotted down his remarks, modifying his characteristic expressions only enough to pass the censor. Thus together we wrote an interesting but far too abbreviated account of his remarkable explorations.

Meyer found the summer months in Washington unbearably warm and grumbled a good deal about the heat, for he was essentially a Northerner. By the time cool weather came again, he was keen to return to China and explore the northwestern Provinces and if possible reach the borders of Tibet and spend considerable time in the Kan-su province. He was such an interesting companion that I hated to see him leave in November on what I realized would be a difficult trip, but which turned out to be both difficult and dangerous.

It was during 1912 that the Montessori Method appeared upon our horizon. Mr. S. S. McClure dined with us one night, just after his visit to Doctor Montessori's school in Rome, and was full of enthusiasm over her method of teaching young children. He was about to publish an account of it in his magazine.

Mrs. Bell had always been quite out of sympathy with the prevailing kindergarten methods. She became so interested in Mr. McClure's description that she went to see the Montessori class which Miss Anne George was conducting on Mrs. Frank N. Vanderlip's estate on the Hudson. The initiative and interest shown by the children, their concentration, orderliness and helpfulness delighted Mrs. Bell and she determined that her grandchildren must have a Montessori education.

As Mrs. Vanderlip's class was to be discontinued, Mrs. Bell brought Miss George and her friend, Miss Roberta Fletcher, to Washington and established a Montessori school in her own house. Our children, Graham and Barbara, and several little Grosvenors were among the members of the first class.

The school soon outgrew Mrs. Bell's big room, and she bought a house on Kalorama Road which she turned over to a Board of Trustees composed of interested parents. This school played an important part in introducing Doctor Montessori's ideas into America. Mrs. Bell remained the godmother of the school as long as she lived, and followed

The World Was My Garden

the careers of all the little "Montessaurs" as we used to call them, with the greatest interest.

In the autumn a new baby was born into our family, Nancy Bell, who was so tiny that only dolls' clothes could fit her, and Marian was very ill for a long time after her birth.

By January Marian was able to leave with me for a fortnight's vacation in Florida. The mango buds were bursting when we arrived at the Brickell Avenue garden, and the hundreds of plants with their explanatory tags showed the excellent care which Simmonds had bestowed upon them.

By this time Miami had begun to attract distinguished visitors, among them William Jennings Bryan, who had rented a house quite near the little garden. When Simmonds and I called on him, he met us at the door, true commoner that he was, and invited us to walk with him to some land which he had purchased and was clearing in the hammock on the road to Coconut Grove. He expressed his interest in the plants which we were introducing and his desire to experiment with some himself, but he had little room, for the native growth of the hammock came close about the house which he was building.

As we walked down the street, a boy on a bicycle brushed past us, striking Mr. Bryan's leg.

"That reminds me of a story illustrating courtesy," said Bryan, turning to me. "An old man was walking down the street and a boy on a bicycle ran into him and knocked him down. He got up and, brushing the dirt off his knees, said, 'My boy, I am sorry. You see, I am so old that I could not get out of your way.'"

Mr. Simmonds was much concerned about a fruit fly (*Toxotrypana curvicauda*) which had recently appeared. He had watched it in the late afternoons as it perched on a young papaya and drove its inch-long ovipositor into them, laying a large number of eggs which soon hatched into maggots, filling the interior of the fruit with their slimy bodies. Wild papayas growing in the hammocks were filled with these disgusting-looking larvæ, and the fly had invaded the plantings in our garden. Some varieties escaped because their fruit flesh—the walls of the "melon" —were so thick that the eggs which the fly deposited could not hatch, since they required more air than they could get in the firm flesh. The maggots could only develop in the open cavity of the fruit among the seeds. Numerous methods for the prevention of this fruit fly were being tried, but as the female merely perched on the fruits to lay her eggs, she seemed unaffected by our sprays, and there was no way of getting at the maggots after they had hatched inside the papayas. For a number of years this pest threatened the papaya industry, and then for some unknown reason, it disappeared, at least as a major factor in the culture of this delightful tropical fruit.

Mr. Charles Deering of Chicago called to see us. He was developing an estate on Biscayne Bay at Buena Vista and had sent for O. C. Symonds, the landscape gardener I had met in Chicago. Mr. Deering seemed much interested in plants and talked of creating an arboretum of tropical trees. His place contained some of the finest mangroves on the coast, an old grove of beautiful coconuts, and a magnificent banyan (*Ficus nitida*) which had spread until it covered an area one hundred feet or more in diameter.

We spent several afternoons together and Mr. Deering took me to see some interesting trees which he had discovered in the district. One was a huge rubber tree (*Ficus elastica*) which is still growing in South Miami. As we stood under the spreading branches of this giant studying its many trunks and pendant, aerial roots, the owner came out to greet us. Mr. Deering asked him where the tree came from.

"A fellow by the name of Fairchild gave it to me about fifteen years ago," he said. "It was a little thing in a pot when I planted it."

One afternoon we called on Professor Charles T. Simpson, who had a hammock jungle on Biscayne Bay near Little River. He was a conchologist and an all-round naturalist, having been connected for many years with the National Museum in Washington. Below his house he kept his shells, thousands of which he had collected himself. Like John Muir and John Burroughs he had gathered about him all kinds of trees and plants, which he loved to handle and classify and study. His charming personality and unfailing generosity towards every one who came for information or plants made his place a general rendezvous.

Professor Simpson and I took a liking to each other from the start, and a friendship developed which lasted to the end of his life. In 1912 he had published an article in the proceedings of the Florida Horticultural Society, in which he gave an account of the tropical plants with which he had experimented. It was the first paper of the kind I had seen, and he gave me one of the few remaining copies.

The World Was My Garden

Mr. Deering agreed with me that such a paper should be widely distributed, and offered to pay for an illustrated publication if I would arrange it. The professor always said that this encouraged him to prepare his first book, *Ornamental Gardening in Florida,* a book which did much to encourage experiments with tropical plants.

Simpson had at that time many rare palms. The African oil palm (*Elæis guineensis*) had already fruited for him as had the sugar palm (*Arenga pinnata*) from the Malay Archipelago, and the gingerbread palm of Egypt (*Hyphæne Schatan*). On his muck land he was growing the Philippine Island Nipa palm which we had sent him. Since it has the ability to grow where the soil is impregnated with salt water, I was delighted to find it thriving. We learned that this palm must be protected against the Florida land crabs which destroyed this specimen before it grew large enough to resist them.

The Professor remembered the freeze of 1907, when a thermometer among his pines had registered twenty-three degrees for most of the night, cutting back his young Lebbek tree to the ground. This experience made him cautious in his recommendations to newcomers, even though his passion for trying new tropical plants led him to import many himself.

I spent most of my time in the experimental garden with Simmonds, and, as I look back over the twenty-five years which have passed since then, it is stimulating to realize how many of the plants have become a part of the landscape of the region. The carissa, for example, is now a common hedge plant. There are sausage trees in private gardens as well as the one on the Homestead highway. The *Eucalyptus alba* that Melchior Treub sent from Timor has grown better than most species of this Australian genus and is represented by stately specimens.

Numerous trees of *Ficus Sycamorus* have reached good size. This was the Tree of Life of the ancient Egyptians and is supposed to be the fig tree mentioned in the New Testament. The *Pithecolobium dulce,* a quick-growing tree from Mexico, now lines the roadways of Coral Gables and grows as far north as Central Florida.

The first Para grass was grown in the experiment garden, and attracted the attention of those who thought to tame the Everglades, for it could survive the floods there, being in fact a water grass. Today I imagine that it is looked upon in the Glades as a native grass. Doctor J. G.

Monsters of the Backyard

Dupuis, who developed one of the largest dairies in the State, told me that the dairy had its first beginning from a patch of Para grass which he grew beside his office in Little River.

This second trip of ours to Florida was brief, but increased our longing to take root ourselves somewhere near Coconut Grove. We returned reluctantly to Washington on February 7th.

The following spring, the long administration of James Wilson came to an end. During the sixteen years that he had been Secretary of Agriculture, the Department had prospered amazingly, particularly in its research bureaus. Also, the Department's functions as a regulatory institution, which those of us engaged in research disliked intensely, had become well organized. When "Tama Jim" returned to civil life, the Assistant Secretary, Willett M. Hays, also resigned. Hays was President of the American Breeders' Association and, although I had not been closely identified with this Association, I was strongly in sympathy with it. Professor Hays had frequently consulted me with regard to the conduct of the magazine which was published by the Association, for, like many such periodicals, it was in grave financial difficulties.

Shortly after my return from Florida, a letter reached my desk addressed to me as President of the American Breeders' Association.

"There's some mistake here," I remarked to my secretary. "Send this letter over to Professor Hays."

Instead, she telephoned to him and, to my amazement, Professor Hays said that he had forgotten to inform me that I had been elected president at the last meeting of the Association. Naturally I demurred, for I had no ambition to take over the responsibility involved, both financial and otherwise. But my friends insisted, in particular O. F. Cook, who sank a shaft of argument deeply into my consciousness when he remarked,

"You can't kick a baby off the doorstep, Fairchild. You'll have to take it in."

Therefore, as Cook, Collins and Kearney gathered around me and Mr. Bell agreed with them about the "baby," we set about the reorganization of the Association's affairs. This resulted in its incorporation and my assuming the obligation to raise a guarantee fund for three years to enable the Association to continue its work.

Mr. Bell, Kearney and I were appointed as a committee to draw up the articles of incorporation. After some persuasion, I succeeded in getting an appointment with Mr. Bell in what he called his Retreat, the little cabin on our place in Maryland.

Kearney and I went down there that evening with the necessary papers, but found Mr. Bell asleeep. We waited for an hour or two until about midnight, when he turned over on his lounge and told us that he would be glad to help us when he was really awake. He talked to us of general matters for another hour or so until both Kearney and I were dead with sleep. Then about two o'clock Mr. Bell's faculties revived and by three he brought his wealth of experience and insight to bear on the problems of the Breeders' Association, in whose future he had much confidence, and kept both Kearney and me busy taking down his dictations. I had seen Mr. Bell in conference many times, but until that night in the woods I had never fully appreciated the deep concentration of which his mind was capable.

This organization, which eventually became the American Genetic Association, has prospered, and now has a membership scattered throughout the intellectual world.

Looking back, I do not know how I found time for any "extra-curricular" activities. I walked into my office every morning expecting some new and interesting event would occur that day, and I was almost never disappointed. There was a concreteness about the conversations which appealed to me, in contrast to the ordinary run of talk speculating about possible events. The men who came in frequently had seeds or plants in their hands and they often took something tangible away with them.

In the summer of 1913, Mr. W. F. Wight appeared one day with seventy-eight distinct varieties of potatoes. He had visited the home of this vegetable in southern Chile, even penetrating into uninhabited portions of the island of Chiloé which could only be reached at low tide, and he had experienced some exciting adventures. As he spread the tubers out before us, our conception of the Irish potato as a conventional, oval vegetable was shattered, for the tubers were of many different forms and colors. The American potato growers were of course delighted to have this new material to use in breeding for high-yielding, disease-resistant varieties.

Silas Mason was at the time investigating the dates of the Libyan oases. His knowledge of the date palm surprised the sheiks of Upper

Egypt, and they found his quiet, courteous manner charming. Mason returned with many ideas for utilizing the leaves of this palm and with a large number of suckers of the famous "Wahi" date, which he had discovered to be identical with the "Saidy" which we already had.

By now, the whole country had become alarmed by the threatened disappearance of the American chestnut. Haven Metcalf and C. L. Shear were investigating *Endothia parasitica,* the virulent fungus disease which attacked the chestnuts, and they were puzzled to know where the causative agent came from. According to certain pathologists, the fungus was a native American species, but its virulence and rapid spread convinced Metcalf and Shear that it was of foreign origin. By 1913, it had already caused a loss of over twenty-five million dollars, but this was insignificant compared with the loss which would ensue when the disease reached the forests of the southern Appalachians. Meyer had already sent in the Chinese chestnut (*Castanea mollissima*), a species distinct from the Amercan, and it had grown rapidly and trees of it had been distributed. I had it growing at "In the Woods" and expected it to bear in a year or so.

One morning in May, Doctor Shear came to see me to discuss his and Metcalf's theory that the chestnut-bark disease must be of foreign origin, probably Chinese. As they had heard that Meyer was in Eastern China, they wondered if he could look for the disease there. Meyer was not a trained pathologist, but I felt confident that he would find the disease if it were common on the Chinese trees. Doctor Shear sent Meyer a fragment of bark with a description of the disease and I instructed him to send back samples of Chinese bark in case he discovered any which seemed infected with the blight.

On June 13, only a short time after my letter reached him, we received a cablegram from Meyer saying that he had discovered a chestnut bark fungus which seemed identical with the American form. In about three weeks, a letter came from him containing a small piece of bark not over two inches square. I looked at it doubtfully, not being familiar with the character of the disease, and wondered why Meyer had cabled with such confidence. When I took it over to Metcalf's office, he lifted the bit of bark in his slow, deliberate way and studied it carefully with his hand lens.

"It looks like it," he said. "The fungus strands are quite characteristic.

However, cultures will soon show. We'll let you know as soon as they mature."

The next week both Metcalf and Shear appeared bringing test tubes containing six cultures. They had inoculated three of them with fungus from the piece of Chinese bark sent in by Meyer, and three from the diseased bark of an American chestnut. The cultures all looked alike to me. Apparently Meyer had found the chestnut blight endemic in China.

Ten days later, Shear found the characteristic ascospores of the fungus on Meyer's cultures, and later in the summer he sent test tube cultures made from Meyer's material out to "In the Woods" and inoculated my Chinese chestnut tree from them. The characteristic cankers were slow in developing, but ultimately disfigured the tree. The Chinese chestnut proved somewhat resistant to the disease, and, although the fungus did a great deal of damage, it did not kill the tree outright as it had the American chestnuts. In less than three months, the fact was established beyond a doubt that chestnut blight was an introduction from the Orient, and probably had arrived on nursery trees some time in the late eighties or nineties.

It is now over fifty years since this pest was introduced. Last summer when I saw the hillsides in the Appalachians dotted with the tragic dead crowns of what had once been magnificent forest trees, I felt like saluting the pathologists who are working to prevent a repetition of this calamity in the future. I regretted any feelings of impatience I may have had towards their quarantines and inspections.

Accounts have been published in *The Journal of Heredity* of the experiments which have been carried on ever since our American chestnuts perished. No less a genius than Walter Van Fleet took a hand in the cross-breeding, and with a certain degree of success too. But the American chestnut tree attained greater size than any of the Oriental trees, and finding a cross to replace it as a forest tree has not been accomplished. The most that can be said is that the Chinese chestnut and some of its hybrids are less susceptible to the blight than the American species, and, as orchard trees, are capable of producing good nuts.

Robert Young had become so immersed in his studies of the dasheen (taro) that he was sent to Hawaii to see it growing there and to study its importance in the economy of the Hawaiians. On his return, he brought some "poi," the favorite food of the Pacific Islanders. He carried on experiments with the ferments which convert taro into poi, but few Americans seem inclined to eat the sour "stuff," nourishing and excellent though it is. After trying to force poi on all of the Office staff, he finally gave up. Had we realized the nutritive value of this food, we would have been more co-operative. Poi contains one of the few alkaline starches —most starches, such as in rice and white flour, being extremely "acid." During the past few years most interesting experiments have been conducted in the Hawaiian Islands. Japanese babies and children brought up on a rice diet in this tropical climate develop serious dental deficiencies. When fed a diet of poi, they have strong teeth and far better health.

About this time we had some particularly good news. Professor Georgesson came in from Sitka with a report that the Petrosky turnip had been a godsend to Alaska, as it was not subject to the turnip root maggot. He also reported the black Finnish oat as a successful variety there.

Much time was devoted that summer to organizing a plant-hunting expedition to southern Brazil which was primarily to investigate the navel orange. A. D. Shamel, of the Citrus Experiment Station at Riverside, California, found that since the seventies, when the Washington navel orange was introduced from Bahia, Brazil, it had developed numerous distinct strains in this country through bud variations. He therefore wanted to visit the original home of the orange.

Dorsett, with his infinite capacity for detail and his talents as a photographer, was selected to head the expedition, and young Wilson Popenoe was selected to go with them. Wilson's enthusiasm and knowledge of all that pertained to subtropical fruits had already made him a valuable man for exploration work.

The preparations were extensive, and card catalogues of all the fruits and other useful plants of Brazil were made. Maps, driers, bales of sphagnum, rolls of oil paper, trunks and cameras were scattered about Dorsett's office for weeks, while Popenoe assiduously studied Portuguese. On October 4, they sailed on the steamer with Theodore Roosevelt and his expedition en route to the River of Doubt, where Roosevelt so nearly lost his life.

CHAPTER XXXI

The Flowering Cherry Trees Are Planted in Washington

THE YEAR 1914 opened with Doctor Galloway exercising greatly enlarged powers and responsibilities in the position of Assistant Secretary of Agriculture. One of his first acts was to plan for a national arboretum, and I found myself associated with Swingle, Cook and Coville on a "National Arboretum Committee."

Since my arrival in Washington in 1889, we had all talked a great deal about an arboretum, but nothing had materialized. There was, it is true, a so-called National Botanical Garden established in 1853 on ten acres near the base of Capitol Hill when that region was virtually a swamp. This garden provided the trees when "Boss" Shepard was planting the streets of Washington and "far out into the country," as it seemed to the taxpayers who hated him and criticised his policy of street extension.

William Saunders was associated with the Scotch gardener, William R. Smith, in Shepard's tree-planting program, and he told me that he and Smith became alarmed at the way trees were being planted in the outskirts of the city, where there were no houses at all, and he went to caution Shepard about the matter. Shepard slammed his fist down on the table, and said most emphatically,

"Damn it all, I want you fellows to plant trees and keep on planting them."

They did, and the trees thus planted have made our National Capital famous for its beauty. Years later Shepard went to Mexico discredited, but on his return was given a royal reception, and his statue now stands on Pennsylvania Avenue.

The Flowering Cherry Trees in Washington

In 1914, William Smith was dead and the professional botanists in the Department resented having the title "National Botanic Garden" applied to a ten-acre park filled with greenhouses not associated in any way with the Department of Agriculture or the Smithsonian Institution. Furthermore, they were annoyed that it was under the management of the Committee of the Congressional Library and got its appropriation through that channel.

The Arnold Arboretum, established in 1872, had an endowment of $150,000 and had already grown into a famous institution under the guidance of Professor Sargent, who raised over $30,000 annually for forty years to maintain it. This Boston institution had more than two hundred acres and a library of over 25,000 volumes. Professor Sargent occasionally came to see me in Washington and generally started the conversation with the facetious remark that he had come to Washington to see what we had in the way of plants. Then I would tell him that he ought to know better, for we had no plants to show him but only people; people, however, who knew a great deal about plants.

Our Arboretum Committee functioned with enthusiasm and I believe helped set the stage for the serious efforts which were made later. This struggle was with Congress and with at least two Presidents, Coolidge and Hoover, who failed to see the situation as we saw it, and who always preached economy when an appropriation was asked for a National Arboretum worthy of the name. There at last seems a probability that success will crown these efforts, and Mrs. Frank B. Noyes of Washington is the person to whom most credit is due. Without her persistent efforts, the fight for an arboretum would have been practically hopeless.

I had first met Frank Noyes through Barbour Lathrop, and I had come to depend on his paper, *The Evening Star,* for publicity of any plant news which I thought advisable.

Some time after the appointment of the Arboretum Committee, Mrs. Noyes asked me if I would address a group of women at her house on plant introduction and our ideas of what an arboretum should be. This seemed a good opportunity to further our cause, and I gladly accepted.

Any public address has always been a matter of anxiety to me, and I spent much time on the preparation of this lecture, and made a special trip from Florida to Washington in order to give it. I was consequently disappointed when I arrived to find only a dozen women informally seated about Mrs. Noyes' drawing room. Of course they were women of importance in their various communities, but nevertheless I could not help bemoaning the amount of time I had spent over the lecture. The lantern was not particularly good and the colored slides did not show up to advantage, but I went through with my address, putting in as much enthusiasm as I could under the circumstances. When I returned home, I remarked to Marian that I would never again waste time lecturing before a lot of sleepy women. However, they were not as sleepy as I thought; at least Mrs. Noyes was wide awake. From the day of that lecture, she threw all her energies into the fight for an arboretum and even besieged the White House during two administrations until her husband told her that she would never be invited there socially again! She saw the Arboretum Bill through Congress, notwithstanding the fact that she was suffering from a serious illness which made it painful for her to carry on a conversation above a whisper.

I cannot refrain from expressing an opinion here which I have held from the beginning. I believe that, with the rapid increase of good roads and motor travel, an arboretum of at least a thousand acres should be acquired far enough in the country to be beyond any possible extension of city streets and away from the fumes and smoke of urban civilization. A tract of several hundred acres in northeast Washington has been secured but I cannot quite feel enthusiastic about it. The Missouri Botanical Garden in St. Louis has recently had to acquire land farther from the city smoke in order to raise many of their plants.

In my last conversation with E. H. Wilson I urged him to persuade the Corporation of Harvard College to acquire a large area of suitable land and gradually to move there the trees and shrubs from the Arnold Arboretum. Sooner or later they will suffer from the effects of the city atmosphere, which has already made its presence felt.

And now the moment has come to tell the story of the Japanese flowering cherry trees planted on the Speedway and around the Tidal Basin in Washington. I am often asked how they came there, for they have become a national institution; they appear on rosy picture postals and each spring they create a special fiesta as increasing thousands come to view their pink perfection.

Back in 1905 when Marian and I bought "In the Woods," one of our

The Flowering Cherry Trees in Washington

chief preoccupations had been to have a place where we could grow Japanese flowering cherry trees, and the first plants to arrive there were cherries from my friend H. Suzuki of the Yokohama Nursery Company. It was all an experiment in those days. Doubt of their hardiness had been expressed by so many horticultural experts that I tried to coddle them by planting them in sheltered spots.

The strange and happy coincidence which sent our gardener Mori to us just at this time, with a letter of introduction from Professor Tomari, was of real benefit to the cherries. Mori cleared places here and there among the cedars and made what he called a "sakura-no" or field of cherries, and later a "sakura-michi," cherry path. Mori also translated the Japanese names into English; there were, for example, the Tiger's Tail, the Milky Way, and The Royal Carriage Turns Again to Look and See.

Curiously enough, there is an important psychological difficulty which attaches to the name "cherry" tree in countries like ours where the cherry is a popular and abundant fruit. This was brought home to me by a question fired at me by Franklin K. Lane, when he was in the Cabinet of President Wilson. He arrived when I was showing our cherry trees to the Japanese Ambassador and his staff, and, with his usual brusque joviality, Mr. Lane remarked, "Cherry trees? Cherry trees? Do they produce good cherries?" In defense of our beloved trees, I retorted, "Must a rose or dogwood produce a fruit for us to eat?" Marian tactfully suggested that if we called them "cherry-blossom trees" we would avoid much of this misunderstanding, and I think she is right.

When it had become evident that the trees would do well, Marian and I wanted to do something towards making them better known in Washington. We therefore sent to Japan for more trees, mainly of the drooping type which seemed to be the most hardy, and the following spring, through Miss Susan B. Sipe, the indefatigable teacher of "Nature Study" in the public schools, we invited a boy from each school in the District of Columbia to come out and get a tree for his particular school yard.

It was a wonderful day in the spring of 1908 when the boys came single file through the woods. The buds on the oaks were red with young growth, and hepaticas and blood-roots and dog-tooth violets were coming up under the dead leaves on the ground. Each boy was shown how to dig and plant his tree; we gave them a little talk on tree culture, and

411

then they went back on the special car which the Street Car Company had provided for their transportation.

At the same time we discussed how the barrier of conservatism in the Department of Parks could be broken through and a "Sakura-no" be made on the Speedway, which had just been created. I was to give an illustrated lecture at the Franklin School the next afternoon after the children had planted cherry trees in the school grounds, and I invited Miss Eliza R. Scidmore, then the most noted writer on Japan, to be present. I secured some views of the new, unplanted Speedway, which I threw on the screen at the close of the lecture saying that the Speedway would be an ideal place for a "Field of Cherries," and quoted Miss Scidmore, the great authority on Japan, who was there as the distinguished guest of the occasion. The next day *The Washington Star* had an item: "Celebrate Arbor Day. Pupils Plant Japanese Flowering Cherry Trees. Mr. Fairchild described the beautiful flora of Japan and aroused the enthusiasm of his audience by telling them that Washington would one day be famous for its flowering cherry trees."

Mrs. Taft, at that time the First Lady of the Land, was much interested in all that concerned the beauty of Washington, so Miss Scidmore took our suggestion to her, with the result that the Park Department was asked to buy any available trees and to plant them along the Speedway. I think that it was the Ellwanger and Barry Nursery who supplied the first lot. Miss Scidmore consulted me regarding the importation of a large number from the Yokohoma Nursery Company. She wanted bigger trees than Marian and I had imported, insisting that it was important to make a show as soon as possible, but I cautioned her not to order large trees, because of the difficulties in making them live as well as the extra expense involved.

Some time later Major Cosby, head of the Office of Public Buildings and Parks, wrote us that he had been notified by the Mayor of Tokyo that he was sending two thousand cherry trees as a gift to Mrs. Taft, to aid in her plans. As Major Cosby had no agent in Seattle through whom to arrange for their entry, I offered the services of our importing agent there, and the Office of Plant Introduction handled the shipment across the continent into Washington.

I had been worried about the trees, fearing that they might prove too large, but I had not dreamed of any difficulty with the Quarantine authorities; this was in the early days of the existence of the Quarantine and it had not yet assumed the rôle which it now plays. The crates arrived January 7, 1910, and immediately came under the inspectors' eyes, with the result that almost every sort of pest imaginable was discovered, and I found myself in a hornets' nest of protesting pathologists and entomologists, who were all demanding the destruction of the entire shipment.

Ghastly as it seems, the trees were all burned.

Public interest had been aroused and much expectancy encouraged, so that the handling of the newspapers was no easy matter, particularly as Commander Hobson chose just this moment to speak on the floor of Congress in a derogatory way of the Japanese. This, added to the disgust of Miss Scidmore, the annoyance of Major Cosby, the criticism of the pathologists and entomologists, and the astonishment of my Japanese friends, combined to give me many sleepless nights. My only comfort was the knowledge that the trees had been so large, and their roots had been so cut, that I felt sure the greater number of them would have perished in the raw soil of the Speedway.

The entire matter was finally hushed up. I wondered, of course, just what kind of impression the affair had made in Tokyo, but it was not until years afterwards, in Geneva, that Miss Scidmore told me of a meeting in Mayor Osaki's office at which it was suggested that if the American public felt towards the Japanese as Commander Hobson's speech seemed to indicate, the matter of the flowering cherries had better be dropped. Fortunately, more generous counsels prevailed, and on February 2, 1912, the Mayor of Tokyo wrote to the Superintendent of Public Buildings and Parks of Washington saying that he was sending another two thousand trees. In this letter he said:

Although a small token of the very high esteem in which the people of this city hold your great country, it gives them boundless pleasure to think that the trees may in a measure add to the embellishment of your magnificent capital. As for the first lot of trees which we sent you three years ago, we are more satisfied that you dealt with them as you did, for it would have pained us endlessly to have them remain a permanent source of trouble. The present trees have been raised under the special care of scientific experts and are reasonably expected to be free from the defects of their predecessors.

When this second shipment arrived in March, 1912, many of the same quarantine inspectors who had examined the previous one were on hand

and Doctor L. O. Howard, Chief of the Bureau of Entomology, stated that no shipment could have been cleaner and freer from insect pests.

It has always seemed a pity to me that the official planting on the Speedway was not made the occasion for an elaborate ceremonial, for I do not know of any greater or more lasting instance of international friendliness than that shown by the Japanese when they sent their favorite trees to flower in the heart of our Capital. The first tree was planted in the presence of Mrs. Taft, the Japanese Ambassador, Major Cosby and Miss Scidmore, but little was published in the papers about the affair.

Several years passed and the cherries around the Speedway came into bloom, adding to our feeling that there should be an attempt on the part of the city of Washington to return the courtesy of the Japanese. About this time a great friend of Count Okuma, Mr. Kuwashima, visited America. Swingle had met him in Japan, and we three were often together when he reached Washington. The thought of some return courtesy was discussed, and finally it was arranged that several hundred dogwood trees, a large quantity of dogwood seed and some plants of our mountain laurel should be sent from the Department of Agriculture to Count Okuma for presentation to the city of Tokyo.

In 1918 I received the following letter from Mayor Tojiri of Tokyo:

The dogwoods which your government sent here in 1915 and 1917 blossomed in the spring quite beautifully with popular admiration as the picture in forwarding by the same mail it seems that they suit to our climate very well and I hope they will do better in the future. I thank you very much for the Kalmias which were sent by your government. Then we shall take the best care for them and I hope they will do just as well as the dogwoods.

On receipt of the photograph of the American pink dogwood flowering in Tokyo, we felt that we should send pictures of the Japanese cherries flowering in Washington. This was done, and at the same time duplicate sets were sent to Mrs. Taft and to Doctor Takamine of New York, who had been an enthusiastic collaborator in the project.

I never dared to imagine the popular enthusiasm which these Washington trees have caused throughout the country. Through them there must have filtered into the consciousness of hundreds of thousands an appreciation of the nobility of a people who can love these trees as do the Japanese. A national holiday is declared in Japan when the cherry trees come into bloom.

At Chevy Chase, Marian wandering in our "sakura-no," field of cherries.

The Mayor of Tokyo sent 2000 flowering cherry trees as a gift from the people of Japan. When they reached Washington, ghastly as it seems, they were all burned but the Mayor sent another shipment.

414A

The single-flowered drooping cherries are the longest lived and among the loveliest of these exquisite flowering trees. The birds scattered the seeds and many lovely seedlings grew wild in our woods. Growing on their own roots, they should live a century or more.

Snowy blossoms of the "Royal Carriage Turns Again to Look and See."

414c

The morphology of the double cherry blossoms is immensely interesting. In one variety, "The Tiger's Tail," I found a pistil which had turned into a leaf with teeth along its margin, and I placed it before my enlarging camera.

414D

The Flowering Cherry Trees in Washington

For keenest enjoyment, I visit the flowers when the dew is on them, or in cloudy weather, or when the rain is falling; and I must be alone or with some one who cares for them as I do. For those who are satisfied with the distant view of a plant, the cherries may not seem as impressive as the hawthorns or flowering crab-apples. The cherries are more delicate and one must stand beneath their branches and see the dainty blossoms against the blue or gray of the sky fully to appreciate them.

I used to roam among our trees at dawn and gaze at the individual flowers through the darkness of my enlarging camera before the dew-drops had vanished from their petals. I seemed a Lilliputian wandering among their soft, velvety surfaces. As normal pictures did not reproduce this feeling, I decided to see whether by enlarging them I could give the impression which the actual flowers made upon my mind. That these enlarged photographs have given something of the spirit of the sakuras I think is evident.

Of course I had my favorites among the twenty-five varieties scattered over our hillside. Alas, their names became involved in a nomenclatorial tangle. In an admirable paper, "Notes on Japanese Cherries," Collingwood Ingram says, "In attempting to create order out of the chaotic confusion of Cherry names that exist in Japan and Europe, I find that I have set myself an almost impossible task!"

However, some of the most beautiful varieties have survived with their names intact. The Murasaki, for example, with its deep purplish-pink, semi-double flowers which covered the ground with a carpet of pink petals as they fell, is still called Murasaki. But my Naden has become Shogetsu, though its extremely double blooms still hang on long flower stems and it still seems to have the most delicate shell-pink color of them all. Choshu, which is now recognized as the Kanzan, has immense double flowers of the deepest pink and was one of the latest of all to bloom; Ussussumi, now Shiro-Fugen, has deep sepia-colored foliage, and the flowers when half-opened are among the most charming blossoms in the world. The Amanogawa, Milky Way, with its strikingly fastigiate form, and its upright branches like garlands of delicate pink stretching upwards into the blue sky, never failed to thrill me, and as a tree for the small garden it deserves to be widely planted. The Asagi and the Ukon, strange, greenish-flowered varieties, were not so impressive on the tree, but Marian loved their strange exotic character and used them often as cut flowers in the house.

The trees I have just mentioned are more or less short-lived, cultivated varieties of the single species *Prunus serrulata,* whereas *Prunus yedoensis,* with its more vigorous growth and longer life, belongs to an entirely different class. I did not get the latter in my first introductions from Japan in 1902 and 1906, but later I ordered several in the collection for the Chevy Chase Land Company, for I had seen a remarkable tree of this species which Professor Sargent was growing in the Arnold Arboretum. These trees grew with amazing rapidity and have charmed all who pass them on Connecticut Avenue, near Chevy Chase Lake. This is the cherry which has been extensively planted in and around Tokyo, and nearly a thousand of the trees sent to Washington from the city of Tokyo were of this variety.

But of all the flowering cherries of Japan, those of the small-leaved species, which Ingram calls the Spring Cherries because they flower earlier than those just mentioned, are the ones which have always fired my imagination as an introducer of plants. There are two distinct forms: the weeping cherry, *Prunus subhirtella pendula,* and the erect type, *Prunus subhirtella ascendens,* either of which may appear when you plant seeds. I can assert this from personal experience, for the birds scattered the seeds of my trees of the weeping type and thousands of seedlings of this lovely tree grew wild in our woods. About one in ten of these seedlings had the weeping habit so characteristic of the parent.

When the drooping cherry trees began to show lavender tips in early April, they seemed like chiffon veils spread over the threadlike tracery of the bare branches. As the flowers opened, the color faded to a ghostly gray. Not only were they the first of the trees to come out in the spring, but from the very start they showed unusual vigor. It is now twenty-five years since little Mori planted the cherries and today the trunks are about a foot and a half in diameter at their bases and fully twenty-five feet in height. Compared with the double-flowered *Prunus serrulata* species, they give the impression of greater hardiness and the promise of far greater age. During the great drought of 1930, when so many trees in Washington suffered, they did not seem affected, although a good many of the shorter-lived cherries either died or were severely injured.

It is a great pity that most of the old specimens scattered throughout the Atlantic States were grafted high on Mazzard (*Prunus avium*) stock and are dying when they should be coming into their glory. When grafted

The Flowering Cherry Trees in Washington

low, beneath the ground, they soon get off onto their own roots, and there is every reason to feel confident that they will live for a century or more in America, as they do in their native country.

The October-bloomer, *Prunus subhirtella* var. *autumnalis,* blooms in the spring as well as in the fall, and is an ideal tree of just the right shape for a large city yard. Its scattering show of bloom in autumn after all other flowers were gone, and the fact that, at any time through the winter, branches taken into the house would bloom, never failed to surprise and charm us. It, too, will probably prove a long-lived tree in America.

Only the barest beginning has been made in a study of these enchanting trees. Hidden away among the ruins of those old feudal gardens of ancient Japan, there are priceless forms as yet unknown to us, while still lovelier and hardier and longer-lived varieties may yet be created to brighten the springtime of our gardens.

As trees, the cherries of Japan are not to be compared with oaks, elms or maples. The double sorts require the treatment of any orchard tree but should not often be pruned. Dead wood must be removed and branches that are too long should be pinched back in June, but otherwise they are best left alone unless attacked by some fungus, when spraying must be resorted to.

I am loath to leave these cherry-blossom trees behind; as loath to leave them in this book as I was to leave them when we sold our place "In the Woods." I will sentimentally close this chapter with a poem written by Frances Hodgson Burnett in Japanese characters on her place-card at dinner one evening. I had told her of my distress that we had sold our cherry trees and would never see them bloom again.

> Only in dreams of spring
> Shall I see again
> The flowering of my cherry trees.

CHAPTER XXXII

Quarantine Increases and War Begins

B Y THIS TIME, it had become evident that the land rented from Mrs. Brickell for the Miami Garden afforded inadequate space to test the number of plants we were bringing in. O. F. Cook and I were deputized to negotiate for another site which would be the property of the government, and from which we could not be ejected by the termination of a lease. Cook had never been in Florida before and was much surprised to find that the conditions in the Miami region were so tropical.

We investigated various locations, including one offered by Carl Fisher on Miami Beach where he had begun to sell the lots he had made by pumping up the sea bottom to fill in the mangrove swamps. This activity had caused much consternation among people who loved the mangrove swamps, crocodile holes and all the wild life of the old sand key which is now Miami Beach.

We did not accept Fisher's offer because Lyman J. Briggs, now chief of the Bureau of Standards, had told me that he had tested the soils of the sand keys and that they were composed almost entirely of broken-up sea shells and contained an amount of lime much greater than even the rock keys or the rocky soils of the mainland. As things have turned out his advice was good, for the range of plants which can be grown satisfactorily on the sand keys is extremely limited.

It is interesting now to remember how afraid both Cook and I were of any site offered us further than a mile or two from the center of Miami. Accessibility was then one of the indispensable factors and we felt that visiting scientists would have difficulty securing transportation back and forth to a garden "in the sticks."

The previous February, Mr. Charles Deering had expressed surprise that the garden should be located on rented land, and had offered to give us a tract near his new Buena Vista estate. This land was by far the best site offered to us, and we finally accepted it. Mr. Deering deeded it to the government with the proviso that in case the government should ever cease to use it as a plant introduction garden, the property should revert to his estate.

One afternoon as we were discussing the garden with Mr. Deering, his brother James came in from the adjoining room where the architect was drawing up the plans for "Viscaya," his magnificent estate on the Bay.

"Charles," he said, "don't let the botanists get all your money. The architects are getting all of mine."

I sometimes think of this remark and realize how differently the dreams of these two wealthy men materialized. Mr. Charles Deering's Buena Vista Estate, where beautiful canals wound through the mangrove forest and beneath overarching coconut palms, contained a collection of succulents planted by Doctor John K. Small and an arboretum of rare trees, many supplied by our Office, aviaries of egrets and cranes, and even an island with monkeys on it; yet it all disappeared overnight in the boom of 1925 and 1926. Even the magnificent *Ficus nitida* tree, the wonder of the Miami region, was torn to pieces and dragged out by tractors as the entire place was turned into a ghastly waste of suburban lots which to this day have not been sold. This wrecked land is a tragic memorial of those crazy, mad days in which the trees of the "hammocks" went down, destroyed by an army of road builders, carpenters and cement mixers.

On the other hand, "Viscaya," James Deering's dream, remains a place of strange, tropical beauty. One winter it was open to the public and thousands of visitors who had never seen a palace or European garden could wander there and imagine they were visiting some Italian Renaissance villa set on the shore of an aquamarine bay where palms and tropical trees grow instead of columnar Italian cypresses.

At this moment, Professor Sargent arrived on what proved to be his last trip to Florida. Together we visited the Cutler hammock, then the finest hammock land remaining in this region. Sargent remarked that this hammock was the best site for an arboretum that he knew of in southern Florida.

He was fond of saying, "*If* you have a million dollars you can start an arboretum."

We did not have a million dollars, and saw no prospect of ever having that much, so we felt elated over the twenty-five acres of land which Mr. Deering had given us.

As we came to the outskirts of Coconut Grove, the seventy-year-old Professor turned to me and said,

"Fairchild, there will be three hundred thousand people down here some day. Nothing can stop this development."

Time has shown that he was correct, both in his estimate of the number of people who would come to this region and in his selection of a site for an arboretum.

Doctor Leo Wallerstein of New York had discovered that a few crumbs of dried papaya juice would clarify a barrel of beer, and I had jokingly insisted that he should come to Florida and set out an orchard of papaya trees. Rather to my surprise, he arrived one morning late in March. He had never seen a papaya plant, for he imported his papaïn from Ceylon. I took him to the garden and led him to one of these odd trees. With his penknife he gently punctured the skin of a green papaya fruit. A stream of milky juice shot out, a foot or more, and sprayed his hands and clothes. It was a spectacular performance and, being completely ignorant of tropical vegetation, he began an attack on other plants near by, imagining that the geyser of sap was common in the tropics and not specific of the papaya.

Cook and I were camping out in the little laboratory which Flagler had given us, and we carried on all sorts of foolish experiments with the juice of the papaya. We found that it would cleanse our hands like soap, and, if squeezed from a ripe fruit, would solidify into a beautiful, golden yellow jelly. A homesteader had set out over five hundred trees and had a magnificent crop of fruits with which Wallerstein experimented. We soon discovered, however, that the cost of collecting papaïn in a land of expensive labor was quite prohibitive. Also, no short-cut mechanical methods for the expression of the juice proved practicable.

A rubber vine (*Cryptostegia madagascariensis*) which we had introduced was growing well in the Everglades and looked promising as a source of first-class rubber. This vine later produced the material for a long series of experiments. Thomas Edison and his friend Mr. Harvey Firestone came to investigate this and also a chance hybrid between this species and another closely related one (*Crypstostegia grandiflora*). The yield of rubber in the leaves of this hybrid proved so high that for a time it ranked among the most promising rubber crops available for the Everglades. With the synthesizing of rubber, which has been made in the chemical laboratories of E. I. du Pont de Nemours & Company, the rubber situation appears to be changing, and I suppose that rubber-producing plants and rubber plantations may disappear like the indigo and the madder.

When we left Miami, Cook and I stopped at the Brooksville garden to see the tung oil trees we were growing for distribution to people who had become interested in its commercial possibilities. The trees were a mass of bloom and the flowers reminded me somewhat of small, white hibiscus, although they are not at all related to the hibiscus.

Mr. William H. Raynes of Tallahassee was the proud possessor of quite an old tung oil tree. He was long past middle life and crippled with rheumatism, yet his greatest delight was in observing everything which took place around him. With the instincts of a real investigator, he had sent me careful notes about his tung oil tree every year, giving a full account of the behavior of his pet. He had ground the nuts into meal, and pressed out the tung oil into bottles which he showed me. He was the first field observer to take an intelligent interest in the tree, and I have always considered him the outstanding pioneer in the culture of the tung oil in America.

The small beginnings of Raynes and the Brooksville garden now seem amateurish compared with the more systematic investigations later carried on by Doctor Wilmon Newell and Doctor Maury of the Florida State Experiment Station, and Mr. B. F. Williamson, the first person to make a large planting, followed by Mr. H. W. Bennett and others whose plantations now represent an expenditure of hundreds of thousands of dollars. Last summer I walked through two thousand acres of tung oil trees on Mr. Bennett's plantation at La Crosse, near Gainesville, and stood with Mr. Williamson before a mountain of the nutshells, which, as a by-product, represent a problem of importance, since the shells contain potash, phosphoric acid and ammonia.

The Paint and Varnish Manufacturers' Association, led by Doctor Henry A. Gardner, also played an important part in the early development of this industry, and Doctor C. C. Concannon of the Department of

The World Was My Garden

Commerce has devoted much time to its commercial aspects. In fact I cannot enumerate the many personalities now concerned, for the industry has entered the commercial stage of exploitation.

A few days after my return from Florida, Dorsett and Wilson Popenoe landed from Brazil. The expedition was by no means an expensive one, but it brought back an amazing amount of valuable material. The men were most enthusiastic about the various plants, and their cuttings and plant materials had all been so carefully packed and stored in the cool room of the steamer that they arrived in beautiful condition. It was thrilling to watch them unpack their treasures in the quarantine house; to witness the first rites connected with their establishment in this country.

Of course the Office force combined to give a banquet to honor the return of the expedition. I was appointed toastmaster, and while I was occupied preparing my speech, the younger members of the staff concocted an amazing assortment of South American beverages to be served at the appropriate moment. Those not in the secret, believed, as I did, that the various gayly colored, sweet flavored drinks had been brought back by the expedition from the interior of Brazil. Before us we beheld concoctions labelled Imbuçada, Grumichamada, Pitombada, and Jaboticabada. Needless to say, these drinks made a tremendous hit and the banquet went down in the history of the Office as a howling success.

But one of the finest accomplishments of this expedition was the completeness and detail of the descriptions and the beautiful photographs of the fruits and plants which it had assembled. We had never had such a collection of pictures, made for the express purpose of illustrating as fully as possible all the characteristics and the uses of the various new plants. I imagine that there must have been at least a hundred photographs of different types of navel oranges in the region around Bahia. They had also assembled a complete collection of budwood of the Bahia navel orange and a full account of its origin, even locating the old negro who had cut the original scions which were sent to Mr. Saunders in Washington.

Their introduction, "capim gordura" (*Melinis minutiflora*), the principal forage grass of the state of Minis Geraes, grew very well, but one of our correspondents wrote us a sarcastic letter saying that the grass grew well but that his cattle refused to eat it; he could not understand why we had introduced the thing. It so happened that the very day this letter arrived

422

The largest Tung Oil tree in this country has grown near Cairo, Georgia, from a seed I imported in 1907. It has borne over 250 pounds of nuts in one year.

Right: Male flowers on a Tung Oil tree. A problem of this new industry is the selection of a variety producing the right proportion of male and female flowers on each tree.

422A

Five-year-old Tung Oil trees in the 2000-acre grove of Mr. H. W. Bennett near
Gainesville, Florida.

422B

Mr. B. F. Williamson examining a mound of tung nutshells near the first successful mill established for the extraction of tung oil from American-grown nuts.

422C

The Grumichama of Brazil, an early spring fruit now growing in South Florida.

422D

a professor from the Agricultural College of Lavras had come in to see me, bringing a motion picture film of a livestock show in Brazil. I asked him what type of hay the cattle were eating, and he replied,

"The capim gordura of course; it is the best hay crop we have down there."

So, the trouble really was not with the grass but with the cattle. They had not been educated to eat it, and they behaved much like many people who refuse to taste a mango, dasheen, or any new fruit or vegetable with which they are not familiar. I was reminded of the difficulties experienced by one ranchman in persuading young range cattle to eat corn. One steer actually died of starvation with a bin full of corn within reach.

All this time, the constant stream of visitors continued. The door of a government official's office must always be open and, when the chairs in his anteroom are filled with waiting visitors, any serious investigation behind that door is a practical impossibility. Carrying on profound research in the hubbub that prevails in most government offices is as difficult and rare as composing good poetry on a noisy street corner.

My visitors were nearly all interesting, particularly those from foreign countries, and it seemed important to help them. As I look back, I am inclined to think that the attention which I gave to their requests was probably as worth-while as the "careful research" I might have given to the plants themselves. It is hard to say.

As I thumb through my notebooks of 1914, I am much surprised to find so little reference to the War. However, since I was writing notes on plants, and not personal diaries, this is perhaps understandable. America was not yet in the War, and the activities of the Office went on much as usual until our actual entrance into the conflict.

It was during the spring of 1914 that the American Breeders' Association was converted, as I have said, into the American Genetic Association. Its *Breeders Magazine* became *The Journal of Heredity,* with Paul Popenoe, now the well-known eugenist, as editor. A campaign for funds was launched for the support of this *Journal,* which represented one of the few avenues then open for the publication of articles on plant hybrids and on the collection of material for the use of the plant breeders of the world.

The Association's affairs absorbed my vacation in Nova Scotia that summer, and Mr. Bell helped me by contributing many ideas. It seems unbelievable, but in those days the terms "breeder" and "breeding" could not be used indiscriminately in mixed audiences. They were, so to speak, taboo, and when Mr. Hays and I were discussing the future of *The American Breeders Magazine,* I remember saying that if it were to have a real future, the name "Breeders" would have to be eliminated, for it smacked too much of the barnyard. Even with the taboo abolished, I do not consider that we made a mistake in changing the name to *The Journal of Heredity.* Now in its twenty-seventh year, *The Journal* has published thousands of original scientific papers and unique photographs dealing with the new science of Heredity. It is the standard publication today in the field, and its able editor, Robert C. Cook, son of my old friend, occupies a high place among the workers in this important new science.

Nothing that Marian and I have ever done has seemed quite so worthwhile as helping to put this publication on its feet. Also, none of our contacts have been more stimulating than those with the members of the Council of the American Genetic Association, every one of whom has done important research in Genetics. Much as I would like to include here some of the interesting experiences which came to me as president of the Association, I shall refrain, since accounts of most of them have been published in *The Journal of Heredity.*

In September, 1914, Dorsett and I seriously took up the propagation and distribution of the beautiful Japanese flowering cherry trees. We had brought in No. 32860 from the island of Oshima, and we used it as a stock for the best of the many varieties in our collection. It was this distribution to nursery firms and private individuals that started the universal planting of Japanese cherry trees throughout the United States; a movement in which the Arnold Arboretum played an important part by establishing E. H. Wilson's collection of named sorts in Boston and duplicating it in Rochester under the supervision of Mr. John Dunbar, an unusual plantsman in charge of the parks there.

The increase of the Plant Quarantine organization and their insistence on more stringent inspection, fumigation and detention of our plants, necessitated the construction of a special quarantine hospital. Dorsett and I drew up elaborate plans for such a structure, but the money for it was not forthcoming. Five thousand dollars was all that could be secured, and with it Dorsett constructed a makeshift building which was inadequate from the start. Nevertheless, for many years it was the chief receiving station for all the plant material coming into the country from abroad.

The failure to provide adequate housing for the important new plant immigrants, notwithstanding the vast sums being expended to control pests already in the country, was always irritating to me, and I tried in vain to remedy the situation. However, the entomologists and pathologists formed a very powerful force, and their leader, Charles Marlatt, was mainly concerned with the protection of our established plant industries.

If the public became alarmed that an important industry might be "wiped out," almost unlimited funds were voted by Congress to combat the invader, even though sane judgment often indicated the futility of the expenditure. Unfortunately, the pathologists were often unable to control the Congressmen who enjoyed the expenditure of Federal monies in their districts, and who voted immense appropriations more to add to the number of their constituents than with any knowledge of or interest in the real problem involved. In other words, the Plant Quarantine, like many excellent government projects, at times approached dangerously close to what is known as a "racket."

Our Office of S. P. I. naturally became a storm center. The Quarantine Act prevented the nursery firms from importing plants themselves, and the only way that they could get them into the country was through the little five-thousand-dollar-quarantine greenhouse on the Mall. There were times when I felt that my old friend Charles Marlatt and his associates would gladly have done away with the introduction of plants from abroad altogether. But I determined that the door should not be shut. As long as I was able, I would keep my foot in the doorway and prevent importation from being entirely forbidden.

I still maintain that it would have been better had a much larger proportion of the millions spent on eradication campaigns been spent to further the invention or discovery of new and better methods of disinfecting living plant material (small plants, scions, cuttings and seeds) so that it could be exchanged between countries without danger. What is more, we might profitably have employed the funds to make a really thorough investigation of the conditions in foreign countries which tend to hold in check the diseases and insect pests which, although of minor importance in those countries, contain threatening potentialities if they reach America.

I also felt strongly that when the Quarantine closed the doors to amateurs, and discouraged private initiative, the government was obligated to develop its own methods of introduction and dissemination for their benefit. With this in mind, I attempted to extend the distribution of new plants to as large a group of capable experimenters as possible. This work of encouraging plant lovers to experiment with our new introductions reached a peak about this time, for our Plant Immigrant *Bulletins* were still being published and carried illustrations of the new things, keeping the experimenters interested in what we were bringing in.

The problem of informative plant labels seemed of great importance. I had learned that the time to acquire the knowledge of a plant is when you stand beside it in the garden, and not later when you are back in the house. This conviction led us to devise fifty-word, explanatory descriptions printed on thin lead labels. These were attached to the plants we sent out and explained just what the plants were and what they might be good for. We had many complimentary letters about these indestructible labels.

With the outbreak of the Great War our lead tags were declared too expensive and the substitutes we were forced to use proved to be short-lived affairs, which generally became illegible long before the small plants we distributed had grown large enough to attract serious attention.

During the spring of 1914, we had made an average of thirteen introductions a day, and had sent 537 shipments to experimenters in foreign countries. The Office had, in fact, become an international plant exchange. I made it a point to fill foreign requests for plant material as promptly as possible, and, in return, foreign correspondents seldom refused to assist us in securing species which occurred within the boundaries of their respective countries.

As the great conflict proceeded and engulfed the whole world, I watched with interest the effect on our traffic in plants and was surprised to find how seldom national antagonisms interfered with our exchange of seeds and plants, beyond the delays connected with all mail and express shipments. Wherever any correspondence was possible, it was carried on during the War with the same degree of freedom as before its outbreak. The volume fell off, but to no such extent as I had anticipated.

But repercussions of the War did of course reach us. Professor Sargent, on his annual trip to Mount Vernon where he was eliminating all the species of exotic trees planted there since Washington's time, dropped in to see me that autumn of 1914. He wanted something done to save Camillo Schneider, a famous Austrian botanist, from being interned in China by the British. Schneider had been collecting in Western China with a British botanist when the War broke out. The two men were great friends, but the events in Europe forced them to part. On the border of Tibet they bade each other farewell. The Britisher walked over into India, and Schneider came down the Yangtzse in an attempt to reach the United States. Instead, he fell into British hands when he reached Shanghai.

Professor Sargent proposed that I employ Schneider in China and keep him there collecting for us until the termination of the War. This plan did not work out, but, to my surprise, Schneider later appeared in Washington where he again barely escaped internment as we ourselves were about to declare war on Austria. His contributions to dendrology since the War have amply repaid the efforts that we made to save him from a concentration camp. I have never forgotten Schneider, or his description of his parting from his British friend in Tibet. Meyer was again far up the Yangtzse and his letters had begun to give an inkling of the emotional situation which surrounded him, far though he was from the great conflict.

CHAPTER XXXIII

Seeds from Afghanistan

WHEN MR. DEERING'S gift of a garden site near Miami became an accomplished fact, Dorsett and I went down in January, 1915, to organize the work there, taking Wilson Popenoe along to study the mango and avocado collections growing in the old Garden. It was delightful to realize that we would have four times as much land as before, but, before we disturbed the native flora, we asked Professor Simpson to make a canvass of the species which occurred on a representative section of land a hundred feet square. He listed twenty-four families and forty-two species in that small space. We then divided the remainder of the land into hundred-foot squares, and blue-printed the whole tract.

Three avocados which had grown to fruiting age in the old garden have played a rather important rôle in the avocado industry. Number 26710, which we named for Doctor W. A. Taylor, proved to be one of the hardiest of the good avocados; from the Collins (# 19058), the well-known Collinson originated; and the Winslow (# 10978), sent in by Admiral Cameron Winslow, was parent to the Winslowson, a delicious variety. The Collinson was a chance hybrid between the hard-shelled Guatemalan Collins and some unknown male parent of the leathery-skinned West Indian type.

During this trip to Florida, I visited the Bright Brothers' Ranch six miles west of Little River where they were keeping several hundred head of cattle on their three hundred acres of Para grass and were also baling Para grass hay which they sold for twenty dollars a ton in Miami. Glenn Curtiss bought this ranch in 1919, but soon cut it up into lots. In fact, upon this tract now stands the town of Hialeah, and Para grass still grows here and there, although few of the inhabitants even know its name.

428

Seeds from Afghanistan

Mr. Deering and I visited a hammock called Paradise Key in the Glades beyond the town of Homestead. Long before we reached there, we could see the magnificent Royal Palms rising among the live oaks and other native trees. We counted sixty-six immense Royals waving their lovely fronds above the untouched jungle, and realized why this was considered the most beautiful hammock in southern Florida. Due to the efforts of some of the women of Florida, this rare spot is now a State Park.

To those who have not seen a "hammock," I should explain that it is a slightly raised area in the Everglades, and is surrounded by water in the rainy season. Being drier land than the surrounding terrain, it becomes the home of the plants and animals which cannot exist in the flooded conditions on the great stretches of saw grass composing the major portion of the vast Everglades. The hammocks are real bits of jungle growth.

This visit to Florida was marred by the discovery in our garden of a virulent bacterial disease called citrus canker. It was caused by the germ *Pseudomonas citri* which had been brought into the country about 1910 on some citrus trees from the Orient, and had gained a foothold on the Gulf Coast and in certain sections of Florida. Before its virulence was recognized, many foci of infection had been allowed to exist in the State, and it finally threatened the complete destruction of the citrus industry in Florida. The infection appeared as small, light-brown spots on both fruit and foliage, and so rapidly did the disease spread that it often killed affected trees in a single season. All methods of pruning and spraying were futile, and it was decided that the only hope lay in burning all infected trees by spraying them with kerosene and setting them on fire. Such a drastic method brought down a storm of criticism and abuse on those proposing it, but a group of forceful, determined men whose names will go down in the history of Florida horticulture—Bird, Krome, Tenney, Stirling, and Rolfe—took up the fight and besieged the Legislature until it passed the State Plant Act of 1915 which later originated the State Plant Board. Sums of money were appropriated to supplement the funds of private individuals and those supplied by the Federal Treasury, until more than two million dollars was spent during the ten years required to rid the State of the disease.

The citrus canker quarantine halted all distribution of plants from our garden and also destroyed practically all the citrus trees in it. Those who waged the war on the canker were so convinced that the disease was carried on the clothing and shoes of any one who visited an infected grove, that special white costumes were worn by the quarantine experts and their boots were sterilized in corrosive sublimate solutions as they left one grove to visit another. In the history of the world no such technique had ever been devised for the control and eradication of a plant disease.

Some time before, I had heard of a California engineer named Jewett who was building a hydro-electric plant for the Amir of Afghanistan. Having the notion that Afghanistan contained a wealth of valuable plant material suitable for California, I wrote to him. But I heard nothing in reply and supposed that my letter had never reached him.

Late one Saturday night, shortly after my return from Florida, I received a telegram signed Jewett, advising me that he was arriving the next day with a collection of seeds which the Amir of Afghanistan was presenting to the American Government.

Early on Sunday, Jewett was at our door, burdened by one of those woven, ruglike saddle bags which the desert Arabs sling across their camel's back.

"I suppose you wonder why I never answered your letter," he said as he dragged the bag in the door. "I'll tell you about it."

We sat in Mr. Bell's library with the saddle bag between us, and he told me that my letter had gotten him into some very hot water. All his letters were censored, and in mine I had asked the simple question as to whether we might place an agricultural explorer in the region and if so, how to go about it. The Amir was very angry at this idea, and since Jewett was the only white man in Afghanistan, and entirely in the Amir's power, he felt distinctly nervous. He realized that the Amir would not hesitate to put him out of the way as he had done before with men who did not please him.

"I had the greatest difficulty in extricating myself from this dilemma," Jewett said. "However, not only did I succeed, but I managed so well that, when the Amir decided to send me on this trip, he conceived the idea of giving me a collection of seeds and dried fruits representing the products of his country to bring as a gift to you. In order to take something in return, I have been on the lookout for devices which might interest the Amir, and yesterday I found one which I know will please him immensely. He is in continual fear of assassination; and I have here a newly invented door-hinge into which a cartridge can be inserted so that when the door is opened the cartridge will explode and frighten the intruder away. I know he will be delighted with it."

We spread out the various seeds which the Amir had sent and found that each packet was accompanied by an elaborate tag printed in Arabic and English.

"Taste these dried white mulberries," Jewett said. "The men who work for me live on them for about eight months in the year. They arrive in the morning with great handfuls of them stuffed in their shirts, and eat them for their lunch with no better accompaniment than water from the irrigation ditches. As the men eat nothing else, the mulberries must be very nourishing."

Like some other mulberries propagated from cuttings, this white mulberry was practically seedless. Our propagator went through the quarts of dried fruit but found only three seeds. From one of these a plant grew, but, as might have been expected, it proved to be a perfectly worthless variety.

Jewett was a fascinating man. He had only one day in Washington, so I took him to the Chevy Chase Club, after we had sorted the seeds and I had written a letter of thanks and arranged to send the Amir some American fruit varieties.

Jewett's picture of conditions in Afghanistan was almost too horrible to believe. Such things as one reads of as happening in the sixteenth century were of daily occurrence there in 1915. The father of the present Amir had put a hundred and eighty thousand people to death during his reign, and it was still a common sight along the highways to see robbers in iron cages hung up to die of starvation.

Jewett related that as he was going along the road one afternoon, he met a woman carrying on her back a bag of stones. She stopped him to inquire where to find the robber who was to be stoned to death as she wanted to take part in the execution. The doctrine of revenge was carried to ghastly extremes, and a mere child not old enough to wield a knife was supposed to revenge his father's death by cutting the throat of the assassin, even though he could not understand what it was all about.

In front of the Chevy Chase Club I pointed to a tree which President Taft had planted. Jewett laughed and said that he was reminded of a tree which the Amir had planted. The Amir was very fond of trees and had given a number to the members of his cabinet with orders to plant them. He suspected that one of his cabinet had failed to take care of his tree, and so went, accompanied by his courtiers, to see for himself. When he found that the minister had not watered the tree, he ordered him to get down on his hands and knees and do so while the Amir and his court pelted the minister with mud.

The captious Amir devised all kinds of strange punishments. His prime minister was so afraid of him that after an interview he had been known to faint from fright. Once he approached while the Amir was at breakfast, bringing some papers to be signed.

"Haven't I told you never to bring me papers while I'm at breakfast?" the Amir bellowed. "I'll teach you a lesson."

And turning to his servants, he said,

"Take that table and follow him wherever he goes *for a week*. I'll teach him." And for a week, the servants dragged the table wherever the prime minister went, even when he went to bed at night.

It is appalling what depths of brutality and cruelty men sink to when given the power of life or death over their fellow creatures. Of course the Amir was later assassinated.

I have always wondered whether he took care of the trees we sent him and whether any of them are still growing in Kabul. I kept in touch with Jewett for years until his relatives wrote me of his death on the island of Tahiti.

Doctor Walter Van Fleet, whose hybrid roses were already attracting attention, wanted a position where he could carry on his plant-breeding work without the necessity of doing the tiresome proof reading and editorial work in which he was then engaged. I was extremely anxious to attach him to the Department, and offered him the only vacant position worthy of his genius, namely, that of superintendent of the Garden at Chico, California. After the green fields of the Atlantic seacoast, neither he nor his wife could endure the arid desert climate with its summer temperatures of 125°, and in the autumn he returned to Washington and was given a position in the Office of Drug Investigations. At heart he was a plantsman, a hybridizer, one of the most remarkable that America has produced, and it was not long before he had established himself in a garden of his own near Bell, Maryland. This place became a center of

432

plant-breeding work, and later was chosen for the site of the permanent plant-introduction garden now known as the "Bell Station."

As early as 1907, Doctor Van Fleet had besought me to secure a certain yellow rose for him to work with. It had been discovered in China by a Catholic priest, Father Hugo Scallan, who sent seeds of it to the British Museum in 1899. At my request, Kew Gardens generously sent several plants of Hugo Scallan's rose, botanically known as *Rosa Hugonis,* and they were propagated and placed in Doctor Van Fleet's hands.

From the very start, this rose attracted my attention. I espaliered one on a trellis near our sleeping room at "In the Woods" and, to our delight, it bloomed each spring before anything but the Japanese flowering apricot and the single flowering cherries came into flower, proving to be one of the earliest of all roses. We distributed this lovely rose to thousands of small home owners, and it has become such a favorite that I wish Father Hugo Scallan were alive to enjoy the success of his charming introduction.

Doctor Van Fleet used this and other species from China to create a host of new and interesting forms, many of which are now obtainable in the commercial nurseries.

On Professor Sargent's last visit to Washington, he and I lunched with the noted oculist, Doctor William H. Wilmer, whose office, curiously enough, was on Eye Street! Doctor Wilmer, a charming, warm-hearted man, was deeply interested in trees and had an "arboretum" of his own in Virginia. Professor Sargent urged us both to spend a Sunday with him at the Arnold Arboretum.

Doctor Wilmer and I therefore travelled to Boston and, notebooks in hand, trailed Professor Sargent as he stalked from tree to tree pounding his cane on the ground as he went. His marvellous memory for names and his interest in every characteristic of his plants made the experience a memorable one for both his guests. I find that I filled two notebooks with the scientific names of the plants we saw that day.

Since we were expected to admire, or at least say something intelligent about, each separate specimen the Professor pointed out to us, our adjectives soon gave out and Doctor Wilmer and I took turns in making trite remarks about the beauty of each one. Professor Sargent led us to one of his numerous varieties of yew which was not perhaps quite happily placed on the landscape. It was Doctor Wilmer's turn to speak and he made the mistake of admiring the tree.

The World Was My Garden

"I planted the thing here twelve years ago," snapped the Professor, "but have felt ever since that I made a mistake," and moved on to another tree. However, maybe he would have snapped as much had we criticised!

Sargent's enthusiasm over trees and shrubs was unlimited. He expected every botanist who visited the Arboretum to know all the plants by name, and by his manner made them feel ashamed to pass under any tree without knowing what it was. As I never had a particularly good memory for names, I received many harsh words from the Professor. But I always carried away something interesting and valuable, not merely in the way of information, but also seeds or plants which he gave freely to our organization and which we propagated and distributed throughout the country.

The grounds of the Professor's interesting old house were landscaped with choice trees and shrubs, but I noticed that one of the most prominent "accents" on the lawn was a stump covered with the native poison ivy. The gorgeous coloring of its autumn leaves appealed to Sargent immensely, and I presume that most of his visitors were familiar with its poisonous character and kept their distance

Mr. Charles Deering and his guest, Major James W. Wadsworth, had asked me what I considered the largest trees in America, exclusive of the Sequoias. I am ashamed to admit that I could only guess at the answer, but suggested that we might find out through *The Journal of Heredity*. Mr. Deering and Major Wadsworth each offered one hundred dollars for photographs of the largest trees in the United States, exclusive of the Sequoias, and we conducted a contest which attracted some attention. In the corn belt near Worthington, Indiana, we located a gigantic American sycamore (more correctly called the plane tree, *Platanus occidentalis*), with a trunk measuring sixty-seven feet in circumference.

The success of this contest led to another prize being offered by Mr. Deering for a photograph of the largest individual tree of the American papaw, *Asimina triloba,* the most neglected native fruit tree in America, and yet the one bearing the largest fruits. This contest proved an especially interesting one, for we not only located a number of large trees but some with fruits which were delicious when properly ripened off the tree.

But this publicity was apparently in vain. There are still no papaw orchards in America, and only fruits of mediocre quality from wild trees find their way into the markets. The only man that I know of who has interested himself in the problem of improving the American papaw is

Seeds from Afghanistan

Doctor G. H. Zimmerman of Piketown, Pennsylvania, with whose experiments I have been in touch for many years.

During 1913 and 1914, Mr. Lathrop had been away on another trip to the Orient and Australia. From Japan he had sent me seeds of the Mitsuba, a vegetable which he liked and which we had failed to get when we were there together. Its young shoots are much esteemed for greens, and its roots are eaten fried. I have a vague recollection that, when the seed came in, Skeels told me that the species *Deringa canadensis,* an umbellifer, occurred wild in America. It had never been cultivated by Americans and I did not know it by sight, so Skeels' remark fell on deaf ears. Our plants grew well, and we added mitsuba to our menu.

Madame Fuji, granddaughter of Prince Ito, was in Washington and one spring morning I drove her out to visit Marian and see our Japanese flowering cherry trees. I also planned to surprise her by showing her our little row of mitsuba. As we passed some woodland, our taxicab had a puncture and, while the driver was repairing it, Madame Fuji and I strolled into the forest. Before we had taken half a dozen steps from the roadside, she knelt beside a plant exclaiming,

"Mitsuba, my favorite vegetable!"

The plant was growing here and there through the woods, and I realized at once that it was the same species which was growing in my nursery. Skeels' remark came back to me, and with it the consciousness that my interest and absorption in foreign plants had left too little time for a study of the plants growing wild in the woods about me. Although the joke was decidedly on me, it was too good to keep and I told her that she had found in the woods the very plant I was taking her to see in my little garden. The only difference was that in Japan it was a favorite vegetable while in America no one knew even that it was good to eat. Alas, notwithstanding our attempts to popularize it, I doubt if the mitsuba is deliberately grown anywhere in America today outside of certain Japanese gardens, although when properly prepared it is a delicate dish.

CHAPTER XXXIV

The Plains of Canada

WHEN FRANK MEYER returned from his third trip to China, I decided to meet him in Seattle, so that we could visit the Chico Garden, where many of his plants were fruiting. On my way west, I decided to go by way of Canada, to study the extensive work of plant introduction and breeding which was going on there.

My first stop was to be Ottawa, but before leaving the United States I went to Boston to see Mr. Lathrop who had just returned from Japan, and we spent a few days motoring through Massachusetts.

"You know, Fairy," he said, "there is a grave up here somewhere which I've always wanted to visit. My ancestor, Captain Thomas Lathrop, was massacred by the Indians and is buried in this part of the world, I think near South Deerfield."

We found the place. It was on the site of the Bloody Brook Massacre where, almost exactly two hundred and forty years before, Indians had massacred some white men who were invading their hunting grounds. The inscription read:

On this ground Captain Thomas Lathrop and eighty-four men under his command, including eighteen teamsters from Deerfield, conveying stores from the town of Hadley, were ambuscaded by about seven hundred Indians and the Captain and 76 men were slain. They were a choice company of young men, the very flower of the county of Essex, none of whom were ashamed to speak with the enemy at the gate. The eighteen of September, 1675.

I had sometimes realized to my cost that Uncle Barbour came of fighting stock, and here was the evidence.

When I reached Ottawa early in September I appreciated for the first

436

time the War and the extent of the tragedy in which Canada was playing so noble a part. In the hotels were sad groups of officers saying farewell to their sweethearts and wives, and already some wounded were to be seen. Faint memories of the Civil War came back to me, stories I had heard at my mother's knee. At Lethbridge I saw my first army camp with its thousands of tents, improvised roads, and soldiers moving about in frantic haste.

In the plant-breeding work of Professor William Saunders and his son Charles, their researches had aimed at two things: the production of earlier maturing wheats which would ripen in the extremely short season of that northern latitude; and the production of hardy apple trees which would not be winter-killed by the excessively long, dry, cold winters which characterize the plains of Central and Western Canada.

As breeding material for his work, Professor Saunders had introduced the Siberian crabapple, *Pyrus baccata,* and some high altitude, short season wheats from India. The fruits of *Pyrus baccata* are not good to eat, but it is so cold-resistant that they hoped to produce hybrids by crossing it with the hardiest types of European apple, *Pyrus Malus.*

The problem of creating a hardier apple proved a difficult one; although some of the hybrids stood the cold well enough, practically none of the fruit was of sufficiently good quality to make it worth while. Further than this, these hybrids came into competition with the Russian apples.

In 1888, a Scotch pioneer by the name of Stevenson, living at Morden, south of Winnipeg, received from Professor Budd of Iowa some of the Russian apple varieties which Budd had collected during a trip to Russia.

I visited Mr. Stevenson and saw the early Russian introductions, which had proved very successful. Growing beside them were the hybrid crabs which Professor Saunders had originated. To my surprise, I found that Mrs. Stevenson had no fancy for the hybrid crabs, even though they had a delicious flavor, and, when I asked her why she did not make cider out of them, she was not in the least interested.

"They are no good commercially," she said. "We can grow good apples here. Why should we waste time with these little ones?"

I learned later than Professor Saunders' enthusiasm about his hybrid crabs had rather tended to blind him to the possibilities of the Russian apples. Some people felt that apple-growing would have been further advanced in Northwest Canada if Saunders had been willing to recognize Stevenson's success and had pushed the hardy Russian and other types instead of his new hybrids.

In Central and Western Canada, Saunders' Marquis wheat was beginning to supplant all other varieties and play its great rôle in the winning of the War. During the most acute wheat shortage of the War, its superior yielding character had created a tremendous amount of Canadian wheat that would not otherwise have been available. The story of the Marquis wheat has often been told, but I am never tired of hearing it, for it illustrates how vast may be the results of a beginning so small that it is represented by a single seed.

The Marquis wheat came from a single wheat kernel, the result of a cross which Professor Saunders made between the Red Fife, a Canadian wheat, and a short-season variety from India which he had imported from Calcutta. The Marquis ripened ten days earlier than any of the other good standard varieties, and during the four years preceding 1915 it had taken prizes wherever it was exhibited.

In Winnipeg, I met Mr. S. A. Bedford who had been Professor Saunders' assistant. He was one of the old pioneers of Canada, having tramped across the prairies before the railroad was completed. He showed me samples of the so-called "club" wheat which was the only kind growing in Canada when he first came out there, and, side by side with it, some heads of the modern Marquis, a giant in comparison. On a hundred acres, one of his friends had raised an average of sixty-five bushels of Marquis per acre, something most unusual on the dry wheat lands of Canada.

I was disturbed and a good deal annoyed to find during this trip through the great wheat fields of Canada that so little recognition had been given to the originators of this marvellous wheat. Several years later, in an address before the Canadian Institute, I called attention to this lack. I had received a letter from Charles Saunders after his father's death, complaining bitterly of the failure of the government to support his unique researches in the improvement of Canadian wheats. Charles was so discouraged at the time that he thought of taking his violin and settling in France. This lack of recognition was later corrected, and I believe the younger Saunders was given the appropriations which he needed and a substantial pension upon his retirement. While this would be nothing compared to the returns which would come to the inventor of a successful invention as widely used as the Marquis wheat, I suppose a pension is all that any plant breeder may expect.

"S. P. I. Number 13," one of our earliest introductions, a sweet corn from Moscow which Hanson had reported was the earliest variety grown in Russia, turned out to be the only variety of corn which matured consistently in the region of Ottawa. It had been crossed with the Squaw corn, a low-growing variety originated by the Indians of the western plains, and from these came a cross of unusual promise, a sweet corn to be grown on the northern limit of sweet-corn culture.

I had heard much about the attempts to grow trees on the plains, and in particular about the forestry work of Mr. N. M. Ross at Indian Head, Saskatchewan, begun in 1901. Since our station at Mandan, North Dakota, was mainly for this purpose, I was delighted to see Ross's experiments, which were on a large scale, and I convinced myself that trees would really grow and make low forests and shelter belts on these treeless wastes. Most of his tests were made with Scotch pine, tamarack, ash, and white spruce.

Ross was growing and distributing millions of windbreak trees and shrubs and sending out "forest mixtures" of deciduous and coniferous seedlings to see which combination would soonest become a satisfactory forest. He had already supplied about forty thousand farmers with windbreak material, and had inspectors on the road all summer instructing the farmers where and how to plant them. Until a man's land was in proper condition to receive these government trees, Mr. Ross would not send him any. This method of distribution also seemed practical for the treeless prairies of the Upper Mississippi Valley.

The Siberian pea, *Caragana arborescens,* had proved an invaluable leguminous shrub in these "shelter belt" experiments. It branches close to the ground and, when planted around small fields, serves to lift the winter winds. The possibilities of this method of wind protection were interesting. However, the function of shelter belts became somewhat misunderstood by the general public, and a ridiculous extreme was reached when a program was advertised of making a single belt a thousand miles long from Canada to the Gulf. Advocates of this plan apparently did not know that tree shelters affect only a small area; the effects are not felt for hundreds of miles on their lee sides, as the proponents of this fantastic proposition claimed.

439

The World Was My Garden

After visiting Saskatoon and Lethbridge, I reached Salmon Arm, B. C., on the western slope of the Great Rockies. The contrast between this valley and the prairie stations I had so recently left was startling. Instead of bare, windswept plains, here was protected, hilly country, covered with forests. The autumn colors of the trees were reflected in the placid waters of many beautiful lakes while hills rose upon hills to the very horizon, misty with the haze of autumn.

I had come to see Mr. Sharp, an old veteran of horticulture, of a type now well-nigh extinct; men who lived for and dreamed only of their orchards and plants. Mr. Sharp was one of the best-informed horticulturists of Western Canada, and was then nearly seventy. He had chosen this spot because of the ideal conditions for fruit growing, and had kept accurate temperature records for a number of years. In this region there appeared to be no February thaws or severe storms; the temperature never went above 95° in the summer and, although it went to 20° below in winter, there was a snowfall of 40 inches and practically no wind to blow it away. Mr. Sharp had among his French pears and delicious European apples a tree which I have rarely seen in this country, the German medlar, *Mespilus germanica*. Like our native persimmon, the fruit of the medlar remains hard and inedible until it is mellowed by frosts. Late in the fall it is harvested, and is then ripened in a cool, dry room. The ripening process is known as bletting. It would be interesting to discover the real reason why this fruit, so popular in parts of Germany, has never appealed to the American taste.

The Experiment Station at Agassiz, B. C., was my next stop, and there I found an arboretum where the imported trees, many of which I had known in their original homes, were behaving in a strange way, showing a remarkable tendency to produce seeds.

Large specimens of Japanese cherry trees and maples were growing with our eastern species of tulip trees, sour gum, sweet gum, and sassafras. Near-by were three species of flowering dogwood; the eastern *Cornus florida,* the northwestern evergreen, *Cornus Nuttallii,* which has enormous white flowers set off by dark green foliage; and the Japanese species, *Cornus Kousa,* which is more or less intermediate in character. I had tried my best to grow the Northwest species at "In the Woods" but it proved too tender, although the Japanese species did well there and fruited abundantly. I wonder that some one has not persisted in an attempt to cross these

440

various species of dogwood to see what would result, for I am not one of those who believe that the wild forms of plants are "invariably better" than the cultivated.

This experimental farm had little money at its disposal, but it had erected two excellent field laboratories, each of which had its own investigators, and there was a spirit of research there which I have never forgotten.

All the Canadian stations impressed me most favorably in one respect. They were much better landscaped than those I knew in the United States, many of which looked as though scientific men felt attractive planting around their buildings to be beneath their consideration. But of course our experimental gardens have a very real alibi to offer. The material care of the many acres of land under cultivation presents administrative problems involving so much time that the man in charge has little left for investigation or beautification. A farmer working his acres has nearly as much leisure for research as the director of an experimental farm of the same size. In fact, an experimental farm can become so complicated that it represents only an administrative unit without accomplishing much research, because its various officers are overburdened with the obvious things which just have to be done.

In crossing the Great Divide to the Puget Sound region, I left behind the arctic winters and reached a land where the winter minimum is twenty degrees above zero instead of twenty degrees below. But in so doing I came into a region where there was too little summer heat to ripen most varieties of Indian corn. In other words, the Puget Sound region resembles England climatically as well as Japan; it has a much heavier rainfall than England, sixty inches as compared with sixteen, but has the same cool summers which characterize the north of England and make it a wonderful region for perennial borders.

On the third of October I reached Victoria, an attractive city located on wooded Vancouver Island. A new experimental farm had recently been started at Bazen Bay, fifteen miles north of Victoria where the lowest temperature on record was nine above zero, and the highest 92. I never saw such quantities of pods as there were on the leguminous shrubs which were assembled in this little garden. They looked as though they were fruiting themselves to death, and I felt that here was a place where the effects of climate on leguminous plants might be studied to advantage.

During my stay, I drove over to Sidney and called upon the great botanist of Canada, Professor Macoun, who was then well in the seventies. He told me of a visit which he had made in 1875 to the famous Peace River region northwest of Edmonton, Alberta. At a mission station five hundred miles from Edmonton, he had found some barley and wheat that had been harvested there. Professor Macoun sent the samples to the Philadelphia Exposition in 1876 and received two bronze medals for them. In 1879, the Government sent Macoun to explore this country. When he returned to Ottawa with his report, it was received with some incredulity and Sir Charles Tupper and Charles Fleming said to him:

"Now Macoun, we are prepared to back you to the limit, only don't draw on your imagination."

It is pleasant to know that he lived to see this Peace River country become a great wheat-growing center.

In Seattle I addressed a luncheon of the College Club, where I met Samuel Hill, son-in-law of J. J. Hill, the railroad magnate. He was an amazing person, a tremendous talker and a good deal of an egoist. For some reason, he took a liking to me and treated me as an intimate friend from the first day of our meeting. His engineer Bowlby had just completed the Columbia River highway, at that time the most perfect automobile drive in the country, the first I believe to be especially designed for automobile travel, with broad sweeping, properly banked curves. Above the rapids of the Columbia, Sam Hill had built an extraordinary country house, designed so that he could drive his car into the top story. He insisted that I should establish a plant introduction garden there, but the site was impracticable.

At the luncheon I also met the head of the Carnation Milk Company, Mr. E. A. Stuart, who was experimenting with the Sudan grass, one of our introductions, and he gave me a photograph of himself standing in a field where the grass came up to his shoulders. I was much interested to find this introduction from Upper Egypt so successful in the far Northwest.

Frank Meyer was surprised to see me on the dock as the boat came in. He stood at the rail, holding in his arms a dwarf cycad in a green porcelain pot. Meyer had good taste, and the gifts he brought back always had beautiful form and color. He loved the work of the old Chinese artists and detested the ordinary run of bric-a-brac. Once, when invited to dine at a lady's home, he astonished his hostess by suddenly rearranging her over-crowded parlor, concealing all of her ornaments except two or three bits of Oriental pottery which he liked and placed appropriately. Fortunately she was pleased by the result!

As the boat approached, Meyer waved to me and began excitedly calling that on his way home he had discovered the chestnut bark disease present in Japan as well as China. This discovery made it appear even more probable that the blight had come into America by way of Japan.

After leaving Seattle, Meyer and I set forth for the Garden at Chico where so many of his early collections were growing. It was a real pleasure to take him through the rows of plants. There were so many interesting introductions that it was bewildering as well as fascinating.

Meyer's dwarf lemon from Peking was producing a high yield; we had heretofore pushed it as a greenhouse curiosity, but it had begun to attract attention as a possible commercial lemon, even though its fruit flesh had an orange tint. The jujubes were making amazing growth; grafted plants only nine months old had dozens of fruits on them. There were his drought-resistant wild almonds from Chinese Turkestan; wild sakura seedlings from Japan to test as stocks for fruiting cherries; the wild olive of Baluchistan, *Olea ferruginea;* long rows of tung-oil seedlings to supply the plantations of North Florida; Doctor Van Fleet's hybrids of the Chinese, Japanese, and American chestnuts; and a curious Chinese walnut beside which I photographed Meyer holding a nut in his hand.

From Chico, we visited Doctor J. H. Barr of Yuba City who was a great believer in new plants. He and his neighbors were much excited about avocados. The first avocado "Association" had just been formed, and one of the burning questions was how hardy the trees were likely to be. The Mexican thin-skinned varieties were proving the hardiest, and the boom days in avocados were just beginning.

Doctor Barr was testing one of Meyer's introductions, *Amygdalus Davidiana,* as a stock for cherries on soil so alkaline that alfalfa could not be grown on it. It bid fair to make cherry-growing possible on soils heretofore considered unsuited. I think that this gave Meyer a greater feeling of the importance of his work than anything else we showed him.

443

Not far from this plantation, Meyer's Tang Hsi cherry was being grown. It was a distinct *species* of Prunus, and not merely an Oriental variety of the European species. This was an unusually interesting discovery of Meyer's; that, in the Chekiang province of China, orchards exist composed of an entirely different botanical species of Prunus from those composing the cherry orchards of the rest of the world.

Leaving Meyer to work over his Chinese collections, I pushed on to Berkeley, with its amazing complex of laboratories, classrooms and personnel. I had always held Doctor Eugene Hilgard, the agricultural chemist, in great esteem. His early work on soil chemistry and physics had become a classic and I welcomed the opportunity of talking to him. I found him sitting quietly in the garden behind his house, for he was approaching the close of life. His eyes were as keen as ever and the enquiring brain which had placed him among the intellectual giants of his day was still active. I suppose that it was natural for him to deplore the trend of superficial and hurried work which he felt was going on around him, for it seems inevitable that in looking back we should forget much of the hurry associated with our own youth, seeing the past as a mirage of placid days filled with study and research.

The Golden Gate Park was developing rapidly under John MacLaren's able management. He introduced me to Miss Alice Eastwood, who was listing the plants growing there. The Office profited considerably from this acquaintance with Alice Eastwood, for she was easily one of the most wide-awake and enthusiastic botanists on the West Coast.

The San Francisco Exposition was in full swing and I was anxious to see the landscaping of the grounds, as John MacLaren had also been responsible for it. The technique which he had developed to produce quick landscape effects was in striking contrast to the methods used at the Chicago World's Fair in 1893, and was responsible for the peculiar beauty which distinguished the San Francisco Fair from all others which I have visited.

The development of the orange groves around Riverside had resulted in the selection of a site for a subtropical station over which Webber was to have charge. As he and I pushed our way through the low bushes and wild grasses, jack rabbits ran to right and left before us over the hills. It made a pretty dry impression on me, as the autumn hillsides of California do to one familiar with the green pastures of the East.

444

Young Nancy Bell Fairchild greatly admired the yellow *Rosa Hugonis* when it first
flowered in Maryland in 1914. (Page 433)

The Mume of Japan, favorite flowering shrub of Japanese poets, blooming in Maryland.

444B

Frank N. Meyer's plant-collecting cart trekking through Chinese Turkestan.

The dark purple fruit of the Gottfried Avocado ripens in Florida in July.

444c

Rows of introduced fruit trees in the Plant Introduction Garden at Chico, California.

The Plains of Canada

It was here, later on, that Clayton Smith carried on his investigations of the virulent bacterial disease known as crown gall which attacks the crowns of plants below the ground. Numerous species are afflicted, including unrelated plants such as beets, daisies, chrysanthemums and orchard fruits. I discussed with Smith the possibility of using the Japanese Mume (*Prunus Mume*) in his experiments. Seedlings of this species had made a remarkable growth at Chico and appeared to indicate possibilities for its use as a new stock. Stock experiments are long and tedious, and liable to be interrupted by the accidents of human life, but when I returned to Riverside again in 1928 Smith showed me peaches grafted on Mume stocks six years old which had proven resistant to the crown gall disease. However, there were signs of uncongeniality between the stock and the scion. There are enough puzzling factors involved in any "stock" experiment to discourage all but the most optimistic and persistent experimenter, yet it is only by means of such experiments, carried on through generations, that suitable stocks for our orchard trees have been discovered. Few fruit industries have reached a high state of development without many such discouragements.

Every tree of the citron grove above Munrovia had been dug out and in their places the new owner had set out avocados. The type of avocados grown in Florida behaved quite differently in California, in fact the Trapp, the leading commercial variety in South Florida, produced small and worthless fruits wherever it was planted in southern California.

At Claremont, I attended a meeting of the local Horticultural Club and listened to addresses by Webber and Wilson Popenoe, who by this time had become quite an authority on the avocado. It seemed to me that the time was ripe and the man at hand to undertake a comprehensive search for avocados in Central and South America, the home of this fruit newly come to our attention. A preliminary study of this region had been made by Cook and Collins and had revealed a great wealth of seedling varieties.

American horticulturists, scattered through Florida and California, were creating a new fruit industry, destined to rival that of the grapefruit which started twenty years earlier and is now measured by thousands of acres and millions of invested capital. Prehistoric avocado culture in Central America reaches back into the Mayan civilization, far beyond the dawn of any historical records, but, so far as I know, nowhere have

445

there been found any evidences that the Mayas had learned to graft or bud the avocado. The first plantings in our country of named varieties propagated by budding date from 1901.

As early as 1887, fruit from unnamed seedling trees scattered through South Florida was sold on the northern markets. However, the opinion prevailed that this tree could not be budded. Seedling avocados probably came into California later than into Florida, but a tree is known to have been planted in 1886 at Hollywood, California, by one Jacob Miller. In 1894, the "Trapp" avocado fruited in the yard of Mr. Trapp at Coconut Grove from a seed of a West Indian avocado brought from Key West. This was a milestone in the development of avocado-growing in America, for I believe that it was the first avocado to be commercially budded and grown as an orchard tree. Mr. George B. Cellon of Miami, Florida, deserves the credit for first developing the technique of avocado bud propagation and first putting on the market budded trees of named varieties such as the Trapp.

It appears that as far back as 1880 Admiral Beardsley carried a few fruits on his flagship from Guatemala to Honolulu, and from these originated a variety of the Guatemalan race which has come to be known as the MacDonald.

As time passed, the avocado increased in popularity. I wish that I had space to include the names and history of the many successful varieties which have been developed. However, this information is to be found in the records of the Avocado Association and our Plant Inventories.

Returning from the Pacific Coast, I stopped over at Reno to get an idea of the climatic conditions which our plants must face in order to succeed in Nevada. I had only a vague conception of the climate there and had not realized that temperatures of seventeen below zero occurred in that part of the world. The people living in the treeless wastes of Nevada were pathetically interested in any shade tree which could be persuaded to grow, and Meyer's Chinese poplars and particularly his dry-land Chinese elm, *Ulmus pumila,* seemed almost built to order. Therefore I arranged to send a hundred seedling elms to the station there for trial.

It was early fall and along the river bottoms the giant poplars (*Populus Fremontii*) had turned to gold. As I stood beside their great trunks

446

and watched the yellow leaves falling about me in the autumn sunshine, I wished that I might have introduced such a tree into a region like Nevada where it would be appreciated as this native tree has been.

President Hendricks of the University invited me to dinner and I learned something about that curious colony of short-term residents for which the town of Reno is famous. In fact, the economic life of the place was entirely colored by the traffic in divorces.

The Utah Agricultural College was the home of the American sugar-beet industry. The industry was said to have been started in 1853 by Charles J. Staymer, a New England dreamer, but it was not until 1890 that the first sugar factory was built in Utah. My familiarity with sugar-cane plantations in the tropics, and with the subsidized beet-sugar industry of Europe, had given me the impression that sugar-cane was superior to the sugar-beet for the production of sugar. I felt convinced that without a State subsidy and a tariff wall against cane sugar, no beet-sugar industry could long exist.

The tremendous over-production of sugar in tropical countries has proved the correctness of this impression; but the problem of providing the American farmer with another crop from which to make a living on this particular type of soil and under these climatic conditions is more complicated than may at first appear. The problem includes the international relations of our country with other countries that are inhabited by races of human beings who can live without the "luxuries" which an American has come to consider as necessities. Some one has said that all life, all chemical activity even, would cease were there no different levels of activity, no differences of potential. Social levels are essential, too, and tariff walls appear to be as necessary to maintain levels of comfort and "culture" as the locks in a canal are essential to maintain the different water levels. At least this seems to be the basis of argument for beet sugar in a world confronted with a sugar surplus produced in many cases by people reduced to starvation because of their dependence on the only crop they have ever learned to grow.

The secondary effect of sugar-beet growing in Utah was excellent, for thousands of water-absorbing roots are left in the ground and they loosen and aerate the soil so that phenomenal yields of wheat are harvested in rotation with the beets. Until some money crop could be found which would produce these soil effects and at the same time pay its way, sugar-beet growing seemed likely to remain a permanent part of Utah agriculture.

Mr. Widstoe, the president of the College, was prominent in the Mormon Church, and his wife was a granddaughter of Brigham Young. They talked freely about the descendants of the great pioneer, and I was particularly interested, as one of Brigham Young's wives was a distant cousin of my mother. Mrs. Widstoe gave me a letter to her aunt, Mrs. Lucy Gates, which I presented when I reached Salt Lake City. Knowing of my connection with *The Journal of Heredity,* Mrs. Gates told me how distressing she found the inaccurate, spectacular stories about the Mormons which had appeared in some of the other magazines. I offered her the columns of *The Journal* in which to present the correct story. She rummaged through her desk searching for some papers and happened to find a most interesting group photograph of eleven daughters of Brigham Young by eight different mothers.

"You know," she said, "I can always tell a female descendant of Brigham Young, for they all have his thin lips. It was one of his striking characteristics, as you can see from any of his photographs."

CHAPTER XXXV

The War and Dried Vegetables

S OME TIME before I left Washington, I had received a rather illiterate letter from a man in Savannah saying that he had heard of my
interest in bamboos (I do not know just how) and that he sold
shoots from a grove near Savannah to a restaurant in town. I
rather doubted the tale—particularly the part about the restaurant, as so
few Americans will eat bamboo anyway.

Having forgotten the episode, I was considerably amazed when a
strange, unkempt individual walked into my office demanding to see me.

"I am just a chore boy," he announced, "but they call me 'Colonel
Dayton.' I'm the fellow who wrote you that letter about the bamboos I've
been selling to Finkner's Restaurant in Savannah."

He laid on my desk some fresh bamboo shoots and described the
grove as located twelve and a half miles south of Savannah on the
Ogeechee Road. I had never heard of the existence of any such grove
and was still skeptical, but Ed Simmonds was leaving for Miami and I
asked him to investigate the place. It proved to be a grove of *Phyllostachys bambusoides,* the most widely cultivated timber bamboo of Japan.
I immediately coveted this bamboo grove but did not have the means to
buy it. Half jokingly I wrote to Mr. Lathrop, asking him if he did not
want to own a bamboo grove. He replied at once that certainly he did
not want a grove himself, but was willing to purchase it and present it
to the Office for experimental purposes.

This was the beginning of the Barbour Lathrop Plant Introduction
Garden on the Ogeechee River near Savannah, which I hope my readers
will visit if they are motoring in that part of the world. To step into
the bamboo grove is almost to step into another world. You suddenly
become a Lilliputian wandering among giant grasses. In the spring it is

449

quite amazing to watch the new shoots come out of the ground. They grow so fast that one can almost see them increase in size from moment to moment. There is also a little museum which houses various products made of bamboo from different parts of the world.

The National Geographic Society was arranging a Peruvian expedition in co-operation with Yale University. Professor Hiram Bingham was to lead it, and Gilbert Grosvenor asked me to sit in on a conference to discuss the personnel of the party. The expedition was to explore the region back of Cuzco, the center of one of the oldest and most highly developed agricultures of the globe. Although the purpose of the expedition was mainly archæological, I felt that any group going into that region would be greatly assisted by having a trained botanist who knew something about agriculture. I was asked to name the man, and immediately telephoned O. F. Cook requesting him to go. Cook joined the expedition and not only made a study of the terraced hillsides of the Andes, and the numerous crop plants of the Incas, but proved by his observations that, contrary to the accepted belief, the stone terraces, which had been so surprisingly preserved in those high mountain regions back of Cuzco, were built for agricultural purposes and not for defense. He also returned with fine photographs of the explorations.

This Peruvian expedition brought back many fascinating things, including the *Canna edulis,* which we grew and experimented with for years. An excellent starch is made from its rootstocks, or rhizomes, and I believe that the best Australian arrowroot on the market today is manufactured from this plant in Australia. They also brought back several tuber-bearing plants such as the Oca, *Oxalis tuberosa,* which is second only in importance to the potato in some districts of Peru; the *Ullucus tuberosus,* grown commonly in the highlands of the country; and the Anyu, *Tropæolum tuberosum.* A cereal plant, the Quinoa (*Chenopodium Quinoa*), was particularly interesting. It is a cultivated "pig weed" and was one of the most widely grown crops of the Inca civilization before the introduction of wheat and barley into the region. Its long heads of small grain are ground and cooked like oatmeal. It is still grown today.

Unfortunately, most of these finds were high-altitude plants, and the high altitudes in the United States are much warmer in summer than the Peruvian Andes, and so much colder in winter that the cultivation of these Andes species has been generally unsuccessful.

The War and Dried Vegetables

Cook's researches made on this expedition revealed the astonishing fact that the agriculture of the Incas had brought into cultivation almost as many species of plants as had the agriculture of all Asia. I have only to mention the maize, the sweet potato, the cherimoya, the avocado, the chayote, the Lima bean, as a few of the well-known plants which the Inca civilization has given to the world.

Recurring colds during the Washington winters began to prove too much for me. I spent many weeks in bed, and finally was ordered off to Florida. Marian and her parents were in Palm Beach and I joined them there early in February.

It was Mr. and Mrs. Bell's first visit to South Florida, and they were enthusiastic about the towns which were beginning to spring up along the coast. We motored down to Ft. Lauderdale, then consisting of a hotel with a few scattered houses and real-estate offices. The evening of our arrival, William Jennings Bryan spoke to a capacity audience in the church, and much impressed Mr. Bell by his gift for oratory and marvellous delivery.

The next day we went to Miami. Mr. Deering was giving a luncheon in honor of Mr. and Mrs. Bryan but, when we arrived, Mr. and Mrs. Bell were too tired to go, for in 1916 the narrow brick road was so rough that the hours bumping along in a motor were most fatiguing. Today this drive takes about forty-five minutes.

I was very anxious to show Mr. Bell the Experimental Garden and the plants which I had been describing to him. However, the weather was warm and I did not want him to visit the garden in the heat of the day, for he was very sensitive to the bright glare of the sun. Also, plants are at their best when the dew is still fresh on them.

"If you insist on my seeing your garden in the early morning," he said, "I will stay up all night, for I certainly could not promise to get up at such an hour. If you want, I will be ready tomorrow at six o'clock."

When I came for him, he was sitting by his window in the Royal Palm Hotel watching with fascinated interest the birds circling about the mouth of the river where a sand bar had formed.

"David," he said, "I have just seen something which I have heard about but never quite believed. I have seen buzzards, there on the sand bar, face the breeze and rise into the air without flapping their wings."

451

The World Was My Garden

As he spoke, I recalled Professor Langley's tower in the Washington Zoo from which he watched the flight of birds. Several years later I learned from Mr. Warner, the telescope manufacturer, that the incident which started Langley on his study of the "Internal Work of the Wind," was a paper on the flight of buzzards, by a Florida man named Lancaster, read in 1888 before the American Association for the Advancement of Science. Lancaster told of lying on his back in a blind built in the top of some cabbage palms, and seeing buzzards remain almost motionless in the breeze a few feet above his head. When Lancaster went on to predict that a machine might be built which could sustain itself in the wind, he was laughed down by his audience. In fact, so little weight was given to his paper that his full name does not appear in the report of the meeting. However, that Langley was much interested in the paper is shown in the opening paragraph of his first contribution on aerial locomotion.

After watching the birds for a few moments more, Mr. Bell went with me to the Garden and became equally fascinated by the plants. Ed Simmonds met us and showed off his treasures to perfection. The slanting rays of the early morning sun shone brightly on a tall papaya tree with forty immense fruits hanging on it, and Mr. Bell stood for some time amazed by the spectacle. Certainly a papaya tree loaded with fruit is one of the sights of the tropical world.

After this visit, Mr. Bell took a keener interest than ever in our work and even advised me to organize a "Plant Utilization Society." I discussed this proposal with Mr. Lathrop, but he was not sympathetic.

Mr. and Mrs. Bell departed for Washington and Marian and I, dissatisfied with hotel life, went to stay with Mr. and Mrs. Simmonds in the Garden. Living there among the plants, we soon realized how fond we had become of Florida and, when Mrs. James Nugent came to see us and said she had heard that we were looking for a place, we did not deny it, but went to look at her property on Biscayne Bay. We drove down a narrow road to the edge of a slight bluff overlooking the blue waters, and stood under a spreading poinciana tree beside the little wooden house. Without a moment's hesitation, Marian said,

"David, I want this place. We just must have it. I will ask Mother to give it to us." Turning to Mrs. Nugent, she added: "Mrs. Nugent, will you give me a ten-day option?"

I suggested that we had not even looked around the place, but my

Alexander Graham Bell stayed up all night rather than rise early to visit the Experimental Garden, but once there, was enchanted by the odd, tropical appearance of the Papaya.

452A

Barbour Lathrop purchased this grove of Bamboo near Savannah and presented it to the Government for a Plant Introduction Garden.

452B

In the Barbour Lathrop Garden there is now a little museum containing articles made of Bamboo.

452C

David Fairchild beside the old stone barn on his place at Coconut Grove—one of the oldest buildings in South Florida. It has been remodelled and is used by him as a laboratory.

words fell on deaf ears. As we drove out of the gate we met a man coming in, and later learned that he had the money in his pocket to purchase the property.

Marian posted off to Washington to talk to her mother, and I went down to Cutler, Florida, to see one of our old correspondents, Doctor S. H. Richmond, to whom we had sent many tropical plants. We checked over his plants, and he warned me that a leguminous tree, *Leucæna glauca,* which we had imported from the island of Réunion, would become a weed in Florida. It has, but as it is a legume, its roots add nitrogen to the soil, while its slender trunks make excellent firewood.

Soon a wire from Marian sent me back to Miami to close the deal with Mrs. Nugent.

Thus it was that The Kampong came into our lives. I stood beside the old poinciana and looked at a group of royal palms growing down near the water. The thought that those stunning gray columns with their waving plumes actually belonged to us gave me a thrill which I have never forgotten. I went down on the shore and put my arms about one. There was also a little stone barn, and both it and the pigsty seemed the most picturesque structures I had seen in South Florida; certainly they were about the oldest. I often recall the impressions of that first afternoon, which I spent all alone among the live oaks, citrus trees, and guava bushes growing around the little wooden house which had so recently become ours. Twenty years have passed but the charm remains. The Kampong is still a dream place to the family.

Midsummer of 1916 saw Wilson Popenoe off for the avocado regions of Guatemala, where he spent the next two years in most productive exploration for the best varieties of this fruit, then new to the Anglo-Saxon world. I wish that Wilson would write the account of this interesting quest. He selected the trees which bore avocados of exceptional quality, tasted and described the fruits minutely, photographed them with great care, and cut budwood which he packed and shipped home. Into the bark of each tree, he cut a number to identify it by his notes.

His budwood sticks arrived in excellent condition and in order to classify the seedlings which resulted, I asked Wilson to give them Mayan names —names which they still bear. Coming in as they did at the commencement of this newest and very promising orchard industry, these varieties contributed materially to its development. A number of them have won places of importance in the California and Florida orchards.

In October, before Meyer left for his fourth trip to China, he decided to make his will. As he came from the lawyer's office he laughingly said, "I've left some money for the Office Staff. It isn't much but I feel grateful to them for all they have done for me." He often spoke of never coming back, and I always tried to stop him from thinking about death.

On June 4, 1918, a cablegram came from the Consul at Nanking. It read, "Frank Meyer, Department Agriculture, disappeared from steamer in this consular district en route Hankow to Shanghai, June 2d." Later another cable reported that a search for his body was being made and a third cable said that it had been found thirty miles above the city of Wuhu.

Meyer had so endeared himself to every one in the force that the shock was a very real one. While we were in our first sorrow over his death, his last letter came. It was from Ichang on the upper Yangtzse where he had been shut in by the fighting troops on the river below.

"At last," he wrote, "I have been able to break through the lines around Ichang and walked to Kingmen, got the stored seeds and baggage and settled the payment for the pear seeds; then marched down to Shasi and took a steamer from there. . . . Of course we passed through villages that had been looted and burned, and food was hard to obtain, but to an old hand out here, like myself, these things have so often been encountered that one is used to them.

"I did not write you from Ichang of late because I was not sure that I really could make the trip. The whole country is so fearfully upset that travel has become a gamble. . . . I am awfully glad that I got away from Ichang; the situation began to depress me. One cannot live for months in an atmosphere of suspension without feeling the effects, and as I had cheerless, uncomfortable quarters and lack of substantial food at times, there were both mental and physical discomforts.

"Well, I just received your very sympathetic letter of February 26. Uncontrollable forces seem to be at work among humanity; and final results, or possible purposes are not revealed as yet, that is, so far as I can look into this whole titanic cataclysm."

There followed a masterly analysis of the Chinese food situation, in which he pointed out that "rice forms three fourths of the total amount of the food of the ordinary people; meat and fish supply a mere fraction, and the rest is in the form of beans, peas, lotus rhizomes, various roots and tubers and in leafy vegetables."

"Concerning substitutes for dairy products," he remarked that "the hundred and one different manufactures of the soy bean supply this protein, but I must admit that it will take some time for the white races to acquire a taste for the large majority of these products."

I had written Meyer that we were putting in three hundred acres of soy beans and he was much interested, as it was the largest area of this bean yet grown in America. He was pleased to hear that his Feicheng peach, a large clingstone variety, which had caused him a great deal of trouble to get, had apparently come true to seed. He closed his last letter with the words "Times certainly are sad and mad, and from a scientific standpoint, so utterly unnecessary."

I have always been at a loss to understand his disappearance, for no evidence of foul play was found. Meyer's death will remain a mystery to his friends.

I called the members of our little staff together and we held a touching memorial meeting. I closed it with the following words:

"Meyer's work is done. Whether his body rests beside the great river of China, or under some of the trees he loved and brought to this country, will matter little to him. He will know that throughout his adopted land there will always be his plants, hundreds of them—on mountainsides, in valleys, in fields, in the backyards and orchards of little cottages, on street corners, and in the arboreta of wealthy lovers of plants. And wherever they are they will all be his."

Then I read them his will. He had left a thousand dollars to be spent by the office force in an outing somewhere in his honor, or, if the force so voted, it could be divided equally among all the members and, as there were a hundred, this would mean ten dollars to each person.

It was the unanimous opinion of the Force that we should have a medal made in his honor, call it the Meyer Medal, and present it for meritorious work in the field of Plant Introduction. My old friend the medalist Theodore Spicer Simson offered to make the medal, and produced what I think is a unique and interesting one. It has already been given to ten investigators in plant introduction.

Since the Office was a government affair, it was against the law for its staff to present a medal but it could suggest to the Council of the American Genetic Association the names of prominent men in Plant Introduction work who would be eligible to receive the medal.

Among the seeds which were found in Meyer's baggage and which his last letter indicates he expected to "write up" when he reached Shanghai, were some of a species of grass, which Piper, the authority on grasses of the Department, named the "Centipede Grass" (*Eremochloa ophiuroides*). This grass has come to be one of the best lawn grasses in the State of Florida. Many thousands of children are already playing on lawns of Meyer's grass, but few know the history of its introduction.

The American Consul brought Meyer's body to Shanghai, where it now rests beside an appropriate gravestone which his great friend and admirer P. H. Dorsett erected. Some day perhaps the beneficiaries of Meyer's life-work will erect a monument to him in America.

The Office was busier than ever. Doctor Galloway, who had left us for Cornell, returned to take a research position in the Office where his wise counsel aided us for twenty years. In Florida, the nematodes had attacked our dasheens and a hot-water method of soaking the corms was being devised to control them. This treatment has since become of wide application and is reflected in the latest Plant Quarantine regulations prescribing the method to be used in treating imported bulbs.

When winter came, we had not yet entered the War, and I went to Florida again, arriving at our new place just before the great freeze of February 3, 1917. This frost devastated the landscape of South Florida and gave us a check on the hardiness of hundreds of plants which we had come to look upon as being perfectly safe there. I set to work to make colored photographs and collect dried specimens of the frozen plants to illustrate the effect of this freeze, the first memorandum of its kind ever made, I believe. These data formed the basis for many of our later decisions and operations, and proved of great value in an argument I presented later to Secretary of War Weeks which induced him to grant us the use of an abandoned air field of eight hundred acres south of Miami which he had advertised for sale after the Armistice. This air field became the large Introduction Garden known as Chapman Field, and the "frozen plant album" became the basis of a bulletin called *Effects of Freezing on Tropical Plants*.

David, Marian, and Nancy Bell Fairchild lunching at the Kampong, with Mangos to left and the Wampi tree to right of them.

456A

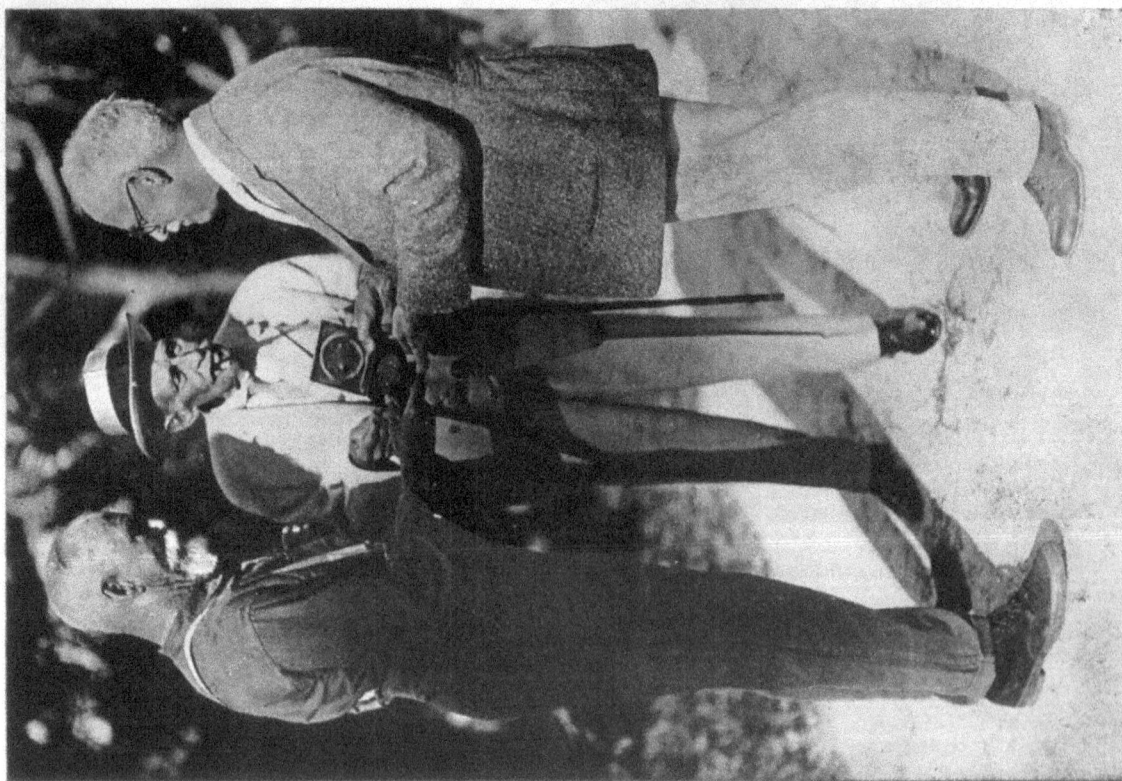

David Fairchild (*right*) presenting the Meyer Medal to Charles Torrey Simpson, with Barbour Lathrop as spectator.

Barbour Lathrop came to Miami every winter and took a paternal interest in the tropical fruit trees at the Kampong.

The War and Dried Vegetables

In March, a wire appointing me to the Committee of National Defense made it necessary for me to return to Washington. On my way North, Mr. Lathrop and I stopped in Savannah so that I might show him the bamboo grove on Ogeechee Road, which he had presented for a Plant Introduction Garden.

Uncle Barbour had continued to be a part of our lives. Of course, after my marriage I had not been available for trips to the ends of the world with him as before, but he had become reconciled to that and fond of Marian too. This had required a little time, and neither she nor I will ever forget her first meeting with Uncle Barbour. I had taken her to his hotel in Boston, and we went up to his sitting-room. When he came in, I said, "Uncle Barbour, this is Marian." He did not speak, merely looked her up and down and then grunted, "Well, she has fine eyes." His glance then travelled to me.

It did not occur to me that there had been any change in my appearance since I had been with Uncle Barbour, who always saw to it that I was correctly clad and kept my hair cut and mustache trimmed. Marian was used to Mr. Bell, who wore his hair rather long, and she did not approve of me if I were too much shorn.

"My God, Fairy," Uncle Barbour ejaculated, "what has happened to your hair? It looks as though it hadn't been cut for months. You had better go right down to the barber shop and get a cut. You are a fright!"

I looked at Marian, who started to say something.

"Oh," said Uncle Barbour to her, "you think you can make him look like your father, do you? But your father has a great leonine head. He can wear his hair long. Fairy can't."

Tears came into Marian's eyes. No one had ever spoken to her with such brutal frankness. She wavered a moment, and then bolted for the door. Uncle Barbour raised his hands and exclaimed,

"Good Lord, what have I done? She's offended——"

I did not wait to hear the rest, but ran after Marian, who was halfway down the hall where she stood sobbing by the stairway. I did not know what on earth to do. I was angry with Uncle Barbour, yet I knew that he did not mean to be brutal, that it was just his manner. Almost before I could say anything, Marian brushed the tears away and started back. Although she did not agree with Uncle Barbour, she consented to my going down to the barber shop, and when I returned I found them both in good spirits and saw that Marian was going to like my old friend and benefactor. "There, you must admit that he looks better," he said. Marian did not argue the matter with him, for which I was deeply grateful.

Once Marian and Uncle Barbour knew and understood each other, they became fast friends. After we purchased The Kampong, he took a great interest in the tropical fruit trees which I planted, for so many of them were the results of our travels together. As he grew older and travelled less, he came to Miami every winter, and was constantly at The Kampong, spending practically all of his afternoons on our little porch chatting and looking out over the Bay. He was a great raconteur and, to the end of his days, surprised us by relating new tales of his personal experiences which we had never heard before.

Washington was seething with excitement. War was declared on April 6, and Dorsett and I were already at work arranging a war budget for the Office. No one can adequately describe the confusion of those early days, nor can I look back at them without a shudder. The wild disorder; the hundreds of futile plans to stimulate food production by increased farm acreage; new methods for prevention of plant disease; the cultivation of drug and oil plants made necessary by the shortage of foreign supplies— all these come back as unpleasant memories.

Almost immediately I found myself in the center of a Dried Vegetable Campaign, and discovered that really nothing definite was known about the drying of even our common vegetables. Frantically I evolved a home-made drier with which Marian and I dried vegetables on wire trays over the kitchen stove. Photographs of this crude apparatus were published by the Department and broadcast over the country, although the resulting product was discolored and poor in quality. The difficulties in persuading people to eat these poorly dried vegetables were almost insurmountable. I took a hand in the propaganda to encourage their use, arranging, among other things, the details of a moving-picture film staged in Mrs. Lansing's back yard, in which five ladies of the Cabinet were shown eating with assumed delight a luncheon of dried vegetables, dried fruits, and dried meats.

I learned much about the food habits of people from this campaign. For example, I discovered that unless people's attention was especially attracted to the vegetable they were eating, they would eat it without dreaming that it was a dried product "brought back" by soaking overnight in water. However, if they were urged to eat dried vegetables, they would vow they could never bear to touch such things.

I had many amusing experiences with the manufacturing firms and canners, as for some time I was their contact with the Departments. They all believed that the boys at the Front ought to be using dried vegetables, but in this they were balked by Herbert Hoover as well as by the officers of the Commissary Departments. Both Hoover and the War Department refused to give dried foods the benefit of the doubt.

The campaign had not been under way very long before E. Clements Horst, the biggest hop grower in California, arrived in Washington with a carload of dried vegetables which he had prepared in his hop kiln. He walked into my office one morning with the declaration that he had come to get President Wilson to eat his dried products, and that he would then launch his propaganda with this news. Knowing something about the President's food habits, I smiled, and told Horst that I thought he would have an uphill job.

When he found that he could not get to the President, he decided to give a big, dried-vegetable luncheon at the Willard, and asked me to assist him in lining up the Bureau chiefs whose conversion he thought desirable. He was particularly concerned that the Food Administration be represented, since its attitude towards dried vegetables had been apathetic, to say the least. Of course Mr. Hoover would not come, I knew that, but I had met Miss Ethel Bagg, who was very active in the organization and who, I thought, had influence with Mr. Hoover; so I persuaded her to represent her chief.

Horst had twenty-six different kinds of dried foods but they bore little relation in quality to the superior products turned out only a year later. In fact they were rather poor stuff. I strongly advised him against serving more than two or three vegetables at his luncheon, but the morning of the affair I saw him marching off towards the Willard Hotel with an armful of packages containing the whole twenty-six varieties. Being dried, they weighed so little that he could easily carry enough to feed his forty guests.

A notable gathering of chemists, Army and Navy men, Red Cross workers and dieticians assembled. From the very outset I was anything but happy, for it seemed to me that the waiters grinned sardonically as they served the first course, although it was really a good vegetable soup.

The reason for their amusement became apparent when they began to pile the various vegetables around the individual steaks which followed the soup. Dried spinach, corn, potatoes, carrots, beets, cabbage, parsnips and all the others. Over this extremely unappetizing array of strange-looking food, was poured an inky fluid as a finishing touch. This was a preparation of Horst's dried mushrooms.

After the luncheon, most of the guests politely lied about it, not wishing to hurt their host's feelings. But some of them disappeared without saying anything.

Horst arrived at my office the next morning glowing with enthusiasm about the success of his luncheon.

"How did Hoover's representative like it?" he asked.

I called Miss Bagg's office on the telephone and was told the devastating news that the representative of the Food Administration had returned to her apartment from the luncheon so nauseated she had thrown up all of Horst's twenty-six dried delicacies.

However, all this took place early in the campaign. Before the end of the War, the big firms were turning out excellent products by using low temperatures and powerful fans to carry off the moisture. So successful were the results that vegetables thus dried could be "brought back" by soaking overnight, until, when cooked and served, they approximated canned or even fresh vegetables.

The closing banquet of this campaign was a very different affair from poor Horst's dismal luncheon. The long table was spread with an amazing variety of dishes which had been prepared with such skill that many of the guests had no realization that they were eating dried vegetables at all. There were creamed spinach, stewed tomatoes, beautiful white mashed potatoes, string beans, and even corn on the cob, none of the vegetables revealing more than faint signs of having had any but the most usual history.

Really, the argument for dried instead of canned vegetables was a logical one. Dried vegetables save space, weigh little, and keep almost indefinitely, reducing the freight and storage problem tremendously. But all these arguments meant little to cooks and housewives who did not want to learn new methods of cooking and were influenced by the newspaper ridicule of the idea. However, I still believe that, had the War lasted longer, a number of the dried products would have become permanently established on the market. As it is, only the mixed vegetables for soups seem to have survived except for campers and expeditions which stock up with desiccated foods to save bulk and weight. Certainly whatever products there are today are as good as they are because of the work and research accomplished during the War.

During the autumn of 1917, the work with dried vegetables took me to Doctor McCullum's laboratory, in Baltimore where he was feeding fifteen hundred rats on a diet similar to that being used by one of the orphan asylums. The diet, largely composed of cereals, had proved so defective that it had produced disastrous effects on the rats. As a result of his researches, McCullum had discovered the presence of a new substance in green, leafy vegetables which was essential to animal life; a *vitamin* which he called "Fat Soluble A." It is also present in butter fat and certain fats of the animal body, and he was already beginning to stress its importance, an importance which has since become universally recognized. McCullum complained that he had no facilities for photographing his half-starved rats in order to show the results of his experiments. I therefore sent Crandall, the photographer of our Office, over to Baltimore and he took the photographs for one of McCullum's earliest publications.

As the German submarines threatened our communications with Europe, it appeared possible that we might not be able to provision our troops adequately from this country. Consequently, it was proposed that sweet potatoes, one of the biggest yielders of all starchy vegetables and one easily propagated, should be grown extensively in southern France or Morocco. A shipload of American farmers familiar with the cultivation of sweet potatoes was to be sent over, equipped with tools and roots.

I was instructed to consult Ambassador Jusserand and find out whether this suggestion were feasible. When I called at the Embassy, I found him much excited, for one of the projectiles of the Big Bertha had landed near his home on the Champs Élysées the day before, and the mystery of the big gun was still unsolved and alarming everybody.

I told him about the scheme to grow sweet potatoes in France. He listened attentively and then remarked,

"You propose to grow these sweet potatoes for the American troops, do you not? For I must tell you that my countrymen will not eat them. One was put on my plate at a dinner in Washington the other night. I ate it. But I would never eat another."

The World Was My Garden

When it was found that our convoys succeeded in passing safely through the U-boat zone, the plans to grow food on the other side were abandoned. However, this revelation of a food prejudice attributed to an entire nation made a great impression on me. It was one of many which were forced upon my attention during the War. Another example was corn-meal. All attempts to get the Food Administration to send corn-meal to the Belgians failed. Mr. Hoover knew their prejudices against it. Doctor L. H. Baekeland, the inventor of Bakelite, was born in Belgium, and tells me that many Belgians consider one of the greatest sacrifices which we Americans made during the War was that we used potato flour and corn-meal so that there would be wheat flour enough for the Belgians.

Another wartime undertaking in which I was involved catered to something more substantially concrete than public taste, in other words, to the chemistry of gases and carbon.

When poison gas was first used, there were of course no gas masks, and the destruction of human life was appalling. Immediately, man's inventiveness was called upon, and the chemists developed the application of a well-known principle called "occlusion." In other words, certain gases are absorbed by the surfaces of porous objects such as charcoal. The gases have the property of adhering to the surfaces of the charcoal, and when the poisonous gases used in the War were breathed through charcoal, it filtered out the poison gases and let the air pass carrying its oxygen to the lungs. At the outset, I believe that any kind of charcoal was used, but it soon was discovered that the finer the grain and the denser the charcoal, the more efficient it proved as a filter. Charcoal made from peach kernels or coconut shells had many small cavities and presented a greater surface area to the gases. At least, this is a crude picture of the process.

This discovery started the gas experts on a search for palm seeds to use in making charcoal, and they turned to us for assistance. We immediately embarked on a wide enquiry based on the palm and other seeds in our seed collection. As our Inventories were completely documented, we knew where the various seeds occurred. Any dense, hard seed of sufficient size made a suitable charcoal. The coconut proved the best, with peach and apricot pits as second choice, and after them some of the rarer palms such as that of the West African oil palm and various palms of the great

The War and Dried Vegetables

Amazon basin. Later O. W. Barrett, who had been a member of our staff, was sent to some of the tropical countries to round up a comprehensive supply of palm seeds for the gas-mask service. This world-wide search for the best seed for charcoal illustrated the importance of our collecting data regarding the plants of the world even before they were as yet on any obvious program.

Nothing connected with our activities during the War was more irritating than the utterly futile and uncalled-for attempt on the part of the War Department to turn agricultural themselves when they wanted castor beans for the Aviation Service. Brushing aside the Department of Agriculture, they undertook to stimulate the growing of castor beans through their own organization. Employing men who might know something about haberdashery but certainly knew nothing of plants, the War Department developed a program to encourage farmers to grow castor-oil plants without ascertaining the requirements, climatologically or otherwise, of their culture. All they knew was that castor-bean oil was essential for the lubrication of the Army aeroplane engines.

I naturally had nothing to do with the matter, but had opportunity to discover the caliber of some of the men in charge. I met one specimen when I was on my way to Moore Haven, a new settlement on Lake Okeechobee, Florida. As the boat pulled out of West Palm Beach, a man stepped on board who had all the earmarks of a ward politician. My companion pointed to him. "There is the man sent here by the War Department to tell the farmers how to grow castor beans," he said. "He is going to make a speech."

This being the case, I thought it behooved me to make his acquaintance. I soon found that he barely knew a castor-bean plant when he saw one. When he discovered that I did, he tried to make me back his program of castor-bean planting in the Everglades. I asked him where the seed he was going to distribute had come from and whether it was adapted to Everglades soil. Apparently the seed was a job-lot from somewhere in India (he did not know where) and he had no evidence that castor beans would fruit satisfactorily on the muck lands of the Everglades even if they grew at all. I soon edged away from him and let him go his uninformed way, since there was little that I could do to prevent what I knew would be a fiasco. I did urge the planters at Moore Haven to investigate the situation carefully, as the outcome was a gamble with the dice loaded for failure.

463

Of the complete collapse of the castor-bean program and the lawsuits which the farmers filed against the War Department for supplying poor seed and causing them to lose time and money, I prefer to say nothing. It was just one of the many ill-advised schemes hatched during the War. (It was a fact, I believe, that the man in charge of the program really was a haberdasher.)

Little occurred in 1917–18 which I take delight in remembering. To make matters worse, the ghastly flu epidemic occurred and Marian spent her days and often her nights heading the Women's Red Cross Motor Corps of Washington. In perspective it all seems a horrible nightmare.

However, the cherries still blossomed at "In the Woods," and others as well as ourselves apparently felt the appeal of the place, for it attracted as tenants the Newton Bakers, followed by the Herbert Hoovers, and brought us into rather intimate relations with both families.

At the outset of the War, President Wilson used to motor out to "In the Woods" to confer with his Secretary of War, and, during one visit, he and Baker strolled through our woods onto a neighbor's place and sat down on an old board fastened between two trees. It broke with a snap, depositing them firmly on the ground, and Secretary Baker told me later that for a moment he thought that they had been bombed! I went to the site of this disaster, and regretted that the historic board was not mine. But my prosaic neighbor felt differently, even complaining that his distinguished trespassers "might at least have replaced the board."

CHAPTER XXXVI

The Allison Armour Expeditions

SOON AFTER THE WAR, Mrs. Bell presented me with five hundred dollars. She had been watching her grandson and had made up her mind that Graham and I ought to go off together on a trip somewhere. In fact, one afternoon she took me to task for not seeing more of my son, and the upshot of this conversation was her offer to pay the expenses of a vacation if I would take fifteen-year-old Graham with me. It was a typical example of the way Mrs. Bell made generous gifts and constructive suggestions to her family.

We settled on Panama as a likely place for a summer holiday because of its jungles and the chance to collect butterflies, this being Sandy's greatest joy in life. As we talked about the trip, I remembered the collapse of Colonel Gorgas' and my plans to establish the mangosteen in Panama and the fiasco of the appointment of Schultz as "Landscape Gardener of the Canal Zone." There were a number of fine mangosteen plants in boxes in the greenhouse, together with some litchi and jaboticaba trees, and I asked Dorsett, who never voluntarily took a vacation, to accompany us as far as Panama and to continue on to California on an inspection trip, as I knew that he was worn out and needed a change. We decided to take some Wardian cases of tropical fruit trees with us, for the Goethals régime had passed and Governor Jay Johnson Morrow was said to be eager to see the Zone developed agriculturally. We felt that perhaps he would help us.

We sailed from Norfolk one August morning after Sandy had helped the sailors hoist our cases of plants onto the deck where they would have light but not too much sun, and where we could care for them—for Wardian cases sometimes fail to transport tropical plants safely, generally because of overwatering.

In an attempt to share with Sandy the training I had received from Uncle Barbour, I promised the boy a quarter for any worthwhile interview

465

he could report. Sandy had already developed excellent powers of observation, but he had as yet no faculty for scraping an acquaintance and eliciting information from the people with whom he came in contact. After a short dissertation on the various methods by which we may secure knowledge, I sent him forth to investigate the possibilities of the ship.

Sandy returned in short order, with eyes like saucers and a tale so astonishing that I felt he had really outdone both himself and the bounds of possibility. He reported interviewing a man who had attended a Gargantuan cannibal feast in the South Seas. Sandy repeated with gusto the details of flavor and fragrance, with sidelights on the etiquette of these affairs thrown in as a colorful background.

Not knowing whether to treat the tale as semi-fiction, or a complete fabrication of the boy's mind, I withheld the quarter and demanded to meet this experienced gourmet.

Sandy's informant proved to be a rather ne'er-do-well type of individual, who readily repeated his tale to me. His statements had the ring of truth in them and I therefore checked up on him with the Captain. Apparently the man had sailed on the Captain's ship before, and the Captain assured me that, to his knowledge, the man had lived among the South Sea Islands for years and had had ample opportunities to see or be at a cannibal feast, either as guest or chief viand.

So I felt that, as a reporter, Sandy was a distinct success and had a real nose for news! Sandy, for his part, realized that I had proved my point that people we meet have much of interest to impart if we will but listen.

I had with me some splendid photographs of mangosteens and, the day after landing, walked up the hill to call on the Governor. On the way I noticed, fruiting beside the road, a relative of the mangosteen, *Garcinia Xanthochymus,* which had come from our shipments of plants to Schultz. I picked some of the fruit and took them along to show the Governor. Remembering the experience we had had with General Goethals, I was keyed up for a battle, but, after looking at my exhibits, the Governor leaned back in his chair and said, "Mr. Fairchild, why can't we grow the mangosteen here?" I was almost speechless with relief. Then he told me that, in the southern Philippines, he had eaten the mangosteen and thought it one of the finest fruits in the world.

I told him that I believed the mangosteen could be grown in the Zone if a suitable place could be found, and a gardener secured who knew how to care for tropical trees. Governor Morrow at once sent for the official in charge of the agricultural lands, and in a few minutes I found that my plans were enthusiastically accepted.

Dorsett and I visited the places which the Governor said could be set aside for a Plant Introduction Garden such as I had pictured to him.

One of the sites lay up the Chagres River some distance from Gatun, at a place called Juan Mina where an earlier attempt to grow citrus fruits had been made. Much to Sandy's delight, we were taken up there and left to camp out for a few days in a little screened house. It was surrounded by dense jungle and Dorsett and Graham spent many hours chasing insects.

I wrote an article for the *National Geographic Magazine* describing Graham's and my experiences there, for while they were not widely different from those which any man and his son might have in the jungles of Central America, they left unique impressions on our minds. Few tropical forests are more beautiful, and surely no one could forget the sight of a Morpho butterfly flashing through the sunlight. Then there was the experience of shooting at something moving in the tree-tops only to find that we had just missed killing a charming little monkey, and the splash of a big iguana falling into the dark water of the Chilibrillo as the Indian paddled us toward the bat caves near the river's source. When I think of Graham, who is now in Brazil, I cannot help feeling that his grandmother's gift had a great deal to do with his career, for he learned not to fear the jungle, to love it in fact, and to see in its changing life a great field for discovery.

The site we finally chose for the Plant Introduction Garden was at Summit, the highest point on the railroad which crosses the Isthmus. A brook flowed through the property, and along its banks the soil seemed wet enough for mangosteens. I was much thrilled years later to visit Summit and pick mangosteens from the trees which Dorsett and I planted on that vacation trip.

The Governor took a real interest in the possibilities of the Plant Introduction Garden, and I was again delegated to select a trained horticulturist who would create an experimental garden in the Zone. It was a satisfaction to find in Holger Johansen such a person, and I was disappointed when he left to join one of the banana companies. The Garden has since developed most satisfactorily under the able management of Mr. J. E. Higgins, and it is now one of the places of special interest in the Canal Zone. Were it not for the quarantine against their entry into the United States, shipments of mangosteens and other fruits would be coming North from there every season.

But this was not all that resulted from that vacation trip. When Gilbert Grosvenor asked me to write it up for the *Geographic,* the article brought me into friendly relations with a noted systematic botanist who had been studying the plants of the Zone and desired very much to write a Flora of that area, Doctor Paul C. Standley of the National Museum. He has given me credit for bringing about his being sent to Panama. This enabled him to publish what is now the only Flora of the region—an admirable work which should be in the hands of all who are curious about that strange and fascinating strip of land which connects the two great oceans and divides North from South America.

But the most interesting link in the chain of circumstances which came from Mrs. Bell's vacation investment was that it brought me into close contact with Doctor Thomas Barbour, one of the outstanding professors at Harvard, and also with Frank M. Chapman, the great ornithologist, and cemented deep friendships which have lasted to the present time.

Hearing of my interest in the Panama forests, Doctor Barbour, Director of the Agassiz Museum, asked me to join his committee and help him create a laboratory of research on an island in Gatun Lake. Frank Chapman was also on the committee.

Of the Barro Colorado Island Laboratory, in sight of which the steamers pass as they make their way through the Panama Canal, I should like to write a volume. I should like to tell of the vision of these great naturalists who, knowing the thrills to be found in tropical jungles, could not bear to see banana and cassava patches destroy the forests on the largest island created by the rising waters of Gatun Lake. They realized that when, in raising the water-level for the Canal, hilltops became islands, a unique opportunity for the study of Natural Science existed, as the wild life thus isolated could be studied as a unit of jungle.

To Tom Barbour, Morton Wheeler, and a number of their friends, C. V. Piper among others, and to Governor Morrow whose alert mind comprehended the wisdom of their proposal, must go the honor for its inception. But the real burden of establishing the laboratory has been borne by

The Allison Armour Expeditions

Doctor Barbour, who selected and has personally directed the small staff, with James Zetek at its head, which carries on the difficult task of conducting a research laboratory in a dense jungle where termites and many other insects and fungi complicate the problem immensely.

The ravine-furrowed island of Barro Colorado comprises six square miles of virgin forest. On it are representatives of the jungle animals and plants of the region, and it appears to be large enough to preserve the original fauna and flora for generations to come. A modest laboratory with living quarters for visiting scientists was built; simple trails were cut across the island; and here and there small huts were placed for observers to stay in when overtaken by the torrential downpours or when they wanted to make night studies of the jungle animals. Means of communication with the mainland, a safe water supply, mosquito and malaria control, patrols to prevent poaching, a library and herbarium, and other details were worked out.

From its beginning, the laboratory has had governmental status, for the committee of management is appointed by the National Research Council. Universities, museums, botanic gardens and other scientific organizations were informed of its establishment, and asked to make annual appropriations for the maintenance of "tables" similar to those in the Biological Station in Naples. These tables not only give partial support to the institution, but make it possible for students to live there with safety in the very depths of a primeval rain forest at a minimum expense. Many of the best biologists of America have been enthusiastic "boarders" at the little laboratory overlooking Gatun Lake. Frank Chapman has made it a second home during the past few winters, and his writings on the jungle birds and animals are familiar to a wide public. The flora and fauna of the island have been described by numerous scientists, and a stream of scientific papers is coming from there every year, covering all sorts of subjects connected with the biology of the tropics.

Some years later, Graham and I spent a memorable, unforgettable month on Barro Colorado with Morton Wheeler and Nathan Banks, two of the world's greatest collectors of insects. I cannot mention those delightful days in the dripping jungle without visions of strange plant and insect life flooding my memory and reminding me how profoundly those intimate glimpses of the unchanged virgin forest have affected my whole philosophy of life.

The World Was My Garden

The courtship of the Calobata flies which Wheeler and I watched on the great broad leaves of a species of *Piper;* the hunt for the queens of the vicious Echiton ants that bivouacked near the laboratory; the air dance of the stingless bees at sunset; the calls of the howler monkeys from the tree-tops at dawn; the two-toed sloths which Curt Richter was studying; and the half-finished laboratory under which I improvised a microscope table where one evening the Indian servant discovered a fer-de-lance— all are pictures in my mind which I shall treasure as long as I live. I feel sure that many other men have carried away as stirring memories as Graham and I, and I can think of no investment quite so well placed as those all too meager sums which have gone into the upkeep of the Laboratory. The images registered in the brain-patterns of some of America's most talented observers will bear fruit for generations and will direct discoveries in the great field of biological things.

The intimate friendships which grew up between the various members of the Barro Colorado Laboratory group led to other events as well. One was a winter visit with Tom Barbour to southern Cuba where I saw the beginning of an important Arboretum at Soledad, the sugar estate of Mr. Edwin Atkins, situated near Cienfuegos on the south coast.

I arrived at Soledad in time to see the building of "Harvard House," a laboratory with herbarium, work tables and sleeping quarters, in all of which Tom was much interested. This event marked the turning point when the splendid collection of tropical trees and plants which Robert M. Gray had gathered together for Mr. and Mrs. Atkins became part of a scientific institution.

A long correspondence with Mr. Gray had kept me in touch with the activities at Soledad, but it was not until this visit that I realized with much pride the rôle which our Office had played in the beginnings of the arboretum. I always tried to make plant introduction an international exchange, and here was an example of the working of this policy. Through the years, we had shared our collections of tropical plants with Soledad, hoping that those which found Florida unsuited to their needs would grow and prosper in Cuba.

But what interested me most was to discover that Tom Barbour, whom I had looked upon as essentially a zoölogist, was in reality just as much interested in the introduction of plants into Cuba as I was in their introduction into the United States.

The Allison Armour Expeditions

Since this first visit, I have made two other trips to Soledad and have watched with increasing interest its development into one of the best tropical arboreta in the Western Hemisphere.

The connection which Mr. and Mrs. Atkins established in the early days with Harvard through Professor Goodale and Professor Oakes Ames and, subsequently, their intimate friendship with Doctor Barbour, have brought about the creation of a Foundation and the setting aside of a large tract of land for arboretum purposes. The interest of Cuban authorities has been encouraged and friendly relations with their forestry experts established, so that today the Harvard Arboretum (Atkins Foundation) at Soledad is a scientific institution where accredited students interested in tropical plants may go for serious study.

At the Office, the six years immediately following the War were hectic but full of romance too. On big maps hung in the office, we followed with absorbing interest the wanderings of our collectors and shared their enthusiasms about the possibilities of each of their new discoveries. It was our job, once the plants were safely landed in America, to see that they were promptly and accurately described; that they were given a fair start towards being grown and propagated; and that they did not get smothered by Quarantine regulations and treatments.

The reception of these foreign plants, however, was only a part of our duties. We gave enthusiastic attention to domestic experimenters who reported that new plants which they were tending were performing well and had such promise that something should be done to let the farmers and gardeners know about them. We were increasingly handicapped by lack of funds to educate the public about the value of proven introductions, and much hampered by the discontinuance of *The Plant Immigrant Bulletin,* which had served this purpose.

Notwithstanding these difficulties, I was determined that the stream of incoming plants should not be allowed to dry up. There is a definite need for new varieties to meet demands which are constantly arising in our swiftly changing agriculture. Neither did I myself propose to turn aside from securing foreign plants. I felt that I had done some of my best work as a Plant Explorer and, on the other hand, had balanced my years of travel by an equal number in an office chair. Our children were reaching ages of some discretion, and Marian and I began to yearn for the open road once more.

It was when we were in this mood, one night in 1923, that Mr. Allison Armour invited us to dine with him on his house-boat anchored near Coconut Grove. The dinner was in honor of Miss Evangeline Booth, and Uncle Barbour and several others were present too. Miss Booth had been speaking at a meeting and was exhausted when she arrived, but when she heard that I was a "plant man" she became delightfully animated and described her garden, which was a gift to her from a friend. She longed for a Japanese flowering cherry to plant there and I later sent her one.

After Miss Booth and the other guests had gone, we sat late with Mr. Armour. He told me of an expedition he had made to Central America with Professor C. F. Millspaugh of Chicago and Professor William H. Holmes, Anthropologist of the National Museum, and of an expedition he had taken to Cyrene. He then turned the conversation to plants.

"I am becoming tired of the purely social life I have been leading lately," he confessed, "and would like to make a long trip exploring for plants. If I fitted up a boat especially for this purpose, would you go with me?"

I told him that of course I would, and we discussed what interesting itineraries were possible for such an expedition. I suggested the Moluccas, islands lying northwest of New Guinea, as I thought that they should yield many valuable plants for Florida. Mr. Armour had been to Borneo with Lord Crawford and knew a little about that part of the world.

"It would all depend on finding the right boat," he said as he bade me good-night, "and I am sure such a one will not be easy to find."

It was all quite vague, but a delightful topic for conversation and thought.

The following winter, Mr. Armour was again at Miami, accompanied by Mr. and Mrs. Jordan Mott, old friends of his. Marian and I went down to the Keys with them in search of a tiny island which Mrs. Mott had heard was absolutely free from mosquitoes because of some plant growing there. We located the island but were surrounded by voracious mosquitoes the moment we landed, and retired defeated.

This excursion, however, gave us an opportunity to know and become very fond of Mr. and Mrs. Mott and to discuss again the possibilities of Mr. Armour's proposed expedition.

Both Marian and I came to admire Mr. Armour immensely; his cour-

Howard Dorsett and David Fairchild (*seated*) spent twenty happy years in close association, trying to increase the number and improve through introductions the quality of the fruits and vegetables of the United States.

On a recent visit to his old place near Chevy Chase, David Fairchild stood beneath one of the flowering cherries which he had imported from Japan as a tiny seedling.

There are thousands of flowers in each cluster of Mango bloom.

472B

The blue waters of Biscayne Bay sparkle beyond the arched entrance of the Kampong at Coconut Grove.

Howard Dorsett showing Marian Fairchild his Regal Lilies which were one of E. H.
Wilson's greatest gifts to America.

472D

tesy towards everybody and his alertness in anticipating the needs of his guests, together with his deep interest in the food plants of the world and their uses, made me conscious that for a second time in my life I had met an expert traveller and a great gourmet who was interested in furthering the cause of plant introduction with which I had been so long associated.

Nevertheless, it seemed beyond the bounds of possibility that he would purchase and refit a commodious cargo vessel, transforming it into a delightful yacht equipped with a laboratory containing microscope tables, special seed-drying apparatus, a dark room, and a library for books of reference. Added to this were launches built for landing on the strands of tropical islands and, what is perhaps more rare, a steward interested in the cooking of any kind of new vegetable or fruit. The whole thing was fairy stuff and I did not allow my mind to believe in it; but, like the vision of Java which Uncle Barbour made come true, the *Utowana* became a reality in a ship-building yard in Sweden.

When I announced in 1924, at a gathering of the Office staff, that Mr. Allison Armour had invited me to go on a plant-collecting expedition in a yacht which he was remodelling, and told them that I had agreed to go with him to the Orient, I could see distinct disapproval in many faces. They felt that I was once more "abandoning the ship," as I had seemed to do in 1898 when I went off with Mr. Lathrop.

I did not realize that evening that I was virtually saying good-bye to my executive days—the office days in Washington. I thought that I saw ahead only a year or two of exploration in the Dutch East Indies and then a return to my desk. But, when we sailed from Montreal that autumn, I virtually burned my bridges, leaving my old friends Howard Dorsett, Peter Bisset, Doctor Galloway, and my young assistant Wilson Popenoe in charge.

For the next six years the Allison Armour Expeditions absorbed most of my attention. An account of three of these is given in *Exploring for Plants,** which was published in 1930. Even after the book was written, the unforgettable days on the *Utowana* were prolonged by a cruise in the summer of 1930 through the Adriatic and Ægean and as far as Constantinople; and expeditions in 1932 and 1933 through the West Indies to the Guianas, in one of which Howard Dorsett joined us and with

*Published by the Macmillan Company, New York.

The World Was My Garden

Harold Loomis and J. H. Toy formed the scientific staff. They were aided by Mrs. Fairchild and our daughter, Nancy Bell. Besides the staff, either Mr. and Mrs. Francis M. Whitehouse or the Jordan Motts accompanied us, inspiring the group with their enthusiasm.

Thus, from 1924 to 1933, the *Utowana* nosed into Mediterranean and African ports, explored the fjords of Norway and the coasts of the Canary Isles, Mexico, South America and thirty of the West Indian Islands. Riding comfortably in the roadstead, she would tug gently at her anchor as she waited for the botanizing scientists scrambling over the rocks and fields ashore. Always she seemed ready and anxious to set sail again, always she seemed a veritable floating home containing good fellowship, consideration and good cheer. Indeed, I cannot think of these rich years without becoming almost overwhelmed by the flood of memories.

There were days spent collecting in the dripping forests, on sandy deserts or along lava or coral-covered beaches; long séances in the laboratory identifying the collections, or cleaning and packing the seeds which had been gathered. I close my eyes and see the constantly swinging beams of sunlight across the laboratory tables as the boat rolled in the trade-winds; and again hear the whir of the fan in the drying cage, and the delightful banter of the volunteer seed cleaners, of whom Francis Whitehouse was the best. When the day's work was over, there were interesting dinner conversations centering upon zoölogical, entomological, or botanical subjects. Even the sudden death on board of Jordan Mott, whom we all loved, has cast no shadow, for it was his often expressed wish to close his life on his old friend's yacht!

It seems incredible that a second dream of my life, that of having a boat at my disposal, did materialize. But, like all things that the fairies bring, it disappeared again into the shadows, not, however, without leaving "stranded" on the shores of Florida, Cuba and elsewhere many seeds which have since grown into tall trees, so large that even the derricks of the *Utowana* would find it difficult to lift them now.

In 1933, the *Utowana's* cruise to the West Indies and Panama included bird, mammal and shell collecting as well as plant gathering. Tom Barbour and his friend James Greenway were the leading spirits that year. To search for a strange, almost extinct species of mammal, confined to a tiny uninhabited island in the Bahamas—East Plana Key—with such a friend as Tom Barbour was an unforgettable experience. Our tastes are very similar, and hours spent together pass all too quickly.

In the interim between 1927 and 1930, Marian and I built a house and a laboratory in "The Kampong," with trellises strong enough to carry the heavy tropical creepers which we admire so much. I had retired by this time from the direction of the Office in Washington, but found abundant scope for my activities in the local problems connected with the increasing collections of plants at the Chapman Field Experiment Garden and on our place in Coconut Grove.

We soon found that our life had readjusted itself, and consisted of a nine months' sojourn in Florida, and a three months' stay at Baddeck, our home in Nova Scotia. Consequently it seemed necessary to sell our flowering cherry trees and "In the Woods." To exclude from our lives such glorious living things as those cherry trees caused a painful wrench of the spirit, but the lure of the many tropical flowering and fruiting trees on our Florida place made us more content to entrust the cherries to Doctor E. A. Merritt of Washington, who has taken excellent care of them and has fulfilled the purpose we had when we set them out, by propagating the best varieties and distributing thousands through the country.

Shortly after my last expedition on the *Utowana,* there followed a long, serious illness and a slow recovery, during which my collection of plants on "The Kampong" played an important rôle in restoring health and life once more.

And now the years pass swiftly, happily, peacefully—years of "simple, quiet, contented *Still-leben"* such as Alexander von Humboldt wrote was, "after all, the highest life that man can have." But the years are only relatively quiet, for every winter brings an increasing number of visitors to "The Kampong." Among these, each year, none is more welcome than Tom Barbour with his amazing store of knowledge in many fields. We search the Keys for shells and the trees for snails. His interest in the Chapman Field plant collections, and his encouragement and support of the tropical garden which Colonel Robert H. Montgomery proposes to name the Fairchild Arboretum, have encouraged me to go along with all of these plans which would tend to stimulate others to engage in this romantic work of introducing new plants.

CHAPTER XXXVII

Aloha

As a boy I was fascinated by the changing patterns of a kaleidoscope with its bits of colored glass tumbling between mirrors. As I look back on my life, vivid scenes pass the eye of memory; a mixture of interesting concrete objects, scenery, personalities and ideas—incomparably more fascinating than any mosaic possible in a toy.

The objects have varied in size from the fine particles comprising the nucleus of a swarm spore, almost beyond the border of visibility, to gigantic volcanoes. At some moments, my life has been expressed in powdery mildews on the lilac leaves and at others expanded to encompass desert sand or the dense, tropical jungle. I have found that the play of electrons in a Crook's tube contains a beauty as enthralling as the sunset on a coral atoll fringed with palms. I have ranged from high, rocky passes in the Andes to strange scenes on the ocean floor among marine sponges and rare seaweeds.

In most lives, men rank above objects. Perhaps because we ourselves are men. In this human field, my life has also contained greater variety than most. I have heard around me voices speaking many languages, including the harsh guttural of Arabic, the soft whisper of Malay, the staccato tongue of southern Italy, and the high, nasal singsong of the Chinese. And I have looked into the eyes of many races of men and wondered what they thought of me; races as distinct as the Fijian cannibal and the tow-headed, blue-eyed Scandinavian.

From the limited fare of my childhood, my menu has expanded to include a range of food stuffs which I never dreamed existed. Furthermore, from my excursions into the field of foods, I have gained the conviction that any one who will sincerely try can learn to enjoy almost

Aloha

any food. The term "not fit to eat" (when applied to things non-poisonous) is indefensible except as a personal opinion of the speaker. For there are no scientific standards in the realm of the palate; in fact, it has no nomenclature comparable to that of the realms of sound or color. When we are new-born babes, the taste buds of our tongue give us the first glimmer of consciousness, and all our lives remain one of our most intimate contacts with the outside world; yet we cannot describe in accurate terms the flavor of a tomato or an apple or any other delicacy further than to say that it is sweet, bitter, sour, spicy, or salty, and continue to ring the changes on these words.

Continuing this discussion, I should add that during this life of mine, my ears have played a part as well as my eyes and palate. They have heard the boom of breakers on many shores, and enjoyed the ripple of brooks in Nova Scotia and roar of waterfalls in lands near the equator. I stand today amid the rustle of autumn leaves in New Jersey and remember the notes of a magic flute on the midnight canals of Venice, the emotion of Wagnerian music in the Berlin Opera house and the immortal voice of Patti singing "The Last Rose of Summer." In the Florida moonlight, the mocking bird has sung to me, and across the sea I have listened to the lark as its sweet song faded in the gray skies of Belgium. In another mood, my ears attune themselves to the interesting voices of the insect world; the shrill song of the cicada in the Sumatra highlands, a hundred times more deafening than the vibrant hum of the seventeen-year locust; the drone of bumble-bees, the higher-pitched note of the honey-bee, the buzz of house-flies on a hot summer afternoon; the cricket on the hearth which has sung to man for thousands of years; and the lesser noises of the woods which must be listened for with understanding ear.

A man who has spent his life surrounded by other people every hour of the day will not understand what I am talking about. To him the personalities of those around him seem everything; the exchange of thoughts, the conversation and doings of the human animal are more exciting than anything which "Nature" can offer. Still, I too have met and mingled with a great variety of people and found them interesting. I have known five presidents and many ambassadors and statesmen. Of artists, sculptors, architects, and city planners, I have known some of the most noted of my time, and learned to believe in their work and appreciate better the great masters of the past because I have met those still living.

The World Was My Garden

With the men of science and invention of my generation much of my life has been spent, together with that choice group of naturalists with whom I have most associated. These explorers of the unknown have filled my imagination as none of the others have, for they are forever pushing on like little children, never losing that thing called curiosity which gives to childhood its fascination and its charm. They seem to carry on zestfully to the very end of life—forever finding new things, new toys to play with. And if their toys break, their theories explode, to find out why they broke seems more interesting to them than the toys themselves. They move in a world so vast and yet so little explored that they have only to enter for a moment some woodland trail to see on every side a thousand unsolved riddles, an ocean of untouched possibilities, for there is no end to the world of natural things once curiosity controls your life. Swingle once said of the work of Nathan Cobb, whose interest was insatiable: "When he shows you what he has discovered it seems like a mosaic, but when you try to pull up any one of the stones of the mosaic, you bring up attached to it something as complicated as an umbrella."

As I near the last of these pages, a fear of being thought immodest, of claiming too much credit for everything in the book, besets me. The eager, wistful eyes of some long gone and others still alive gaze at me from the past. They ask, somewhat eagerly, what credit I have given them for their years of toil which have brought no word of recognition save the pay check from Uncle Sam. Poor Stuntz, whom we buried in the snow, and Wester, who died in the Philippines, and Barrett, still wandering in the tropics, and Beagles and Klopfer, who labored in the Chico garden, and Morrow, who still carries on there.

"After all the years we worked together, the ungrateful fellow never mentioned me in his book!" they say. There is much warrant for such a feeling in one who drudges year after year to help build up an organization, whether inside or outside the government service. Surely he is as deserving of mention as the figure-heads of the concern.

In its last analysis, the matter comes down to the utterly illogical, unreasoning use of words. Why should one great organization bear the name of a man, a personality, and another not? The word "Ford" brings to mind a gray-haired, slender man, while "General Motors" creates the picture of wheels and whirring machinery. The United States Department of Agriculture is a rather nebulous term; so, to remedy its vagueness,

Aloha

the public substitutes the name of the political head of the Department. Convenience and a certain feeling of picturesqueness play important parts here, but it cannot be denied that these lead to great abuses, for as Frederick Adams Woods has said, there is "the unearned increment of fame."

If I mentioned everybody I should like to, the book would become a mere list of names. So this is my excuse to the gallery of personalities which I feel around me as I write.

But I must at least mention with feelings of real appreciation that group of loyal women whom I have always designated as "The Girls of the S. P. I." It has been their agile fingers on the keyboards which have accumulated a mass of information regarding useful plants and plant experiments, such as the world has never known before.

It is with regret that I close before bringing the book down to date, but additional volumes would be required to describe the wide extension of the activities of the organization, which is now known as the Division of Plant Exploration and Introduction, with Mr. Ben Y. Morrison as its head.

A year ago, in New Jersey, in the delightful library of Alfred and Elizabeth Kay—which was once an old barn of "Hidden River Farm"—at their insistence, I began writing this biography. Now, after twelve months of work in Florida and Nova Scotia, I have come back to Hidden River Farm to finish the tale.

On the shelves beside me are the autobiographies of many great men, and as I study their writing, curious to learn their feelings as they approach the close of their narratives, I like best the lines of John Hay:

I really believe that in all history I never read of a man who has had so much and such varied successes as I have had, with so little ability and so little power of sustained industry. It is not a thing to be proud of, but it is something to be grateful for.

Still, I should not like to close this book looking backward, for the vision of the future in store for those who shall live on is too alluring. The vistas opened by the discoveries of science are bewildering in their possibilities. I have enjoyed each moment of each experience of my life and have found that each experience has given me greater interest and powers of understanding, so that I stand each day more ready for the next. Why should men collect tangible assets which may be swept away in a moment? Marian and I have a treasure-trove of knowledge and memories which are ours as long as we shall live.

The leaves are reddening on the trees along the river here and wood asters and golden rod are scattering their seeds in the wind. A cold rain of autumn will soon come to drop the leaves to the ground and close the careers of the flowers.

A spirit of melancholy came over me as I entered the forest this morning. Memories of the walks Uncle Byron and I used to take through the woods of Iowa came back to me. He taught me to enjoy every detail of field and forest and to feel the relative unimportance of the man-made things of life.

And what a contrast there is between man's civilization and Nature's! I walked today until the forest engulfed me on every side. The rough dark trunks of the red oaks and the smooth-barked birches shut me in.

It is early autumn and the beech and maple leaves on the ground make patterns more gorgeous than ever the Persians dreamed of for their carpets. Near by, at the base of a rotten stump, a graceful Amanita mushroom with its slender ringed stem warns me that within my reach is a poison as deadly as the Gambians ever used in warfare. A fallen log covered with cup fungi brings back my student days when Harper and I went digging for truffles in the deep beech woods of western Germany. It seems deathly still, as still as in the dense jungle, but yet leaves are falling here and there. Now crickets' voices break the stillness, then the tapping of a woodpecker in the distance, and suddenly I am startled by the chatter of a squirrel in a near-by tree-top. A crow caws across the Hidden River, and I hear the sound of falling water at the dam.

My feet feel the soft leaf-mold beneath them and I imagine that it is the deeper, moister leaf-mold of the tropical forest until the asters and golden rod remind me that these are northern woods. However, the feeling of being alone and far away is the same. I am amazed how quickly my thoughts change as I stand where there is no reminder of human personalities—no interruption of a spoken word to bring me back from my dream of solitude where nothing matters and things go on forever.

Standing here alone, the scenes from childhood to age of a very happy human life pass on the screen before me; of a life spent wandering among the myriad plant forms of the globe; and I sense again the pleasure of seeing them, and gathering their seeds to grow in other lands than those which they call their home.

But the autumn chill reminds me that the picture is drawing to a close; both the romance of plant introduction, and the romance of a happy marriage to the most charming Gypsy I met with on the trail, have now reached the moment when visions of another generation mingle with recollections of the past. The eldest of our next generation is returning from South America to marry before continuing his research in the tropics; Barbara is now the mother of our two delightful grandchildren; while Nancy Bell is developing her natural faculty for music in order to charm us with her songs.

Moving in the background of the picture, like scenery in the cinema, are pastures, grain fields, orchards, flowering trees and vines in many people's gardens; the other children of my wanderings.

I return through the woods. Soon fences and plowed fields and a notice warning trespassers bring back the complications of a man-made world.

Index

Index

Page references followed by a letter indicate illustrations.

Index

Index

Index

Index

489

Index

Index

Index

Index

493

Index

494